Tsunami

The Underrated Hazard

Between 1990 and 2000, over ten major tsunami events have impacted on the world's coastlines, causing devastation and loss of life. These events have made scientists aware that tsunami have been underrated as a major hazard, mainly due to the misconception that they occur infrequently compared to other natural disasters. Evidence for past great tsunami, or "mega-tsunami", has also recently been discovered along apparently aseismic and protected coastlines, such as those of Australia and Western Europe. These mega-tsunami are caused either by huge submarine landslides or the impact of meteorites and comets with the ocean. With a large proportion of the world's population living on the coastline, the threat from tsunami cannot be ignored. Were a mega-tsunami to occur today the death toll would be in the tens of thousands, while the damage would exceed that of the largest disasters of the past century.

Tsunami: The Underrated Hazard comprehensively describes the nature and process of tsunami, outlines field evidence for detecting the presence of past events, and describes particular events linked to earthquakes, volcanoes, submarine landslides, and meteorite impacts. While technical aspects are covered, much of the text can be read by anyone with a high school education. The book will appeal to students and researchers in geomorphology, earth and environmental science, and emergency planning, and will also be attractive for the general public interested in natural hazards and new developments in science.

Edward Bryant is Associate Professor and Head of the School of Geosciences at the University of Wollongong, Australia. He has published over fifty papers in international journals on topics ranging from sea level, beach erosion, coastal evolution over the last 200,000 years, climate change, tsunami dynamics, and catastrophic agents of coastal evolution. He is author of two previous books with Cambridge University Press: *Natural Hazards – Threat, Disaster, Effect, Response* (1991) and *Climate Process and Change* (1997).

The remnant plug at the centre of a vortex bored into granite at the front of a cliff at Cape Woolamai, Phillip Island, Victoria, Australia. The ridges in the foreground indicate that flow around the
plug was counterclockwise and consisted of a double helix. A catastrophic tsunami wave produced the vortex as it overwashed the cliff.
©John Meier

Tsunami

The Underrated Hazard

EDWARD BRYANT

University of Wollongong

CAMBRIDGE
UNIVERSITY PRESS

PUBLISHED BY THE PRESS SYNDICATE OF THE UNIVERSITY OF CAMBRIDGE
The Pitt Building, Trumpington Street, Cambridge, United Kingdom

CAMBRIDGE UNIVERSITY PRESS
The Edinburgh Building, Cambridge CB2 2RU, UK
40 West 20th Street, New York, NY 10011–4211, USA
10 Stamford Road, Oakleigh, VIC 3166, Australia
Ruiz de Alarcón 13, 28014 Madrid, Spain
Dock House, The Waterfront, Cape Town 8001, South Africa

http://www.cambridge.org

First published 2001

Printed in the United Kingdom at the University Press, Cambridge.

Typeface Utopia 10/13.5 pt.   *System* QuarkXPress® [GH]

A catalogue record for this book is available from the British Library.

Library of Congress Cataloging-in-Publication Data
Bryant, Edward, 1948–
    Tsunami : the underrated hazard / Edward Bryant.
        p.   cm.
    Includes biographical references and index.
    ISBN 0-521-77244-3 – ISBN 0-521-77599-X (pb)
    1. Tsunamis. I. Title.
  GC221.2 .B78 2001
  551.47′02 – dc21   00-046761

ISBN 0 521 77244 3 hardback
ISBN 0 521 77599 X paperback

To the Memory of

J Harlen Bretz

Contents

List of Illustrations

List of Tables

List of Symbols in Formulae and Greek Symbols

a	length of a boulder (m)
a_o	initial acceleration of a submarine slide (m s^{-2})
A_x	horizontal bottom orbital diameter of wave motion (cm s^{-1})
b	the intermediate axis or width of a boulder (m)
b_i	the distance between wave orthogonals at any shoreward point (m)
b_o	the distance between wave orthogonals at a source point or in deep water (m)
b_l	intermediate diameter of largest boulders (m)
B_r	the breakdown parameter of a wave (dimensionless)
c	thickness of a boulder (m) or
	thickness of a submarine slide (m) or
	soil cohesion (kPa)
C_m	added mass coefficient of a submarine slide (dimensionless)
C	wave speed or velocity of a wave (m s^{-1})
C_d	the coefficient of drag (dimensionless)
C_f	the De Chézy friction coefficient (dimensionless)
C_i	wave speed at any shoreward point (m s^{-1})
C_l	the coefficient of lift (dimensionless)
C_o	wave speed at a source point or in deep water (m s^{-1})
d	water depth below mean sea level (m) or
	flow depth (m)
d_o	water depth at a source point or in deep water (m)
d_i	water depth at any shoreward point (m)
D	diameter of an impact crater (m)
$D\,(\text{ø})$	direction of tsunami propagation relative to an observer (degrees)
F_d	the drag force (N)

F_l	the lift force (N)
F_r	the Froude number (dimensionless)
F_{res}	the restraining force (N)
g	gravitational acceleration (9.81 m s^{-1})
h	wave height above the seabed (m)
H	crest-to-trough wave height (m)
Ha_o	the Hammack number at termination of a submarine slide (dimensionless)
H_b	wave height at the breaker point (m)
H_o	crest-to-trough wave height at the source point or in deep water (m)
H_r	tsunami run-up height above mean sea level (m)
$\bar{H}_r$	mean tsunami run-up height above mean sea level (m)
H_{rmax}	maximum tsunami run-up height above mean sea level (m)
H_s	wave height at shore or the toe of a beach (m)
H_t	tsunami wave height above mean sea level (m)
$\bar{H}_{tmax}$	mean maximum tsunami wave height along a coast (m)
i_s	Soloviev's tsunami intensity scale (dimensionless)
k	coefficient (dimensionless)
K_r	refraction coefficient (dimensionless)
K_s	shoaling coefficient (dimensionless)
K_{sp}	coefficient of geometrical spreading on a sphere (dimensionless)
l	length of a submarine slide parallel to the slope (m)
L	wavelength of a tsunami wave (m)
L_b	length of a bay, basin, or harbour (m)
L_s	bedform wavelength (m)
L_r	the length of fault rupture (km)
m	mass of a meteorite (kg)
m_{II}	tsunami magnitude, Imamura–Iida scale (dimensionless)
m_s	mass of a submarine slide (kg)
m_o	mass of water displaced by a submarine slide (kg)
M_m	mantle moment scale (dimensionless)
M_o	seismic moment measured (N m)
M_s	surface magnitude of an earthquake on Richter scale (dimensionless)
M_t	tsunami magnitude (dimensionless)
M_w	moment magnitude scale (dimensionless)
n	Manning's roughness coefficient (dimensionless) or an exponential term
r	the radius of the crater made by a meteorite impact in the ocean (m)
r_m	radius of meteorite (m)

R_e the shortest distance from a location to the epicentre of an earthquake (km)

R_t the distance a tsunami travels from the centre of a meteorite impact (m)

s_o travel distance of a submarine slide (m)

S_p density correction for a meteorite impact (g cm^{-3})

S_t area of seabed generating a tsunami (m^2)

t_o duration of the slide(s)

t time(s)

T wave period(s)

T_s wave period of seiching in a bay, basin, or harbour(s)

u_{max} maximum bottom orbital velocity (cm s^{-1})

$u(x, y, t)$ depth averaged velocity under a wave at right angles to shore (m s^{-1})

U velocity of a submarine slide (m s^{-1})

U_i the initial velocity of a submarine slide (m s^{-1})

U_∞ the terminal velocity of a submarine slide (m s^{-1})

v flow velocity (m s^{-1})

$\bar{v}$ mean flow velocity of water (m s^{-1})

$v(x, y, t)$ depth-averaged velocity under a wave parallel to shore (m s^{-1})

v_m impact velocity of meteorite (m s^{-1})

v_{min} minimum flow velocity of water (m s^{-1})

v_r velocity of tsunami run-up (m s^{-1})

w width of submarine slide (m)

W kinetic energy of a meteorite impact (kilotons of TNT)

x distance at right angles to bottom contours or the shoreline (m)

x_{max} limit of tsunami penetration landward (m)

y distance parallel to bottom contours or the shoreline (m)

GREEK SYMBOLS

α_i the angle a wave crest makes to the bottom contours at any shoreward point (degrees)

α_o the angle a wave crest makes to the bottom contours at a source point (degrees)

β slope of the seabed (degrees)

β_w slope of the water surface (degrees)

Δ angle of spreading on a sphere relative to direction of wave travel or angle between a location and the epicentre of an earthquake (degrees)

ΔC correction on tsunami magnitude, Mt, due to source (dimensionless)

Δp_{max} maximum change in pressure on the seabed (dyne cm^{-2})

Δu_{max} maximum asymmetry in onshore versus offshore bottom velocities (cm s^{-1})

Σ	surf scaling factor
ξ	pore water pressure (kPa)
η	wave height above mean sea level (m)
ρ_e	density of material ejected from an impact crater (g cm^{-3})
ρ_m	density of meteorite (g cm^{-3})
ρ_s	density of sediment (g cm^{-3})
ρ_w	density of sea water (g cm^{-3})
$\emptyset$	the azimuth of an observer relative to the direction of fault rupture (degrees)
π	3.141592654
ω	radian frequency of a wave (s^{-1})
t_s	the shear strength of the soil (kPa)
σ	the normal stress at right angles to the slope (kPa)
ϕ	the angle of internal friction or shearing resistance (degrees)
υ_R	velocity of an earthquake rupture (m s-1)
μ	Coulomb friction coefficient (dimensionless)

Preface

Before 10 A.M., 18 March 1989, I was a process geomorphologist who had dabbled into the coastal evolution of rock platforms and sand barriers along the New South Wales coastline of Eastern Australia. I was aware of tsunami, and indeed had written about them, but they were not my area of research expertise. No one had considered that tsunami could be an important coastal process along the East Coast of Australia. On that March morning in brilliant sunshine, with the hint of a freshening sea breeze, my life was about to change. I stood with my close colleague Bob Young, marvelling at a section of collapsed cliff at the back of a rock platform, at Haycock Point south of Merimbula. We saw a series of angular, fresh boulders jammed into a crevice at the top of a rock platform that did not appear to be exposed to storm waves. Unlike many before us, we decided that we could no longer walk away from this deposit without coming up with a scientific reason for the field evidence that was staring us in the face. After agonising for over an hour and exhausting all avenues, we were left with the preposterous hypothesis that one or two tsunami waves had impinged upon the coast. These tsunami were responsible, not only for jamming the rocks into the crevice, but also for the rock-fall that had put the rocks on the platform in the first place. We did not want a big tsunami wave, just one of about 1–2 m depth running about 5–6 m above the highest limits of ocean swell on the platform. Over the next eight years that wave grew immensely until we finally found evidence for a mega-tsunami overwashing a headland 130 m above sea level at Jervis Bay along the same coastline. Subsequent discoveries revealed that more than one wave had struck the New South Wales Coast in the last 7,000 years, that mega-tsunami were also ubiquitous around the Australian Coast, and that the magnitude of the field evidence was so large that only a comet or meteorite impact with the Earth could conceivably have generated such waves. From being a trendy process geomorphologist wrapped in the ambience of the 1960s, I had descended into the abyss of catastrophism dredged from the dark ages of geology when it was an infant discipline. Bob Young subsequently retired in 1996, but his clarity of thinking about the larger picture and his excellent eye for the landscape are present in all of our publications and reflected in this textbook. There was not a day in the field with Bob that did not lead to excitement and discovery.

Since 1995, I have worked closely with Dr. Jon Nott from James Cook University in Cairns, Queensland. Bob Young trained Jon, so I have lost none of Bob's appreciation for landscape. Jon has enthusiastically continued field research with me in remote locations, and has uncanny luck for being able to obtain funding for a strange topic in an age where economic rationalism and blinkered adherence to the safe academe of the 1960s dominates. To stand with Jon at Point Sampson, Western Australia, and both realise simultaneously that we were looking at a landscape where a mega-tsunami had washed inland 5 km – not only swamping hills 60 m high, but also cutting through them – was a privilege. Few geomorphologists who have twigged for the first time to a catastrophic event have been able to share that experience in the field with anyone else. Jon, Bob, and I formulated the signatures of tsunami described in Chapter 3. David Wheeler did the fieldwork that first identified the dramatic tsunami chevron-shaped dunes at Steamers Beach, Jervis Bay. All of us have withstood the rebuff of peers that goes with ideas on catastrophism in an age of "minimal astonishment." I hope that this book conveys to some the excitement of our discoveries about tsunami.

It is difficult to write a book on tsunami without using equations. The relationships amongst tsunami wave height, flow depth at shore, boulder size, and bedform dimensions were crucial in our conceptualisation of mega-tsunami and their role in shaping coastal landscapes. I have avoided some of the more elaborate mathematical equations used to characterise tsunami and apologise to the master artisans for debasing or simplifying some of their formulae. Wherever I have used equations, I have tried to appease the nonexpert by including figures and photographs that illustrate the formulae. Terms used in equations are only defined once where they first occur in the text, unless there could be confusion about their meaning at a later point in the book. For reference, all terms and symbols are summarised at the beginning of the text. Many dates are only reported by year. Where ambiguity could exist, the prefix A.D. (Anno Domini) or the suffix B.C. (Before Christ) is used. If there is no ambiguity, then the affix is dropped and the year refers to A.D. Units of measurement follow the International System of Units except for the use of the terms kilotons and megatons. In order to convey viewpoints and arguments, unobstructed by copious referencing, strict adherence to formal, academic referencing has been relaxed. Usually, each section begins by listing the relevant journal articles or books that either have influenced my thinking or are central to the topic. Again, I apologise to anyone who feels that I have ignored their crucial work but the breadth of coverage precluded a complete review of the literature on many topics. All references to publications can be found at the end of the book. Some articles and data were acquired from the Internet. The Internet addresses in these cases are also referenced. Such material may not be readily available because the addresses have changed or because of the lack of an archival tradition for this new resource medium. Where material is not available in the literature or through these forums, it has been acknowledged at the beginning of the text.

Acknowledgments

A number of people and organisations should be acknowledged for their information about tsunami. First is the U. S. government, which has a policy of putting all its information in the public domain. Many photographs used throughout this book and detailed information on events were obtained from U.S. government agencies and their employees. None of these accepts liability nor endorses material for any purpose. I have acknowledged these sources in this book, but mention here the National Geophysical Data Center (NGDC) at **http://www.ngdc.noaa.gov/cgibin/ seg/m2h?seg/haz_volume3.men+MAIN+MENU** and the Pacific Marine Environmental Laboratory at **http://www.pmel.noaa.gov/tsunami-hazard/** for their excellent sources. Many of the maps in the text are based upon Generic Mapping Tools (GMT), an online software package developed by Paul Wessel, School of Ocean and Earth Science and Technology, University of Hawaii at Manoa, and Walter H. F. Smith, Geoscience Laboratory, NOAA. The package can be found at **http: //www.aquarius.geomar.de/omc/about_gmt.html.**

Background information on tsunami in Chapter 1, and reference to individual events throughout the book, were obtained through the Tsunami Laboratory run by Dr. Viacheslav Gusiakov at the Institute of Computational Mathematics and Mathematical Geophysics, Siberian Division Russian Academy Of Sciences, Novosibirsk, Russia. His web address is **http://omzg.sscc.ru/tsulab/.** I am also indebted to Slava for his comments on a cosmogenic source for the New South Wales mega-tsunami. Dr. Efim Pelinovsky, Institute of Applied Physics, Russian Academy of Sciences, Nizhny Novgorod, Russia, provided encouragement and information on Caspian Sea tsunami used in Chapter 1. Dr. Edelvays Spassov, formerly of the Bulgarian Academy of Sciences, provided information on Black Sea tsunami. Edelvays, thank you for making me aware that tsunami are significant.

Milena Mascarenhas, a second-year mathematics student, Wollongong University, helped formulate spreadsheets for the calculations of tsunami dynamics used in Chapter 2. Dr. Vasily Titov, U.S. Department of Commerce, National Oceanic and Atmospheric Administration, Pacific Marine Environmental Laboratory, Seattle, kindly permitted the computer-simulated tsunami wave train generated by the 1996 Andreanov earthquake to be reproduced in Figure 2.11. This simulation was downloaded from the Internet at **http://corona.pmel.noaa.gov/~tsunami/Titov/ show/andr1.mpeg.** It is from the Maui High Performance Computing Center, Kihei, Hawaii, which is funded by the United States Department of Defense and the University of New Mexico. Dr. Steven Ward, Institute of Tectonics, University of California at Santa Cruz, kindly provided a preprint of a paper on tsunami that gave a novel view of the mathematical treatment of tsunami. Concepts from this paper, especially the term *tsunami window,* are included in the discussion of tsunami dynamics. Dr. Ward also gave his kind permission to use the timelapse simulation of tsunami generated by a meteorite impact in a deep ocean in Figure 8.6 and provided information on the tsunami generated by the Nuuanu slide in Hawaii. Mr. Alan Rodda of Toowoomba, Queensland, provided the details of freak waves off the coast of Venus Bay east of Melbourne, Victoria. Mark Bryant provided the description of the freak wave at North Wollongong Beach in January 1994.

Prof. Toshio Kawana, Laboratory of Geography, College of Education, University of the Ryukyus, Nishihara, Okinawa, provided the photograph of boulders on the Ryukyu Islands used in Chapter 3. Prof. John Clague, Professor and Shrum Chair of Earth Sciences, Simon Fraser University, Burnaby, British Columbia, provided Figure 3.4, showing a sand layer deposited by a tsunami and sandwiched between peats. Susan Fyfe, a Ph.D. student in the School of Geosciences, University of Wollongong originally drew the outstanding sketches of s-forms and bedrock sculpturing used in Chapters 3 and 4. The *Journal of Geology* graciously allows these and other figures published in their journal to be used here. The helical plug that forms the frontispiece, also a bedrock-sculptured feature, is by John Meier, who is a landscape photographer working out of Melbourne.

Information on specific earthquake-generated tsunami and the summary of warning systems presented in Chapters 5 and 9 respectively originate from TSUNAMI!, the web site of the Department of Geophysics, University of Washington, at **http://www. geophys.washington.edu/tsunami/.** The names of individual authors for these pages are not published on the web and could not be properly referenced in the relevant sections. Information on the affects of the Alaskan tsunami of 27 March 1964 came from **http://www.tsunami.gov/64quake.htm.** Details about the Papua New Guinea tsunami of 17 July 1998 were provided by Dr. Philip Watts, Department of Chemical Engineering, California University of Technology, Pasadena, and by Prof. Hugh Davies, Department of Geology, University of Papua New Guinea, Port Moresby. Prof. Davies also provided much of the background information for the story used in Chapter 1. This information is now available at **http://corona.pmel.noaa.gov/ ~tsunami/PNG/Upng/index.html.** Sediment information on the Papua New Guinea event was taken from the U.S. Geological Survey Western Region Coastal and Marine Geology web page at **http://walrus.wr.usgs.gov/tsunami/PNGhome.html.** Figure 5.1, of a tsunami approaching the Scotch Cap lighthouse, Unimak Island, Alaska, was obtained from the web site of Lieutenant Alan Yelvington, U.S. Coast Guard at **http:// www.teleport.com/~alany/uscg/ltsta.html.** These figures are the property of the U.S. Government and are in the public domain. Figure 5.22 of the sediment splay at Arop was prepared especially for this book by Dr. Bruce Jaffe and Dr. Guy Gelfenbaum, U.S. Geological Survey.

Additional information for tsunami generated by earthquakes, submarine landslides and volcanoes originated from the web pages of Dr. George Pararas-Carayannis, retired director of the International Tsunami Information Center (ITIC), at **http:// www.geocities.com/CapeCanaveral/Lab/1029/.** Specific details about the Lituya Bay landslide of 9 July 1958 also originated here, at **http://www.geocities.com/ CapeCanaveral/Lab/1029/Tsunami1958LituyaB.html.** As well, details about the International Tsunami Warning System were gleaned from these pages, from the web pages of the International Tsunami Information Center at **http://www.nws.noaa.gov/ om/tsunami.htm,** and from the National Oceanic and Atmospheric Administration (NOAA) at **http://vishnu.glg.nau.edu/wsspc/tsunami/HI/Waves/waves00.html.** Additionally, particulars on the Alaskan Warning System were taken from the West Coast and Alaska Tsunami Warning Center Internet home page at **www.alaska.net/ ~atwc.**

Dr. Simon Day of the Greig Fester Centre for Hazards Research, Department of Geological Sciences, University College, London, provided information on submarine landslides and their possible mega-tsunami – especially for the Canary Islands. Dr. Day also provided unpublished material and correspondence for Figure 6.8 and the descriptions of tsunami deposition on Fuerteventura and Gran Canaria. Dr. Barbara Keating, School of Ocean and Earth Science and Technology at the University of Hawaii, Manoa, provided the locations of landslides associated with volcanoes plotted in Figure 6.2 and detailed descriptions of historical tsunami in Hawaii related to landslides. Barbara also passed on her comments criticising the tsunami origin of the boulder deposits on the island of Lanai referred to in Chapter 6. Figure 6.10 is taken from Figure 6 in Bondevik et al. (1997b) and is used with the permission of Blackwell Science Ltd. in the United Kingdom. Figure 7.1 is copyrighted and provided by Lynette Cook, who is an astronomical artist/scientific illustrator living in San Francisco.

The following people gave information about Near Earth Objects (asteroids, comets, and meteorites), the characteristics of these objects impacting with the Earth, and the effect of such impacts on human history: Prof. Mike Baillie, Palaeoecology Centre, School of Geosciences, Queen's University, Belfast; Dr. Andrew Glikson, Research School of Earth Science, Australian National University; Dr. Peter Snow, Tapanui, New Zealand; and Dr. Duncan Steel, Spaceguard Australia P/L, Adelaide, South Australia. Michael Paine's unofficial Spaceguard Australia web page at http://www1.tpgi.com.au/ users/tps-seti/spacegd.html provided information on comets, asteroids, and impact events. The Cambridge Conference Network (CCNet) – an electronic newsletter published by Dr. Benny Peiser, School of Human Sciences, Liverpool John Moores University, Liverpool, United Kingdom at http://abob.libs.uga.edu/bobk/cccmenu.html – also was a source of further information. Figure 8.1 appeared originally in Alvarez (1997) and is reprinted by permission. Dr. David Crawford of the Sandia National Laboratories kindly gave permission for the simulations of an asteroid hitting the ocean and the resulting splash tsunami to be used in Figure 8.4. Complete reference to his work can be found at http://sherpa.sandia.gov/planet-impact/asteroid/. Finally, information about the 1886 Charleston earthquake and subsequent events in the region used in Chapter 10 were taken from the United States National Earthquake Information Center, World Data Center A for Seismology at http://wwwneic.cr.usgs.gov/neis/states/south_carolina/south_carolina_history.html.

Ted Bryant

PART ONE

Tsunami as Known Hazards

CHAPTER ONE

Introduction

1.1 The Hollow of the Deep-Sea Wave off Kanagawa (Kanagawa Oki Uranami), a grey-scale print from a colour woodcut, No. 20 from the series *Thirty-Six Views of Fuji*, circa 1831, by Katsushika Hokusai, a famous late eighteenth- and early nineteenth-century Japanese artist. Textbooks and many web sites depict this wave as a tsunami wave, but in fact it is a wind-generated wave. It has a special shape called an *N*-wave, characterised by a deep leading trough and a very peaked crest. Some tsunami, such as the one that struck the Aitape coast of Papua New Guinea on 17 July 1998, emulate this form close to shore.

INTRODUCTION

Tsunami are water waves generated by the disturbance associated with seismic activity, explosive volcanism, a submarine landslide, a meteorite impact with the ocean, or in some cases meteorological phenomena. These waves can be generated in oceans, bays, lakes, or reservoirs. The term *tsunami* is Japanese and means harbour (*tsun*) wave (*ami*), because such waves often develop as resonant phenomena in harbours after offshore earthquakes. Both the singular and plural of the word in Japanese are the same. Many English writers write the plural of tsunami by adding an *s* to the end of the singular form. The Japanese usage will be adhered to throughout this text.

In the 1990s, fourteen major tsunami events struck the world's coastlines. While other disasters over this period have caused more deaths and greater economic destruction, these tsunami events have made scientists aware that the tsunami hazard is pervasive. Before 1990, the public perceived tsunami as originating primarily from large, distant, underwater earthquakes – mainly in the Pacific Ocean. The fear of tsunami was allayed by the knowledge that an early warning system existed to prevent loss of life. Recent tsunami have occurred as near-coastal events – generated by small earthquakes or even submarine landslides – and have occurred in many cases with minimal warning to local inhabitants. In addition, evidence for mega-tsunami has been discovered along the apparently aseismic and protected coastline of eastern Australia. These tsunami have run up heights exceeding the largest earthquake-generated tsunami documented anywhere in the world over the past 5,000 years. These events have not only been repetitive, but also recent. They are novel only in that they have not occurred in a country with a long, scientifically based, written history. Aboriginal legends, however, orally record their occurrence. Ongoing research indicates that such mega-tsunami are more widespread. Their signature not only dominates the Australian Coast, but also that of New Zealand and Eastern Scotland. The generation of these mega-tsunami is contentious but most likely due to either great submarine landslides or the impact of meteorites and comets with the Earth's oceans.

These recent occurrences and discoveries have serious implications when it is realised that Western Civilisation is unique in its settlement of the shoreline and its development of great coastal cities. If a submarine landslide generated a near-coastal tsunami off the coasts of Sydney, Los Angeles, Tokyo, Honolulu, Chennai (formerly Madras), or any of a dozen other large cities, the death toll would be in the tens of thousands, while the damage would exceed that of the largest disasters of the 1990s. This book describes tsunami as an underrated hazard and summarises some of these recent discoveries. It presents for the first time a comprehensive coverage of the tsunami threat to the world's coastline.

FIVE STORIES

1 An Aboriginal legend
(Peck, 1938; Parker, 1978)

It was a stifling hot day, and all the Burragorang people lay prostrate around their camp unable to eat. As night approached, no one could sleep because of the heat and the mosquitoes. The sun set blood red and the moon rose full in the east through the haze. With just a remnant of red in the western sky, the sky suddenly heaved, billowed, tumbled, and then tottered before crumbling. The moon rocked, the stars clattered, and the Milky Way split. Many of the stars – loosened from their places – began to fall flashing to the ground. Then a huge ball of burning blue fire shot through the sky at enormous speed. A hissing sound filled the air, and the whole sky lit like day. Then the star hit the Earth. The ground heaved and split open. Stones flew up accompanied by masses of earth followed by a deafening roar that echoed through the hills before filling the world with complete noise. A million pieces of molten fire showered the ground. Everyone was awestruck and frozen in fear. The sky was falling. Smaller stars continued to fall throughout the night with great clamouring and smoke. The next morning, when all was quiet again, only the bravest hunters explored beyond the campsite. Great holes were burnt into the ground. Wherever one of the larger molten pieces had hit, it had piled up large mounds of soil. Many of these holes were still burning with flames belching out. Down by the sea, they were amazed. Fresh caves lined the cliffs.

Soon stories reached them from neighbouring tribes that not only had the sky fallen, but also the ocean. These neighbours began talking about a great ancestor who had left the Earth and gone into the sky, and who had travelled so fast that he had shot through the sky. The hole he had made had closed up. This ancestor had tried to get back through the sky by beating on top of it, but it had loosened and plummeted to the Earth, along with the ocean. Before anyone could discuss this story, it began to rain – rain unlike anything anyone had seen before. It rained all day and all night, and the rivers reached their banks and then crept out across the floodplains. Still the rain came down, and the people and all the animals fled into the hills. When the water rose into the hills, the people fled to the highest peaks. Water covered the whole land from horizon-to-horizon unlike anything anyone had ever seen before. It took weeks for the water to go down, everyone got very hungry, and many people died. Nothing was the same after the night that the sky fell. Now, whenever the sea grows rough and the wind blows, people know that the ocean is angry and impatient because the ancestor still refuses to let it go back whence it came. When the storm waves break on the beach, people know that it is just the great ancestor beating the ocean down again.

2 The Kwenaitchechat Legend, Pacific Northwest
(Heaton and Snavely, 1985; Satake et al., 1996; Geist, 1997a).

It was a cold winter's night along the Cape Flattery Coast of the Pacific Northwest. At Neeah Bay, the Kwenaitchechat people had eaten and settled into sleep. Then the ground began to shake violently. The land rolled from west to east and jerked upwards, leaving the beach exposed higher above the high-tide line than anyone had seen it before. Everyone ran out into the moonless night and down to the beach, where there was less chance in the dark of being flung into trees or the sides of huts by the shaking. As they fled onto the beach, the adults began to sink into the sand as if it were water. The old people were the last to get to the beach, and when they did, they were yelling frantically for everyone to run to higher ground. The young men laughed at them, saying that it was safer in the open. Suddenly the water in the bay began to recede, far beyond the limit of the lowest tide, further than anyone had seen it go. Everyone paused and stared at the ocean as if for eternity. Then the water began to come back. There was no sound except for the loud rushing of water swallowing everything in the bay. As one, all the tribespeople turned and began to run back to the village – to the canoes. Few got back. Those that did flung themselves, children, and anything else they could grab in the dark into the canoes. Then they were all picked up and swept north into the Straits of Juan de Fuca and into the forests. The water covered everything on the cape with only the hills sticking out. When the waters finally receded, many had drowned. Some canoes got stuck in the trees of the forest and were destroyed. Some people without any means of paddling the canoes got swept onto Vancouver Island beyond Nootka. In the light of day, all trace of the village in Neeah Bay was gone. So were all the neighbouring villages. No sign of life remained except for the few survivors scattered along the coast and the animals that had managed to flee into the hills.

On the other side of the Pacific Ocean, in Japan, ten hours later, the residents of villages along the coast at Miyako, Otsucki, and Tanabe had finished their work for the day and had gone to sleep. It was cloudy but calm along the coast. Then at around nine in the evening, without any preceding earthquake, the long waves started coming in – 3 m high at Miyako, 2 m at Tanabe. All along the coast, the sea suddenly surged over the shore without warning into the low-lying commercial areas of the towns and into the rice paddies scattered along the coastal plains. The merchants, fishermen, and farmers had seen such things before – the small waves that came in like tsunami but without any earthquake. They were lucky, because if there had been an earthquake, many people would have died. Instead, only a few lost their possessions. The events of that night were just a nuisance thing, of no great consequence.

3 Krakatau, 27 August 1883
(Myles, 1985; Bryant, 1991)

Van Guest was sweating profusely as he climbed through the dense jungle above the town of Anjer Lor. He stopped to gasp for breath, not because he was slightly out of shape, but because the sulphurous fumes burned his lungs. He looked down at the partially ruined town. There was no sign of life although it was nearly 10 o'clock. His head pounded as the excitement of the scene and the strain of the trek sped blood through his temples. He did not know if he felt the thumping of blood in his head or the distant rumbling. Sometimes both were synchronous, and it made him smile. This was the chance of a lifetime. No one was paid to do what he did or had remotely thought to climb to the top of one of the hills to get the best view. Besides, most of the townspeople had fled into the jungle after the waves had come through yesterday and again in the early morning. As he neared the top of the hill he looked for a spot with a clearing to the west, reached it, and turned. Beyond lay purgatory on Earth, the incredible hell of Krakatau in full eruption.

As a volcanologist for the Dutch colonial government, Van Guest was aware of the many eruptions that continually threatened Dutch interests in the East Indies. Tambora in 1815 was the worst. No one thought that anything else could be bigger. He had seen Galunggung go up the previous year with over a hundred villages wiped out. Krakatau had had an earthquake then, and when it began to erupt in May, the governor in Batavia had ordered him to investigate. He had come to this side of the Sunda Strait because he thought he would be safe 40 km from the eruption. Van Guest tied his handkerchief over his nose and mouth, slipped on the goggles to keep the sting from his eyes, and peered through his telescope across the strait, hoping to catch a glimpse of the volcano itself through the ash and smoke. Suddenly the view cleared as if a strong wind had blown the sky clean. He could see the ocean frothing and churning chaotically. Only the Rakata peak remained, and it was glowing red. The smallest peak, Perboewatan, had blown up at 5:30 that morning. Danan, which was 450 m high, had gone just over an hour later. Each had sent out a tsunami striking the coastline of Java and Sumatra in the dark. That is what had cleared out the town in the early hours.

As he glanced down at the abandoned boats in the bay, Van Guest noticed that they were all lining up towards the volcano. Then they drifted quickly out to sea and disappeared in the maelstrom. Suddenly, a bolt of yellow opened in the ocean running across the strait to the northwest and all the waters in the strait flooded in. Instantly, a cloud of steam rose to the top of the sky. As Van Guest stood upright, awestruck, a blast of air flattened him to the ground and an incredible noise deafened him. The largest explosion ever heard by humans had just swept over him. The shock wave would circle the globe seven times. When he gained his feet, Van Guest thought he was blind. The whole sky was as black as night. He stumbled down the slope back towards the town. It took him nearly 30 minutes to get down to the edge of the town through the murk. Just as he approached the outskirts of Anjer Lor, he could see the telegraph master, panic-stricken, racing up the hill towards him, silhouetted against the sea – or what Van Guest thought was the sea. It was hilly and moving fast towards him. The sea slowly reared up into an incredible wave over 15 m high and smashed through the remains of buildings next to the shoreline. Within seconds it had splintered through the rest of the houses in the town and was closing fast. The pace of the telegraph master slowed noticeably as he climbed the hill. The wave crashed through the coconut palms and jungle at the edge of the town. Tossing debris into the air, it sloshed up the hill. The telegraph master kept running or stumbling towards Van Guest, then collapsed into his arms with only metres to spare between him and the wave. It had finally stopped. Both men had just witnessed one of the biggest volcanic eruptions and tsunami ever recorded.

4 Burin Peninsula, Newfoundland, 18 November 1929
(Cox, 1994; Whelan, 1994; Dawson et al., 1996)

It was just after five in the afternoon on a cold autumn evening when the residents of outports along the Burin Peninsula of Newfoundland felt the tremors. Windowpanes rattled and plates fell out of sideboards. It was so unusual that one by one people poked their heads out of their pastel clapboard houses to see if anyone else had noticed – fishermen and their families at Taylors Bay, Point au Gaul, Lamaline, Lord's Cove, and thirty-five other communities nestled into the narrow coves along one of the most isolated coasts in North America. Isaac Hillier – who was just 18 at the time – went outside and saw an elderly French man gesturing excitedly to a group of his neighbours. When the man stooped and put his ear to the ground, Isaac's curiosity got the better of him and he went closer to hear what was going on. The old man began to wave his arms and shout that the water would come. Those gathered around him turned to each other and asked, "How would he know that?" One by one they went back to their evening chores before the storm set in. Isaac, although curious, did likewise.

The young children were put to bed upstairs in the wood houses shortly after their evening meals. At Lord's Cove, three-year-old Margaret Rennie was one such child. The excitement of the earthquake was beyond her comprehension, and she only wanted to get into bed to keep warm. Towards 7:30 P.M., seven-year-old Norah Hillier could hardly keep awake any more. Her father had come back with the news that the telegraph line to St. John's was broken. He had gone out again to see if he could do anything before it got colder. Norah heard a loud roar and glanced out the window to the sea only a few metres away. "Oh!" she cried out, "All the sheep!" All she could see were thousands of white sheep riding a mountain of water that was getting higher and higher, and louder. Within seconds, the foaming water was in the house. Her oldest sister bolted for the door and pushed against it. They were up to their waists in water, and the house began to move.

Lou Etchegary had never seen cars before, but with beams of moonlight breaking through the cloud and shining on its crest, the tsunami looked like a car with its headlights on – driving fast up the harbour. Within seconds a wall of water 3 m high was smashing crates off the wharf and lifting fishing dories and schooners 5 m high as if they were matchsticks. Anchors snapped, and all the boats either surged on the crest of the wave or raced belly-up to the pebble beach at the back of the cove. No one had a clue in the dark what was happening. At Taylors Bay, Robert Bonnell heard the wave coming and, grabbing his two children, raced for the hills. He tripped in the dark, fell down, and watched helplessly as the water dragged his children back into the maelstrom. Margaret Rennie slept as her house was swept into the pond out back. Rescuers raced to the house in the dark and smashed in the windows to get into the rooms. Margaret was found unconscious and still lying on her bed. Her mother, Sarah, and three brothers and sisters were found drowned downstairs in the kitchen. Norah Hillier's dad raced back to the house as soon as he saw water flooding his house. The only thing he could think of was to grab his soaking wet girls and drag them through the peat bog to the hills. He could already see a number of bonfires being lit by his neighbours who lived further inland.

Isaac Hillier stood in disbelief. How did that old man know that the water would come? Before he could think further, another wave flooded in. It picked up the remaining boats and pieces of houses, and thrashed them across the beach. Isaac could also see the barrels of flour, molasses, and salted fish, stored on the wharves for the coming winter, floating in the mess. Before it was over, two more waves smashed into the debris stacking it 2 m high in places. Not only was there no food or shelter, their lifeline to St. John's – the boats – was also gone. Stunned, Isaac froze in shock as shivers swept up and down his spine. Stumbling towards the bonfires, he became acutely aware of the shouts of rescuers and the crying, and then of the snow and the bitter cold.

5 Papua New Guinea, 17 July 1998

It was a perfect tropical evening along the Aitape Coast of Papua New Guinea. Here on the narrow sand barrier that ran for 3 km in front of Sissano lagoon, life was paradise – sago trees, coconut groves, white beaches, and the ever-present emerald waters of the Bismarck Sea. It was the dry season, and as the sun set, people in the villages were busying themselves preparing their evening meals. The men had had a good day fishing in the ocean; the women, good returns from their nets set in Sissano lagoon. Children and young people, many who had come home from Port Moresby for the school holidays, played along the beach. Ita glanced at her watch. It was ten to seven – still plenty of time before the sing-sing. She glanced at her two babies who were lying beside her and smiled. These holiday periods when all the children were home were the happiest of times.

She bent over to check her cooking, and that is when she first noticed the earthquake. The water in the pot began to shimmer. Then the ground began to roll. It came in from the north, from the sea. Everyone in the village froze in their tracks. The region often experienced earthquakes but they were always small. How big was this one going to be? Ten seconds, thirty seconds, a minute, two. Then the shaking stopped. Ita looked around. She lived at the back of the lagoon, only 75 m from the ocean. She saw some of the older people gathering around a cluster of buildings closer to the sea. They were talking frantically. She would never forget the look on their faces; it was one of sheer panic. Some of the younger men joined the group. One old man began pointing at the ocean. He was yelling, and Ita could just catch his words. He talked about "leaving the village", "the wave was coming", and "everyone must run". She thought how foolish. The village was on a barrier between the ocean and the lagoon. There was nowhere to run. One of the young men in the group put his arm around the old man's shoulders, smiled, and then began to laugh – not at him, not with him, but in that reassuring way that went with the nonchalant attitude of a people comfortable with a relaxed, carefree lifestyle.

Ita heard a rumbling like thunder, and as she glanced through the trees to the ocean, she noticed that the tide was going out, further than she could remember. By now, some of the children had run up from the beach. One of them said that they had seen the ocean splash tens of metres into the air on the horizon just after the ground shock. They were now asking their parents to come down to the beach. It was full of cracks. Within minutes, everyone was talking about the earthquake. It had not been a big one. The houses built on stilts were still standing. Some people had wandered down to the beach; but the older people were more distraught then ever. Then someone yelled, "Look", and pointed to the ocean. Ita strained to view the horizon in the twilight. A thousand lights from phosphorescence began to sparkle in the water, which had now retreated several hundred metres from shore. Then she noticed that the horizon was moving; it was getting higher and higher. Abruptly a second earthquake jolted her. This time it rolled in from the southeast. As she turned and looked east along the coast, she saw a large wave breaking – not really breaking, but frothing and sparkling. Everyone instantaneously began yelling "Run", but the roar of the wave cut off the shouts. Like a jet plane landing, it engulfed the night. Ita turned, grabbed her two babies beside her, raced the few steps to the canoe, and jumped in. Before the wave hit with a thud, a blast of air knocked her flat to the bottom. The canoe was tossed several metres into the air and then flung like a surfboard into the lagoon and across to the swamp on the other side. At Sissano, Warapu, Malol, Arop, and a half dozen villages all along the Aitape Coast, the scene was the same. A 10- to 15-m-high tsunami swamped the coast. At some places the wave raced along the shore; at others it just reared from the ocean and ran straight inland through buildings at 10–15 m s^{-1}. Everywhere, people were knocked into trees behind the beach or flushed into the lagoon.

It was now night. One could only hear the noise of the wave as it crossed Sissano lagoon and the screams of people as they gasped for air or tried to swim in the turbulent water. A putrid odour filled the air. In all, three waves swept one on top of the other across the coast. From the beginning of the first earthquake to the last wave, it was all over within half an hour. The villages were gone; debris was everywhere. As a surreal mist rose from the lagoon and crept into the silent swamps, the feeble cries of survivors, grunts of foraging pigs, and isolated barks from hungry dogs looking for an evening meal were the only sounds of songs to be heard along the Aitape Coast that night.

In all 2,202 people died, 1,000 were injured, and 10,000 were made homeless. Many of the dead died from their injuries as they clung to trees in the impenetrable mangrove forest on the other side of the lagoon, waiting days for rescue, on a remote coastline thousands of kilometres from nowhere. Ita? She survived. Her canoe was caught in the mangrove. After two days she was rescued with her two babies and reunited with an overjoyed husband.

SCIENTIFIC FACT OR LEGENDS?

All of these stories have elements of truth, yet only two are reliable – those of Krakatau and the Burin Peninsula. The description of the tsunami generated by the eruption of Krakatau in 1883 is based upon historical scientific records – mainly the diary of Van Guest, the colonial volcanologist. The Burin Peninsula story is linked to the Grand Banks earthquake and tsunami of 1929. This event is known more for the breaking of telegraph cables on the seabed between New York and Europe than for the deadly tsunami that struck the Southeast Coast of Newfoundland. Both stories also have elements of fabrication. While the individual experiences of Van Guest and the telegraph master are true, the descriptions of their feelings and eventual meeting at the end have been embellished to produce a more colourful story. The Krakatau eruption and tsunami are probably one of the best-documented tsunami events in the scientific literature. At least four articles about it have been written in *Nature* and two in *Science*. However, there is still scientific debate whether or not the largest tsunami that reached Anjer Lor and other locations in the Sunda Strait was the result of the eruptions in the early morning or the one at 9.58 A.M. If witnesses had not been wiped out by the earlier tsunami, they more than likely did not see the last one because they had fled inland to safety. Thick ash also obscured the last event turning day into night. Field surveys afterwards could not discern the run-up of individual waves, but only the highest run-up elevation of the biggest wave.

Readers may also be willing to accept the story from Papua New Guinea because it, like the Newfoundland one, is based upon interviews with eyewitnesses. Scientists have cobbled the story together from newspaper reports and interviews. Unfortunately, both stories are unreliable because interviewers, unless they apply structured qualitative methodology, can be prone to exaggeration. In essence, both the Papuan New Guinea and Newfoundland stories represent the early phases of an oral tradition or folklore about a tsunami event that is being passed on by word of mouth, or in the twentieth century supported by written documentation. When there are no witnesses to a notable event left alive and no written records, then all these stories become legends. Legends have an element of truth, but often the exact circumstances of the story cannot be verified. The tsunami story by the Kwe-naitchechat native people is a legend. However, when the most likely source of a documented tsunami in Japan on 26 January 1700 was evaluated – using computer modelling – as being a giant earthquake off the coast of Washington State, the legend suddenly took on scientific acceptability and received front-page coverage in *Nature*. None of the other events has ever achieved that status.

The Aboriginal story incorporates numerous published legends in the southeast part of Australia. One story actually uses the colloquial word *tidal wave* for tsunami. Scientific investigations along the Southeast Coast of Australia now indicate that the Aboriginal stories are not myths, but legends of one or more actual events. While no Aborigine at the time thought of writing up a description of any of these tsunami events and publishing it as a scientific paper in *Nature* or *Science*, the legends are just as believable as any newspaper article or scientific paper. They are just briefer and less specific. The sky may have fallen in the form of an asteroid or meteorite shower with large enough objects to generate tsunami tens of metres high. There is geomorphic evidence along the Southeast Coast of New South Wales, the Northeast Coast of Queensland, and the Northwest Coast of Western Australia for mega-tsunami. Certainly the coastal features are so different in size from what historical tsunami have produced anywhere in the world in the past 200 years that a comet or meteorite impact with the ocean must be invoked. The subsequent floods also have veracity. Meteorite impacts with the ocean put enormous quantities of water into the atmosphere either as splash or vapourised water from the heat of the impact. That heated vapour condenses and falls as rain because it is not in equilibrium with the preexisting temperature of the atmosphere. Research is beginning to indicate that rivers and waterfalls across Australia have flooded beyond maximum probable rainfalls – the theoretical highest rainfall that can occur under existing rain-forming processes. Meteorite impacts with the ocean may explain not only some of the evidence for mega-tsunami, but also this mega-flooding.

The single thread running through all five stories is tsunami. The stories have been deliberately selected to represent the different causes of tsunami. The Aboriginal legend refers to the impact of a meteorite and the associated airburst. The historically accurate Krakatoan story recounts a volcano-induced tsunami, while the Kwenaitchechat legend undoubtedly refers to a tsunami, generated by an earthquake of magnitude 9.0 along the Cascadia subduction zone of the Western United States in January 1700. The Newfoundland tale refers to the Grand Banks earthquake and submarine landslide of 1929 – the only well-documented tsunami to affect the East Coast of North America. Finally, and more worrisome, the origin of the Papua New Guinea event is still being debated. The event is worrisome because the wave was too big for the size of the earthquake involved. The combination of downfaulting close to shore and slumping of offshore sediments on a steep, offshore slope may have caused the exceptionally large tsunami. Many countries have coastliness like this. The event is also disconcerting because our present scientific perception and warning system – especially in the Pacific Ocean – is geared to earthquake-induced waves from distant shores. Certainly very few countries, except Chile and Japan, have developed a warning system for nearshore tsunami. The stories deliberately cover this range of sources to highlight the fact that, while earthquakes are commonly thought of as the cause of tsunami, tsunami can have many sources. Our present knowledge is biased. Tsunami are very much an underrated, widespread hazard. Any coast is at risk.

CAUSES OF TSUNAMI
(Wiegel, 1964; Iida, 1963; Bryant, 1991)

Most tsunami originate from submarine seismic disturbances. The displacement of the Earth's crust by several metres during underwater earthquakes may cover tens of thousands of square kilometres and impart tremendous potential energy to the overlying water. Tsunami are rare events, in that most submarine earthquakes do not generate one. Between 1861 and 1948, only 124 tsunami were recorded from 15,000 earthquakes. Along the West Coast of South America, 1,098 offshore earthquakes have generated only 20 tsunami. This low frequency of occurrence may simply reflect the fact that most tsunami are small in amplitude – and go unnoticed – or the fact that most earthquake-induced tsunami require a shallow focus seismic event with a surface magnitude, M_s, greater than 6.5 on the Richter scale. Earthquakes as a cause of tsunami will be discussed in fuller detail in Chapter 5.

Submarine earthquakes have the potential to generate landslides along the steep continental slope that flanks most coastlines. In addition, steep slopes exist on the sides of ocean trenches and around the thousands of ocean volcanoes, seamounts, atolls, and guyots on the seabed. Because such events are difficult to detect, submarine landslides are considered a minor cause of tsunami. The 17 July 1998 Papua New Guinea event renewed interest in this potential mechanism. A large submarine landslide or even the coalescence of many smaller slides has the potential to displace a large volume of water. Geologically submarine slides involving up to 20,000 km³ of material have been mapped. Tsunami arising from these events would be much larger than earthquake-induced waves. Only in the last thirty years has coastal evidence for these mega-tsunami been uncovered. Submarine landslides as a cause of tsunami and mega-tsunami will be discussed in Chapter 6.

Tsunami can also have a volcanic origin. Of ninety-two documentable cases of tsunami generated by volcanoes, 16.5% resulted from tectonic earthquakes associated with the eruption, 20% from pyroclastic (ash) flows or surges hitting the ocean, 14% from submarine eruptions, and 7% from the collapse of the volcano. A volcanic eruption rarely produces a large tsunami, mainly because the volcano must lie in the ocean. For example, the largest explosive eruption of the past millennium was Tambora in 1815. It produced a local tsunami 2–4 m high because it lay 15 km inland. In contrast, the 27 August 1883 eruption of Krakatau, situated in the Sunda Strait of Indonesia, produced a tsunami with nearby run-up heights exceeding 40 m above sea level. The wave was detected at the Cape of Good Hope in South Africa 6,000 km away. The atmospheric pressure pulse generated water oscillations that were measured in the English Channel thirty-seven hours later on the other side of the Pacific Ocean at Panama and in San Francisco Bay, and in Lake Taupo in the centre of the North Island of New Zealand. Probably the most devastating event was the Santorini Island eruption around 1470 B.C., which generated a tsunami that must have destroyed all coastal towns in the Eastern Mediterranean. The Santorini crater is five times larger in volume than that of Krakatau, and twice as deep. On adjacent islands, there is evidence of pumice stranded at elevations up to 50 m above sea level. The initial tsunami waves may have been 90 m in height as they spread out from Santorini. Volcanoes as a cause of tsunami will be discussed in Chapter 7.

There has been no historical occurrence of tsunami produced by a meteorite impact with the ocean. However, this does not mean that they are an inconsequential threat. Stony meteorites as small as 300 m in diameter can generate tsunami over 2 m in height that can devastate coastlines within a 1,000-km radius of the impact site. The probability of such an event occurring in the next fifty years is just under 1%. One of the largest impact-induced tsunami occurred at Chicxulub, Mexico, 65 million years ago at the Cretaceous–Tertiary boundary. While the impact was responsible for the extinction of the dinosaurs, the resulting tsunami swept hundreds of kilometres inland around the shore of the early Gulf of Mexico. Impact events are ongoing. Astronomers have compiled evidence that a large comet encroached upon the inner solar system and broke up within the last 14,000 years. The Earth has repetitively intersected debris and fragments from this comet. However, these encounters have been clustered in time. Earlier civilisations in the Middle East were possibly destroyed by one such impact around 2350 B.C. The last rendezvous occurred as recently as A.D. 1500; however, it happened in the Southern Hemisphere, where historical records did not exist at the time. Only in the last decade has evidence become available to show that the

Australian coastline preserves the signature of mega-tsunami from this latest impact event. One of the main themes of this book is the exposition of this evidence. The geomorphic signatures of tsunami will be presented in Chapters 3 and 4, while meteorites as a cause of mega-tsunami will be discussed in detail in Chapter 8.

Finally, meteorological events can generate tsunami. These tsunami are common at temperate latitudes where variations in atmospheric pressure over time are greatest. Such phenomena tend to occur in lakes and embayments where resonance of wave motion is possible. Resonance and the features of meteorological tsunami will be described in Chapter 2.

DISTRIBUTION AND FATALITIES

Accounts of tsunami extend back almost 4,000 years in China, 2,000 years in the Mediterranean – where the first tsunami was described in 479 B.C. – and about 1,300 years in Japan. However, many important tsunamigenic regions have much shorter documentation. For example the Chile–Peru coastline, which is an important source of Pacific-wide tsunami, has records going back only 400 years to 1562, while those from Alaska have only been documented since 1788. Tsunami records in Hawaii, which is a sentinel for events in the Pacific Ocean, exist only from 1813 onwards. Few records exist along the West Coast of Canada and the contiguous United States. The Southwest Pacific Ocean records are sporadic and almost anecdotal in reliability. Only in the last ten years have records been compiled from Australia and New Zealand, with historical documentation extending back no further than 150 years.

The regional distribution of major tsunami is tabulated in Table 1.1. Only the South Atlantic appears to be immune from tsunami. The North Atlantic coastline also is virtually devoid of tsunami. However, the Lisbon earthquake of 1 November 1755, which is possibly the largest earthquake known, generated a 15-m-high tsunami that destroyed the port at Lisbon. It also sent a wall of water across the Atlantic Ocean that raised tide levels 3–4 m above normal in Barbados and Antigua in the West Indies. Tsunami also ran up and down the West Coast of Europe, and along the Atlantic Coast of Morocco. The Spanish port of Cádiz and Madeira in the Azores were also hit by

TABLE 1.1 Percentage Distribution of Tsunami in the World's Oceans and Seas

Location	Percentage
Atlantic East Coast	1.6
Mediterranean	10.1
Bay of Benga	0.8
East Indie	20.3
Pacific Ocea	25.4
Japan–Russi	18.6
Pacific East Coas	8.9
Caribbean	13.8
Atlantic West Coast	0.4

Source: From Bryant, 1991.

waves 15 m high, while a 3- to 4-m-high wave sunk ships along the English Channel. The continental slope off Newfoundland, Canada, is seismically active and has produced tsunami that have swept onto that coastline. The Burin Peninsula tsunami described earlier reached Boston, where it registered a height of 0.4 m. By far the most susceptible ocean to tsunami is the Pacific Ocean region, accounting for 52.9% of all events. The following sections describe the more noted areas for tsunami.

Mediterranean Sea

(Kuran and Yalçiner, 1993; Tinti and Maramai, 1999)

The Mediterranean Sea has one of the longest records of tsunami. Over three hundred events have been recorded since 1300 B.C. Large tsunami originate in the Eastern Mediterranean, the Straits of Messina of southern Italy, or southwest of Portugal. About 7% of known earthquakes in this region have produced damaging or disastrous tsunami. Around Greece, 30% of all earthquakes produce a measurable seismic wave, and seventy major tsunami have been recorded. Around Italy, there have been sixty-seven reliably reported tsunami over the past 2,000 years. The majority of these have occurred in the last 500 years, as records have become more complete. Of these, forty-six were caused by earthquakes and twelve by volcanoes. By far the most destructive tsunami followed an earthquake on 28 December 1908 in the Messina Strait region. A small proportion of the 60,000 people killed during this event were drowned by the tsunami, which flooded numerous coastal villages and reached a maximum run-up exceeding 10 m in elevation.

Caribbean Sea

(National Geophysical Data Center and World Data Center A for Solid Earth Geophysics, 1984; Lander and Lockridge, 1989; Schubert, 1994)

The Caribbean – including the South Coast of the United States – is particularly prone to tsunami as the Caribbean plate slides eastward relative to the North American plate at a rate of 2 cm yr^{-1}, producing strong seismic activity in the Puerto Rico Trench. Unfortunately, the threat here has been overshadowed by the more frequent occurrence of tropical cyclones or hurricanes. The record of tsunami is one of the longest in North America, beginning on 16 April 1690 with an event near St. Thomas in the Virgin Islands. Here the sea retreated vertically 16–18 m. The devastating Port Royal, Jamaica, tsunami, which drowned 3,000 people, followed this event two years later in June 1692. An earthquake that sent much of the city sliding into the sea triggered the tsunami. Ships standing in the harbour were flung inland over two-storey buildings. An earthquake in the Anegada trough between St. Croix and St. Thomas produced another significant tsunami on 18 November 1867. The resulting tsunami reached 7–9 m at St. Croix, 4–6 m high at St. Thomas, 3 m at Antigua, and 1–6 m in Puerto Rico. Run-ups of 1.2–1.5 m were common elsewhere throughout the southern Caribbean. Other notable events have occurred in 1842, 1907, 1918, and 1946. Of these, two bear mention – the tsunami of 25 October 1918 and 4 August 1946. The former event had a maximum run-up height of 7 m at Frederiksted, St. Croix, and was recorded at Galveston, Texas. The latter event followed a magnitude 8.1 earthquake off the Northeast Coast of the Dominican Republic. Locally, the tsunami penetrated several kilometres inland and drowned about 1,800 people. It also was observed at Daytona Beach, Florida.

Pacific Ocean

(Cornell, 1976; Iida, 1983, 1985; National Geophysical Data Center and World Data Center A for Solid Earth Geophysics, 1984, 1989; Lockridge, 1988b; Gusiakov and Osipova, 1993; Howorth, 1999; Intergovernmental Oceanographic Commission, 1999)

Figure 1.2A plots the distribution of 1,274 observations of tsunami reported along the coastlines of the Pacific Ocean since 47 B.C. The size of the circles is proportional to the number of observations per degree square of latitude and longitude. The data are biased in that the same event can be recorded at more than one location. The map excludes 217 observations that cannot be precisely located. The distribution of all observations is tabulated by region in Table 1.2. Because some countries have better observation networks than others, smaller events are overemphasised. This is true of the West Coast of North America, which is overrepresented in the modern record, despite having records of tsunami for only the last 200 years. Some countries are underrepresented in Figure 1.2A. For example, over a hundred observations from Australia are not included. The coastline of Japan has the longest historical record of tsunami, with 22.1% of all events originating here. Two other regions also stand out as having a high preponderance of tsunami – the coast of South America with 18.6% of events and Indonesia with 12.3%. A few small areas are highly prone to tsunami. These areas include Northern California, Hawaii, Southwest Chile, and the Chile–Peru border region. Destructive tsunami have inundated the Chilean Coast at roughly thirty-year intervals in recorded history.

Tsunamigenic earthquakes with surface magnitudes greater than 8.2 on the Richter scale affect the entire Pacific Ocean once every twenty-five years. Figure 1.2B plots the source region of oceanwide events. For completeness, the map also includes some events, such as the eruption of Krakatau in 1883, not technically lying in the Pacific Ocean. Major events are also listed in Table 1.3. Significant events have increased in frequency in the twentieth century. Earthquakes in southern Chile, Alaska, and the Kamchatka Peninsula have the greatest chance of generating oceanwide tsunami in the Pacific. The West Coast of the United States provides a fourth source; however, the last Pacific-wide event originating here before European settlement occurred 300 years ago on 26 January 1700.

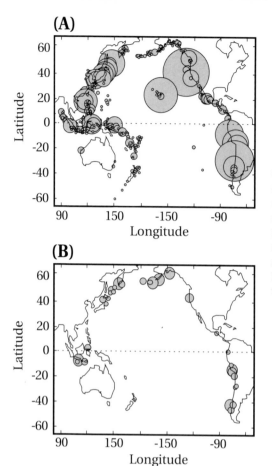

(A)

(B)

1.2 Location of tsunami in the Pacific Ocean region: **A.** Location of 1,274 tsunami since 47 B.C. Size of circle increases proportional to number of events per degree square of latitude and longitude. **B.** Source of significant distant (teleseismic) tsunami. Size of circle increases proportional to area affected and magnitude of the event (based upon Lockridge 1985, 1988b; and Intergovernmental Oceanographic Commission, 1999).

TABLE 1.2   Origin of Tsunami by Region around the Pacific Ocean

Region	Number of Tsunami Generated	Percentage
Japan	329	22.10
South America	278	18.60
Indonesia	184	12.30
New Guinea–Solomon Islands	129	8.70
Philippines	128	8.60
Kamchatka–Kuril Islands	95	6.40
New Zealand–Tonga	82	5.50
North America	78	5.20
Alaska–Aleutian Islands	77	5.20
Central America	72	4.80
Hawaii	39	2.60
TOTAL	1,491	100.00

Source: Based on Intergovernmental Oceanographic Commission, 1999.

TABLE 1.3 Occurrence of Significant Tsunami in the Pacific Region Documented since 47 B.C.

Date	Source	H_{rmax} (m)	Distant Areas Affected
26 January 1700	Washington State	unknown	Japan
8 July 1730	Concepción, Chile	unknown	Japan
7 November 1837	S. Chile	8	Hawaii
9 July 1854	N. Kuril Islands	unknown	Japan
13 August 1868	Arica, Chile	21	Peru, Japan, Hawaii, New Zealand, Australia, Fiji, United States
20 January 1878	Aleutian Islands	3	Hawaii
10 May 1877	Arica, Chile	24	Peru, Hawaii, California, New Zealand, Australia
27 August 1883	Krakatau, Sunda Straits	42	India, Australia
31 January 1906	Ecuador	5	California, Hawaii
17 August 1906	Chile	3	Hawaii, Japan, California
15 August 1918	Celeves Sea	12	Indonesia, Philippines
7 September 1918	S. Kuril Islands	12	Japan, Hawaii
11 November 1922	Atacama, Chile	12	Samoa, Hawaii, Columbia, Japan, New Zealand
3 February 1923	Kamchatka Peninsula	8	Hawaii
2 March 1933	Sanriku, Japan	29.3	Hawaii
1 August 1940	West Hokkaido	3.5	Russia, Korea
1 April 1946	Unimak Is. Alaska	35	Hawaii, California
4 November 1952	Kamchatka Peninsula	20	Hawaii, Sanriku Japan, Samoa, Peru, Chile
9 March 1957	Aleutian Islands	16.2	Hokkaido, California, Hawaii, El Salvador
6 November 1958	S. Kuril Islands	5	Japan
22 May 1960	S. Chile	25	South America, Central America, North America Hawaii, Japan, Marquesas Islands, Pitcairn Is., Samoa, Easter Is., Samoa, Kuril Islands, Johnston Atoll, Christmas Is., Taiwan, Fiji, New Zealand, Australia
20 November 1960	Peru	9	Japan
28 March 1964	Alaska	67.1	U.S. West Coast, Canada, Hawaii, Japan
4 February 1965	Aleutian	10.7	Japan, Hawaii
11 August 1968	SE Hokkaido	5	USSR
10 June 1975	Kuril Islands	5.5	Hokkaido Japan
19 August 1977	S. Sumbawa	15	Australia
3 March 1985	Central Chile	3.5	Hawaii, Alaska, Japan, Ecuador
2 June 1994	Java	13.9	Australia

Note: One event may be recorded at more than one location.
Source: Based on Lockridge, 1988b, and Intergovernmental Oceanographic Commission, 1999.

The 22 May 1960 Chilean event is the most significant historical tsunami. This event will be described in detail in Chapter 5. It is a benchmark for tsunami in the twentieth century. A series of tsunami waves spread across the Pacific over a period of twenty-four hours, taking over 2,500 lives. The tsunami significantly affected such diverse places as Hawaii, Pitcairn Island, New Guinea, New Zealand, Japan, Okinawa, and the Philippines.

The most tsunamigenic coastline in the world is that of the Kamchatka Peninsula, Russia. Between 1737 and 1990, the region experienced almost 8,000 earthquakes of which 96 generated localised tsunami. Volcanic eruptions here have also produced six tsunami, while four events have an unknown source. During the same period, the region was subject to fifteen tsunami from distant sources. A significant tsunami floods alluvium plains along the peninsula every 12.3 years. The most destructive tsunami, penetrating up to 10 km inland, occurred in 1737, 1841, 1923, 1937, 1952, and 1969. The largest event followed the Great Kamchatka Earthquake of 17 October 1737. Tsunami run-up heights reached 60 m above sea level in the North Kurile Islands. The second largest event occurred on 4 November 1952, with run-up heights of 20 m in the same area. This latter tsunami was also a Pacific-wide event.

Local tsunami are also common on the islands of the South Pacific; however, no significantly sized, earthquake-induced tsunami has propagated outside this region. Because many islands drop off into deep water, locally generated tsunami can travel at their maximum velocity, up to 1,000 km hr^{-1}, and reach the adjacent shore in 5–10 minutes. The Papua New Guinea–Solomon Islands region has experienced seventy-eight tsunami in the period between 1768 and 1983. Volcanism has caused one eighth of all tsunami. The largest event occurred on 13 March 1888, when the Ritter Island volcano off the North Coast of Papua New Guinea collapsed, generating a 15-m-high tsunami. The most recent event of significance occurred in the early evening of 17 July 1998 along the Sissano–Aitape Coast of Northern Papua New Guinea. One of the stories earlier referred to this event. It will be described in more detail in Chapter 5. Eleven tsunami have struck Fiji in the one-hundred-year period 1877–1977, averaging one tsunami every ten years. Tsunami are a more frequent hazard here than tropical cyclones. Many small islands in the South Pacific are vulnerable to tsunami because populations are concentrated around coastlines and perceive the hazard as being rare.

Over the past 2,000 years there have been 462,597 deaths attributed to tsunami in the Pacific region. Of these deaths, 95.4% occurred in events that killed more than a thousand people each. The number of deaths are tabulated in Table 1.4 for each of the main causes of tsunami, while the events with the largest deathtolls are presented in Table 1.5. Tectonically generated tsunami account for the greatest death toll, 84.5 %, with volcanic eruptions accounting for 11.2% – mainly during two events, the Krakatau eruption of 26–27 August 1883 (36,417 deaths) and the Unzen, Japan, eruption of 21 May 1792 (14,524 deaths). The number of fatalities

TABLE 1.4 Causes of Tsumani in the Pacific Ocean Region over the Last 2,000 Years

Cause	Number of Events	Percentage of Events	Number of Deaths	Percentage of Deaths
Landslides	65	4.6	14,661	3.2
Earthquakes	1,171	82.3	390,929	84.5
Volcanic	65	4.6	51,643	11.2
Unknown	121	8.5	5,364	1.2
TOTAL	1,422	100	462,597	100

Source: National Geophysical Data Center and World Data Center A for Solid Earth Geophysics, 1998, and Intergovernmental Oceanographic Commission, 1999.

TABLE 1.5 Largest Death Tolls from Tsunami in the Pacific Ocean Region over the Last 2,000 Years

Date	Fatalities	Location
22 May 1782	50,000	Taiwan
27 August 1883	36,417	Krakatau, Indonesia
28 October 1707	30,000	Nankaido, Japan
15 June 1896	27,122	Sanriku, Japan
20 September 1498	26,000	Nankaido, Japan
13 August 1868	25,674	Arica, Chile
27 May 1293	23,024	Sagami Bay, Japan
4 February 1976	22,778	Guatemala
29 October 1746	18,000	Lima, Peru
21 January 1917	15,000	Bali, Indonesia
21 May 1792	14,524	Unzen, Ariake Sea, Japan
24 April 1771	13,486	Ryukyu Archipelago
22 November 1815	10,253	Bali, Indonesia
May 1765	10,000	Guanzhou, South China Sea
16 August 1976	8,000	Moro Gulf, Philippines

Source: National Geophysical Data Center and World Data Center A for Solid Earth Geophysics, 1998, and Intergovernmental Oceanographic Commission, 1999.

has decreased over time and is slightly concentrated in Southeastern Asia, including Japan. The biggest tsunami of the twentieth century occurred in Moro Gulf, Philippines, on 16 August 1976, where 8,000 people died. The largest total death toll is concentrated in the Japanese Islands where 211,300 fatalities have occurred. Two events affected the Nankaido region of Japan on the 28 October 1707 and 20 September 1498, killing 30,000 and 26,000 people respectively. The Sanriku Coast of Japan has the misfortune of being the heaviest populated tsunami-prone coast in the world. About once per century, killer tsunami have swept this coastline, with two events striking within a forty-year time span between 1896–1933. On 15 June 1896 a small earthquake on the ocean floor, 120 km southeast of the city of Kamaishi, sent a 30-m wall of water crashing into the coastline, killing 27,122 people. The same tsunami event was measured 10.5 hours later in San Francisco on the other side of the Pacific Ocean. In 1933, disaster struck again when a similarly positioned earthquake sent ashore a wave that killed 3,000 inhabitants. Deadly tsunami have also affect Indonesia and the South China Sea. In the South China Sea, recorded tsunami have killed 77,105 people, mainly in two events in 1762 and 1782. Indonesia has experienced a comparable death toll (69,420 deaths) over this period – the largest following the eruption of Krakatau in 1883.

New Zealand and Australia
(de Lange and Healy, 1986; Bryant and Nott, 2000)

Australia and New Zealand are not well represented in any global tsunami database. This is surprising for New Zealand, because it is subject to considerable local tectonic activity and lies exposed to Pacific-wide events. At least thirty-two tsunami have been recorded in this latter country since 1840. The largest event occurred on 23 January 1855 following the Wellington earthquake. The run-up was 9–10 m high within Cook Strait and 3 m high at New Plymouth, 300 km away along the open West Coast. However, the highest recorded tsunami occurred following the Napier earthquake of 2 February 1931. The earthquake generated little in the way of a tsunami; however, it triggered a rotational slump in the Waikare estuary that swept water 15.2 m above sea level. The most extensive tsunami followed the 13 August 1868 Arica earthquake in Chile. Run-up heights of 1.2–1.8 m were typical along the complete East Coast of the islands. At several locations, water levels dropped 4.5 m before rising an equivalent amount. Subsequent earthquakes in Chile in 1877 and 1960 also produced widespread effects. New Zealand has the distinction of recording two tsunami generated by submarine mud volcanism associated with diapiric intrusions. These occurred near Poverty Bay on the East Coast of the North Island. The largest wave had a run-up of 10 m elevation.

In Australia, forty-three tsunami events have been recorded, beginning with the 1868 Arica, Chile, event. The closest sources for earthquake-generated tsunami lie along the Tonga–New Hebrides trench, the Alpine Fault on the West Coast of the South Island of New Zealand, and the Sunda Arc south of the Indonesian islands. The Alpine Fault is an unproven source because it last fractured around 1455, before European settlement. It has the potential to produce an earthquake with a surface magnitude, M_s, of 8.0 on the Richter scale, with any resulting tsunami reaching Sydney within two hours. The largest tsunami to be recorded on the Sydney tide gauge was 1.07 m following the Arica, Chile, earthquake of 10 May 1877.

However, the Chilean tsunami of 22 May 1960 produced a run-up of 4.5 m above sea level. In Sydney and Newcastle harbours, this tsunami tore boats from their moorings and took several days to dissipate. The Northwest Coast is more vulnerable to tsunami because of the prevalence of large earthquakes along the Sunda Arc, south of Indonesia. The largest run-up measured in Australia is 6 m, recorded at Cape Leveque, Western Australia, on 19 August 1977 following an Indonesian earthquake. Waves of 1.5 m and 2.5 m height were measured on tide gauges at Port Hedland and Dampier respectively. Another tsunami on 3 June 1994 produced a run-up of 4 m at the same location. The Krakatoa eruption of 1883 generated a tsunami run-up in Geraldton, 1,500 km away, that obtained a height of 2.5 m. This tsunami moved boulders 2 m in diameter 100 m inland and more than 4 m above sea level opposite gaps in the Ningaloo Reef protecting the Northwest Cape. South of the Northwest Cape, tsunami heights decrease rapidly because the coastline bends away to the east. At present, Indonesia is the only known source for tsunami in the Indian Ocean.

Bays, Fjords, Inland Seas, and Lakes
(Bryant, 1991; Kuran and Yalçiner, 1993; Camfield, 1994; Ranguelov and Gospodinov, 1995; Altinok et al., 1999)

Tsunami are not restricted to the open ocean. They can occur in bays, fjords, inland seas, and lakes. The greatest tsunami run-up yet identified occurred at Lituya Bay, Alaska, on 9 July 1958. The steep slope on one side of the bay failed following an earthquake, sending 0.3 km^3 of material cascading into a narrow arm of the bay. A wall of water swept 524 m above sea level on the opposite shore, and a 30- to 50-m-high tsunami propagated down the bay, killing two people. Steep-sided fjords in both Alaska and Norway are also subject to similar slides. In Norway, seven tsunami-genic events have killed 210 people. The heights of these tsunami ranged between 5 and 15 m, with run-ups surging up to 70 m above sea level.

Inland seas are also prone to tsunami. There have been twenty observations of tsunami in the Black Sea in historical records. In Bulgaria, maximum probable run-up heights of 10 m are possible. One of the earliest occurred in the first century B.C. at Karvarna. In A.D. 853, a tsunami at Varna swept 6.5 km inland over flat coastal plain and travelled 30 km up a river. Last century, on 31 March 1901, a 3-m-high tsunami swept into the port of Balchik. Bulgarian tsunami originate from earthquakes on the Crimean Peninsula or from the eastern shore in Turkey. Submarine landslides are also likely sources because the Black Sea is over 2,000 m deep with steep slopes along its eastern and southern sides. The Anatolian Fault Zone that runs through Northern Turkey and Greece has produced many tsunami in the Black Sea and the Sea of Marmara to the west. At least ninety tsunami have been recorded around the coast of Turkey since 1300 B.C. A tsunami flooded Istanbul on 14 September 1509, overtopping seawalls up to 6 m high. At least twelve major tsunami have occurred historically in the Sea of Marmara, mainly in Izmit Bay. The most recent occurred on 17 August 1999. This tsunami appears to have been caused by submarine subsidence during an earthquake. Maximum run-up was 2.5 m along the northern coast of the bay and 1.0–2.0 m along the southern shore. Ten tsunami generated by earthquakes or landslides have been recorded in the Caspian Sea. Seven of these occurred on the West Coast and three on the East Coast. However, the risk is small, as run-ups for the 1:100 event do not exceed 1 m.

Finally, tsunami can be generated even in small lakes. The Krakatoa eruption of 27 August 1883 sent out a substantial atmospheric shock wave that induced a 0.5-m-high, 20-minute oscillation in Lake Taupo situated in the middle of the North Island of New Zealand. Burdur Lake in Turkey has had numerous reports of tsunami although the lake is only 15 km long. On 1 January 1837 an earthquake-generated tsunami swept its shores and killed many people. Tsunami have washed up to 300 m inland around this lake.

This chapter has presented stories on the human impact of tsunami and alluded to the numerous mechanisms that can generate this hazard. In doing so, terms such as *wave height, run-up,* and *resonance* have been used without any explanation about what they mean. While such terms can be described simply, they are usually treated mathematically in the tsunami literature. The next chapter deals with the mathematical description of tsunami waves and their dynamics. While photographs and drawings illustrate the concepts, most of the material presented in this book can be comprehended without any detailed understanding of these formulae.

CHAPTER TWO

Tsunami Dynamics

2.1 An artist's impression of the tsunami of 1 April 1946 approaching the five-storey-high Scotch Cap light house, Unimak Island, Alaska. The lighthouse, which was 28 m high, stood on top of a bluff 10 m above sea. It was completely destroyed (see Figure 2.12). The wave ran over a cliff 32 m high behind the lighthouse. Painting is by Danell Millsap, commissioned by the United States National Weather Service.

INTRODUCTION

The approach of a tsunami wave towards shore can be an awesome sight to those who have witnessed it and survived. Figure 2.1 represents an artist's impression of a tsunami wave approaching the coast of Unimak Island, Alaska, early on 1 April 1946. Similar artists' impressions of breaking tsunami will be presented throughout this text. The impressions are accurate. Whereas ordinary storm waves or swells break and dissipate most of their energy in a surf zone, tsunami break at shore. Hence, they lose little energy as they approach a coast and can run up to heights an order of magnitude greater than storm waves. Much of this behaviour relates to the fact that tsunami are very long waves – kilometres in length. As shown in Figure 2.1, this behaviour also relates to the unusual shape of tsunami wave crests as they approach shore. This chapter describes these unique features of tsunami.

TSUNAMI CHARACTERISTICS

(Wiegel, 1964; Bolt et al., 1975; Shepard, 1977; Iida and Iwasaki, 1983;
Myles, 1985; von Baeyer, 1999; Ward, 2000)

The terminology used in this text for tsunami waves is shown schematically in Figure 2.2. Much of this terminology is the same as that used for ordinary wind waves. Tsunami have a wavelength, a period, and a deep-water height. They can undergo shoaling, refraction, and diffraction. Most tsunami generated by large earthquakes travel in wave trains containing several large waves that in deep water are less than 0.4 m in height. Figure 2.3 plots typical tidal gauge records or *marigrams* of tsunami at various locations in the Pacific Ocean. These records are taken close to shore and show that tsunami wave heights increase substantially into shallow water. Tsunami wave characteristics are highly variable. In some cases, the waves in a tsunami wave train consist of an initial peak that then tapers off in height

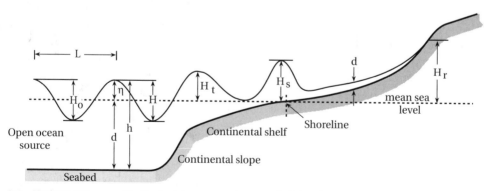

2.2   Various terms used in the text to express the wave height of a tsunami.

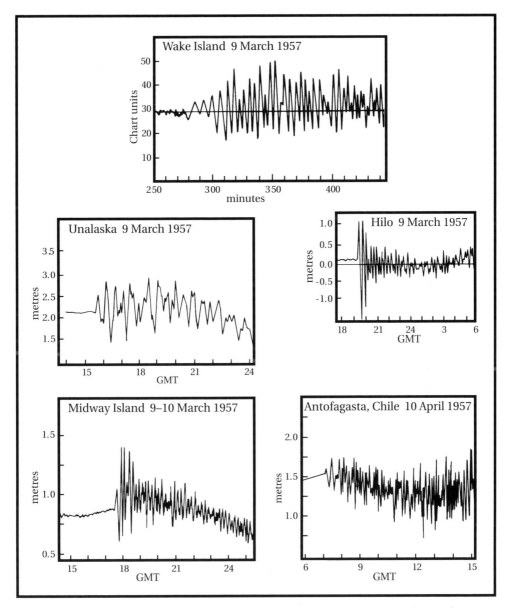

2.3 Plots or marigrams of tsunami wave trains at various tidal gauges in the Pacific region (based on Wiegel, 1970).

exponentially over four to six hours. In other cases, the tsunami wave train consists of a maximum wave peak well back in the wave sequence. The time it takes for a pair of wave crests to pass by a point is termed the wave period. This is a crucial parameter in defining the nature of any wave. Tsunami typically have periods of 100–2,000 seconds (1.6–33 minutes), referred to as the tsunami window. Waves with this period travel at speeds of 600–900 km hr^{-1} (160–250 m s^{-1}) in the deepest part of the ocean, 100–300 km hr^{-1} (30–85 m s^{-1}) across the continental shelf, and 36 km hr^{-1} (10 m s^{-1}) at shore. The upper limit is the speed of a commercial jet aeroplane. Because of the finite depth of the ocean and the mechanics of wave generation by earthquakes, a tsunami's wavelength – the distance between successive wave crests – lies between 10–500 km. These long wavelengths make tsunami profoundly different from swell or storm waves.

The simplest form of ocean waves is sinusoidal in shape and oscillatory (Figure 2.4). Water particles under oscillatory waves transcribe closed orbits. Hence there is no mass transport of water shoreward with the passage of the wave. Oscillatory waves are described for convenience by three parameters: their height or elevation above the free water surface, their wavelength, and water depth (Figure 2.3). These parameters can be related to each other by three ratios as follows:

$$H:L, H:d, L:d \qquad\qquad\qquad\qquad\qquad 2.1$$

where H = crest-to-trough wave height (m)
       L = wavelength (m)
       d = water depth (m)

In deep water, the most significant factor is the ratio $H:L$, or wave steepness. In shallow water it is the ratio $H:d$, or relative height. Sinusoidal waves fit within a class of waves called cnoidal waves: c for cosine, n for an integer to label the sequence of waves, and *oidal* to show that they are sinusoidal in shape. The shape of a wave or its peakiness can be characterised by a numerical parameter. For sinusoidal waves this parameter is zero. While tsunami in the open ocean are approximately sinusoidal in shape, they become more peaked as they cross the continental shelf. In this case, the numerical parameter describing shape increases and nonlinear terms become important. The wave peak sharpens while the trough flattens. These nonlinear, tepee-shaped waves are characterised mathematically by Stokes wave theory. In Stokes theory, motion in two dimensions is described by the sum of two sinusoidal components (Figure 2.4). Water particles in a Stokes wave do not follow closed orbits, and there is mass movement of water throughout the water column as the wave passes by a point. As a tsunami wave approaches shore, the separation between the wave crests becomes so large that the trough disappears and only one peak remains. The numerical parameter characterising shape approaches one and the tsunami wave becomes a solitary wave (Figure 2.4). Solitary waves are translatory in that water moves with the crest. All of the wave form also lies above mean sea level. Finally, it has been noted that a trough that is as high as the wave crest precedes many exceptional tsunami waves. This gives the incoming wave a wall effect. The Great Wave of Kanagawa in Figure 1.1 is of this type. These waves are not solitary because they have a component below mean sea level. Such waveforms are better characterised as *N*-waves. This

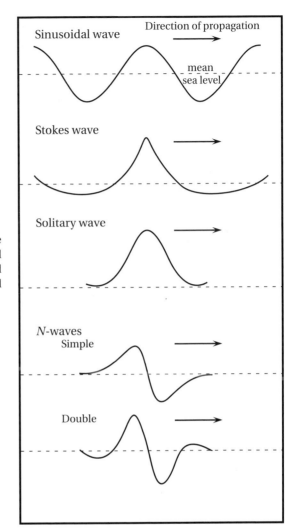

2.4 Idealised forms characterising the cross section of a tsunami wave (based on Tadepalli and Synolakis, 1994, and Geist, 1997b). Note that the vertical dimension is greatly exaggerated.

chapter uses features of each of these wave types: sinusoidal, Stokes, solitary, and *N*-waves to characterise tsunami.

Tsunami Wave Theory

(Wiegel, 1964, 1970; Kastens and Cita, 1981; Okal, 1988; Pickering et al., 1991; Satake, 1995; Pelinovsky, 1996; Trenhaile, 1997; Komar, 1998)

The form of a sinusoidal wave is analogous to a sine curve (Figure 2.4), and its features can be characterised mathematically by linear, trigonometric functions known as Airy wave theory. This theory can represent local tsunami propagation in water depths greater than 50 m, although in many cases linearity is only violated near the shore, where wave breaking is suppose to occur. In this theory, the three ratios presented in Equation 2.1 are much less than one. This implies that wave height relative to wavelength is very low – a feature characterising tsunami in the open ocean. The formulae describing sinusoidal waves vary depending upon the wave being in deep or shallow water. Shallow water begins when the depth of water is less than half the wavelength. As oceans are never more than 5 km deep, the majority of tsunami travel as shallow-water waves. In this case, the trigonometric functions characterising sinusoidal waves disappear and the velocity of the wave becomes a simple function of depth as follows:

$$C = (gd)^{0.5} \qquad\qquad 2.2$$

where C = wave speed (m s^{-1})
g = gravitational acceleration (9.81 m s^{-1})

While Equation 2.2 indicates that the full range of tsunami periods in a tsunami wave train travels as shallow-water waves, not all of the individual waves travel at the same speed. Longer-period waves outrun the very short ones, so that a tsunami wave train after travelling across an ocean tends to reach shore with regular long-period waves followed by shorter ones. This phenomenon is known as dispersion. The wavelength of a tsunami is also a simple function of wave speed, C, and period, T, as follows:

$$L = CT \qquad\qquad 2.3$$

Equation 2.3 holds for linear, sinusoidal waves and is not appropriate for calculating the wavelength of a tsunami as it moves into shallow water. Linear theory can be used as a first approximation to calculate changes in tsunami wave height as the wave moves across a shelf and undergoes wave shoaling and refraction. The following formulae apply:

$$H = K_r K_s H_o \qquad\qquad 2.4$$

$$K_r = (b_o b_i^{-1})^{0.5} \qquad\qquad 2.5$$

$$K_s = (d_o d_i^{-1})^{0.25} \qquad\qquad 2.6$$

where K_r = refraction coefficient (dimensionless)
K_s = shoaling coefficient (dimensionless)
b_i = distance between wave orthogonals at any shoreward point (m)
b_o = distance between wave orthogonals at a source point or in deep water (m)
d_i = water depth at any shoreward point or in deep water (m)
d_o = water depth at a source point (m)

Note that there are a plethora of definitions of wave height in the tsunami literature. These include wave height at the source region, wave height above mean water level, wave height at shore, and wave run-up height above present sea level. The distinctions between these expressions are presented in Figure 2.2. The expression for shoaling – Equation 2.6 – is known as Green's Law. Because tsunami are shallow-water waves, they *feel* the ocean bottom at any depth and their crests undergo refraction or bending around higher seabed topography. The degree of refraction can be measured by constructing a set of equally spaced lines perpendicular to the wave crest. These lines are called wave orthogonals or rays (Figure 2.5). As the wave crest bends around topography, the distance, b, between any two lines will change. Refraction is measured by the ratio $b_o : b_i$. Simple geometry indicates that the ratio $b_o : b_i$ is equivalent to the ratio $\cos\alpha_o : \cos\alpha_i$, where α is the angle that the tsunami wave crest makes to the bottom contours as the wave travels shoreward (Figure 2.5). Once this angle is known, it is possible to determine the angle at any other location using Snell's Law as follows:

$$\sin \alpha_o \, C_o^{-1} = \sin \alpha_i \, C_i^{-1} \qquad\qquad 2.7$$

where α_o = the angle a wave crest makes to the bottom contours at a
                   source point (degrees)

          α_i = the angle a wave crest makes to the bottom contours at any
                   shoreward point (degrees)

          C_i = wave speed at any shoreward point (m s^{-1})

          C_o = wave speed at a source point or in deep water (m s^{-1})

This relationship is straightforward close to shore, but for a tsunami wave crest travelling from a distant source – such as occurs often in the Pacific Ocean – the wave path or ray must also be corrected for geometrical spreading on a spherical surface. Equation 2.4 can be rewritten to incorporate this spreading as follows:

$$H = K_r K_s K_{sp} H_o \qquad\qquad 2.8$$

where K_{sp} = $(\sin \Delta)^{-0.5}$

       K_{sp} = coefficient of geometrical spreading on a sphere (dimensionless)

       Δ = angle of spreading on a sphere relative to the direction of wave
               travel

In a large ocean, bathymetric obstacles such as island chains, rises, and seamounts can refract a tsunami wave such that its energy is concentrated or *focussed* upon a distant shoreline. These are known as teleseismic tsunami because the effect of the tsunami is translated long distances across an ocean. Japan is particularly prone to tsunami originating from the West Coast of the Americas, despite this coastline's lying half a hemisphere away. On the other hand, bottom topography can spread tsunami wave crests, dispersing wave energy over a larger area. This process is called defocussing. Tahiti, but not necessarily other parts of French

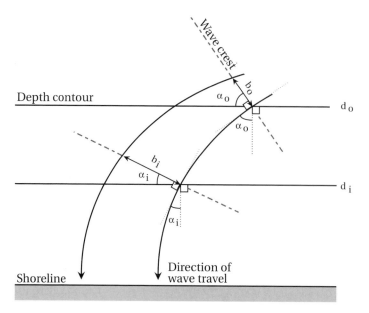

2.5   Refraction of a tsunami wave crest as it approaches shore.

Polynesia, is protected from large tsunami generated around the Pacific Rim because of this process.

Figure 2.6 shows how efficient the refraction of a tsunami wave can be once it reaches the continental shelf. In the figure the degree of refraction is plotted for various initial angles to bottom contours, from the continental shelf break to shore. As the wave crest travels only short distances, geometrical spreading has been ignored. No manner what the original angle of approach of a tsunami wave to the edge of a continental shelf, even for a shelf steeper than 0.5°, the crest of the wave will tend to refract or approach shore at an angle of less than 10°. Refraction of tsunami generated by linear faults close to shore tends to focus the tsunami wave energy onto a narrow stretch of coastline. Within the last 5 m depth of water, this angle will often be reduced to only 3–4°. Hence, tsunami waves tend to rush directly onto coasts rather than run alongshore. This is one of the reasons why the tsunami that struck the North Coast of Papua New Guinea on 17 July 1998 was so unusual. Many locals reported that the wave ran along the coast. For this to happen the source of this tsunami had to be close to shore.

Close to shore and around islands, tsunami follow classic diffraction theory. Diffraction is a process whereby energy leaks laterally along a wave crest. The best example of this effect is the spreading out of a wave into a harbour after it passes through a narrow entrance. Diffraction, together with refraction and geometric spreading, decreases wave energy and reduces the amplitude of a tsunami. In addition, tsunami in shallow water – such as those on the continental shelf – lose energy through frictional dissipation with the seabed. The frictional coefficient

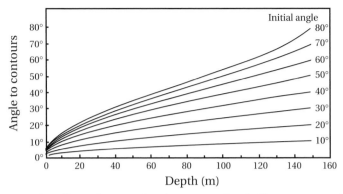

2.6   The effect of wave refraction across the shelf upon tsunami waves of different angles of approach to the continental shelf edge.

used to determine this dissipation is a function of the grain size on the seabed and the amplitude of water motions under the tsunami wave at the bottom. In hydraulics, two coefficients are often used: the De Chézy friction coefficient, C_f, and Manning's roughness coefficient, n. Manning's n in coastal waters typically has a value 0.03. The De Chézy coefficient can be related to either wave velocity or to Manning's n using the following formulae:

$$C_f = g^{0.5} C^{-1} \hspace{5cm} 2.9$$

$$C_f = g \, n^2 \, (d + H_t)^{-0.333} \hspace{4cm} 2.10$$

where C_f =   De Chézy frictional coefficient
$\quad\quad\quad\; n$ =   Manny's n
$\quad\quad\quad\; H_t$ =   tsunami wave height above mean sea leavel (m)

Because tsunami waves are long waves, frictional attenuation is independent of the wave period. The De Chézy coefficient under normal conditions in shallow seas has a value of 0.0024–0.0028. For solitary waves such as tsunami the value may be greater, 0.01. On dry land, the De Chézy coefficient may take on values between 0.04 and 0.09. If Manning's n has a value of 0.03, then the De Chézy coefficient is equal to 0.0023 in waters depths of 50 m and 0.01 just before reaching shore.

Table 2.1 tabulates the amount of frictional attenuation of typical tsunami waves as they cross the shelf. For comparison, this value has also been calculated for ocean swell with a wave period of 10 seconds. This latter wave period is typical of east-coast swell environments in the Pacific Ocean. Frictional dissipation of wave energy under both types of waves becomes a function of the shelf width. This distance can be represented by various slope angles. The highest slope angle, 0.6°, used in Table 2.1, is characteristic of steep continental shelves such as those found along the Southeast Coast of Australia. Here the shelf break lies 12–14

TABLE 2.1   Total Amount of Frictional Attenuation of Tsunami and
Wind-Generated Waves of Various Heights across Shelves of Differing
Slope

Slope Angle	Tsunami Wave Height at Source (percentages)			Wave Heights of Ocean Swell, T = 10 s (percentages)		
	1.0 m	2.0 m	3.0 m	1.0 m	2.0 m	3.0 m
0.6°	2.1	2.0	1.9	3.1	2.8	2.4
0.5°	2.5	2.4	2.3	3.8	3.3	2.9
0.4°	3.1	3.0	2.9	4.7	4.1	3.6
0.3°	4.1	4.0	3.8	6.6	5.3	4.7
0.2°	6.0	5.9	5.7	9.6	7.8	6.9
0.1°	12.2	11.9	11.5	20.0	16.8	14.0
0.05°	23.5	24.2	24.6	38.9	34.5	30.0

km from the coastline. The lowest slope angle, 0.045°, is characteristic of coast-lines with wide shelves. Here the shelf break lies 165 km or more from the coast. This is typical of the East Coast of the United States. On steeper shelves of 0.4° or greater, the total frictional attenuation of tsunami wave height is less than 3.1%. These values are similar to those for ocean swell. As shelves become shallower, fric-tional attenuation becomes more significant, increasing to over 20% on shelves such as those found off the East Coast of the United States. These attenuation effects are less than for ocean swell. Thus, the coasts most prone to the full impact of tsunami are coastlines with narrow shelves such as the East Coasts of Japan or Australia.

The change in wave height as a tsunami crosses the continental shelf is shown in Figure 2.7. This figure assumes that refraction effects are minimal. Most earth-quake-generated tsunami are initially less than 2 m in height. Values as high as 5 m are plotted to take into account possible tsunami generated by submarine slides and meteorite impacts with the ocean. Again, the heights plotted in this fig-ure are dependent upon the depth of water and independent of slope. About 60% of the increase in wave height occurs within the last 20 m depth of water. It is a very noticeable phenomenon that ships anchored several kilometres out to sea hardly notice the arrival of a tsunami wave. However once the wave approaches shore, it very rapidly shoals and reaches its maximum height.

While the heights presented in Figure 2.7 are mathematically correct, they may not be realistic because tsunami close to shore behave as solitary waves. Solitary waves do not have a wave trough, so the wave height lies completely above mean sea level. Many researchers overcome this problem for tsunami by simply halving

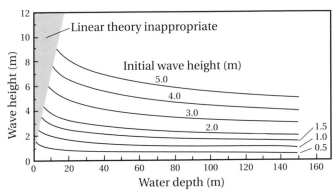

2.7 Amplification of wave height as tsunami of various initial wave heights cross the continental shelf.

the values shown in Figure 2.7. Solitary wave theory may also not be realistic because many observations of tsunami approaching shore note that water is drawn down before the wave crest arrives. This feature can only be generated by the accumulation of nonlinear effects that produce a trough in front of the wave. Solitons or *N*-waves mimic these features. Wind-generated waves are limited in Stokes wave theory by depth. A Stokes wave will break when the height-to-water depth ratio exceeds 0.78. Tsunami in most cases do not break, but surge onto shore at speeds of 5–8 m s^{-1}. However, nonlinear effects can produce wavelets on the wave crest or in some cases result in the tsunami overriding the shoreline as a bore. Whether a tsunami wave breaks or not can be determined using the following formula:

$$B_r = \omega^2 H (g \tan^2 \beta)^{-1} \qquad\qquad 2.11$$

where B_r = is the breakdown parameter of a wave (dimensionless)
β = slope of the seabed (degrees)
ω = $2\pi\, T^{-1}$ (radian frequency)

If B_r is greater than one, the wave breaks down. Note that, in contrast to storm waves, a tsunami wave is more likely to break on a steeper coast. The popular media often portray this latter aspect as a plunging tsunami wave breaking over the coast. Equation 2.11 also indicates that a tsunami wave is more likely to undergo breaking if its wave height increases or its wave period decreases. Hence small tsunami with long wave periods will more than likely surge over a coast, while big tsunami with low wave periods of a few minutes are more likely to break down before reaching shore. While tsunami crests may have the same energy as storm waves, the two types of waves differ in that storm waves break in a surf zone and dissipate most of their energy before reaching shore. On the other hand,

2.8 Sequential photographs of the 9 March 1957 tsunami overriding the backshore at Laie Point on the Island of Oahu, Hawaii. An earthquake in the Aleutian Islands, 3,600 km away, with a surface magnitude of 8.3 generated the tsunami. Photograph Credit: Henry Helbush. Source: United States Geological Survey, Catalogue of Disasters #B57C09-002.

tsunami reach shore usually without breaking and bring tremendous power to bear on the coastline (Figure 2.8).

Effect on the Seabed

(Kastens and Cita, 1981; Pickering et al., 1991; Trenhaile, 1997)

Three other attributes of waves are important in terms of sediment transport: maximum bottom orbital velocity, asymmetry in these velocities, and the maximum change in pressure on the seabed. Both the magnitude of the velocity under a wave and the bias in the onshore versus offshore velocity – termed asymmetry – controls the magnitude and direction of bottom sediment transport, and the type of bedforms that form on a shelf. This asymmetry, which is independent of any resulting backwash or undertow that may develop, can be calculated using Stokes wave theory. All waves also force a pressure pulse through bottom sediments that has the potential to liquidise the seabed. This pressure pulse affects a large mass of material. Generally, as a wave shallows, the time of increased pressure under the wave crest exceeds the decrease in pressure under the passage of the trough. Under ideal conditions, pore water pressure can build up steadily with the passage of two or more waves until the shear strength of sediment is exceeded. Erosion is much more effective when this occurs, and the potential even exists for submarine landslides on steeper slopes. These three parameters for long waves can be calculated as follows:

$$u_{max} \quad = \quad 0.5 \, (gd)^{0.5} H \, d^{-1} \qquad\qquad 2.12$$

$$\Delta u_{max} \quad = \quad 2.5 \, u_{max}^{2} T \, L^{-1} \qquad\qquad 2.13$$

$$\Delta p_{max} \quad = \quad 0.5 \, \rho_{w} g H \qquad\qquad 2.14$$

where u_{max} = maximum bottom orbital velocity (cm s^{-1})
Δu_{max} = maximum asymmetry in onshore vs. offshore bottom velocities (cm s^{-1})
Δp_{max} = maximum change in pressure on the bottom (dynes cm^{-2})
ρ_{w} = density of sea water (~1.024 g cm^{-3})

Maximum orbital velocity, onshore–offshore velocity asymmetry, and maximum pressure under swell, storm, and tsunami waves are presented in Table 2.2 for a steep shelf of 0.6° – typical of the Southeast Coast of Australia. The values for nontsunami have been calculated just before wave breaking. The values for tsunami are taken from similar depths because tsunami usually do not break. The results include frictional attenuation effects, which amount to no more than 3%. The storm wave data include a height of 7 m, which is the highest wave height measured for the storm of the century along the Southeast Australia Coast on May 25, 1974. The 10-m storm wave height has been exceeded at least five times in the Sydney region since then. This wave height is also the largest measured off the East Coast of the United

TABLE 2.2 Maximum Bottom Orbital Velocity, Onshore–Offshore Velocity Asymmetry, and Maximum Pressure under Swell, Storm, and Tsunami Waves on a Slope of 0.6°

Wave Period(s)	Initial Wave Length (m)	Initial Wave Height (m)	Maximum Bottom Orbital Velocity (m s⁻¹)	Asymmetry in u_{max} (m s⁻¹)	Max Pressure Pulse (kdyne cm⁻²)
OCEAN SWELL					
10	156	1	1.5	1.3	64
		2	2.0	1.8	111
		3	2.3	2.1	152
STORM					
15	351	4	2.8	2.5	228
		7	3.5	3.1	350
		10	4.0	3.7	462
TSUNAMI GENERATED AT SHELF EDGE					
900	34,520	1	2.1	2.0	128
		2	2.7	2.5	220
		3	3.3	3.2	308
		4	3.6	3.4	385
		5	3.9	3.5	458

Notes: Values for swell and storm waves are calculated at the breaker point, while those for tsunami are shown for similar depths.

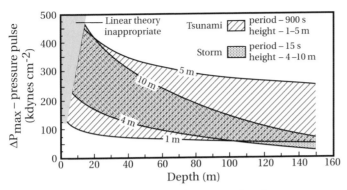

2.9 Maximum pressure pulse under tsunami and storm waves of various heights across the shelf. Compacted muds on the seabed fail at around 280 kdyne cm^{-2}.

States. The combination of high wave heights and wave periods as long as 15 seconds is rare because long wave periods require more wind energy. The lower tsunami wave heights are typical of waves generated by large earthquakes. Heights in excess of 2 m are typical of volcanic eruptions, submarine landslides, and meteorite impacts with the ocean.

The values for tsunami, while greater than the average swell wave striking a coast, are similar to those produced under storm waves. For example, the 1 : 100-year storm along the Southeast Australian Coast, with a deep-water wave height of 7 m generated bottom orbital velocities of 3.5 m s^{-1} at the breaker point. This value exceeds those for small tsunami less than 3 m in height. Even on the shelf, in water as deep as 40–100 m, storm waves produce greater shear force on the seabed and greater onshore sediment transport of sand-sized particles than do small tsunami waves of 1–2 m height. On the outer shelf, tsunami are unlikely to entrain sediment coarser than very fine sand. In 20 m depth of water, large storms can induce onshore bottom drift on the order 2.0 m s^{-1} – in contrast to small tsunami, which can only generate values of 1.0 m s^{-1}. Both storm waves and tsunami tend to flatten the seabed because any bedforms that form will typically have wavelengths as great as 50–100 m and maximum amplitudes of only a few centimetres.

The 1 : 100-year storm wave on the Australian Southeast Coast can also generate a pressure pulse of 350 kdyne cm^{-2} at the breaker point. This value is bigger than that generated by most tsunami in similar depths. Loosely compacted muddy sediments such as those making up the Mississippi River delta will fail under a pressure pulse of 280 kdyne cm^{-2}. The largest storm waves can trigger failure in such sediments near shore, but earthquake-generated tsunami appear incapable of doing so. To show further the difference between tsunami and storm waves, the maximum pressure pulse is plotted in Figure 2.9 out to the shelf edge for both storm waves and tsunami. It is often assumed that the pressure pulse from a storm wave cannot reach to the seabed at the outer shelf edge. However, at long wave periods, a 15-second-period storm wave can generate a pressure pulse equal to or greater than a small tsunami at the shelf edge. Only at wave heights above 2 m is the pressure pulse under a tsunami wave dominant over storm waves on the shelf. On the seabed, there is very little difference between large storms with long wave periods and the effects of ordinary tsunami generated by earthquakes in terms of hydrodynamics and morphology on the shelf.

Clearly, other attributes of tsunami waves must be important, otherwise there would not be the prominent differences in the coastal features generated by storms and tsunami. The main and substantial difference is the length of time that the pressure increases and shoreward velocities exist under a tsunami. For example, shoreward velocities can exist for no more than 7.5 seconds under a 15-second storm wave. They can last for up to 7.5 minutes under a 15-minute tsunami, although skewing in the shape of the wave as it crosses the relatively shallow shelf ensures that this value is in fact less. This large difference in the duration of shore-ward flow between tsunami and any other wave becomes very important towards shore, especially under run-up.

Edge Waves

(Wiegel, 1970; Geist, 1977b; Trenhaile, 1997; Komar, 1998)

The morphology of a coastline also controls the degree of dissipation of wave energy that can occur. In the coastal literature, this relationship is embodied in the term morphodynamics. Morphodynamics can be characterised by a surf scaling parameter defined as follows:

$$\Sigma \ = \ H_b\, \omega^2 (2g\, \tan^2\beta)^{-1} \qquad\qquad 2.15$$

where Σ = surf scaling factor
H_b = wave height at the breaker point (m)

While this parameter is calculated at the breaker point, it can also be used to char-acterise the morphodynamics of a coastline before a wave breaks. Note the simi-larity of Equation 2.15 to Equation 2.11. If the surf-scaling factor has a value greater than 20, wave energy will be fully dissipated. For wind-generated waves, this occurs along flat coastlines and the wave is guaranteed to break before reach-ing shore. This type of morphology is termed dissipative. If the surf scaling factor has a value less than 2.5, wave energy will be reflected from the coastline. For wind-generated waves, this occurs along steep coasts. If waves reach shore with-out breaking, then there is a high potential for wave energy to be reflected seaward off topography. This energy usually becomes trapped alongshore. This type of morphology is termed reflective. Because tsunami have such long wave periods, the morphodynamics of a coast under a tsunami wave is always reflective no mat-ter what the beach slope or wave height. Tsunami thus expend energy at the shore and are highly reflective over the continental shelf. This process tends to change the tsunami waveform, with energy leaking towards the production of edge waves that run along the shore rather than travelling towards it.

Edge waves are standing waves that oscillate up and down along a shoreline. The direction of oscillation is at right angles to the direction of travel of the incoming tsunami wave crest. The height of an edge wave decreases exponentially offshore. The amplitude of tsunami-generated edge waves can reach 70% of the amplitude of the tsunami wave. Usually edge waves have a periodicity that is twice that of the incoming tsunami. This produces variability in the run-up height of a tsunami along a coast over several kilometres distance, depending upon the amount of reflection and the wavelength of the tsunami. At one point, tsunami run-up may simply be a function of the incoming tsunami wave height, whereas a short distance along the coast, the tsunami wave may be superimposed upon the crest of an edge wave that could conceivably have equivalent amplitude. Standing waves are also one reason for drawdown of water along a coastline before the approach of a tsunami. Drawdown is not ubiquitous, but simply a function of shoreline topography and the timing of the arrival of the tsunami wave at shore relative to the longer periodicity in the oscillation of the standing wave.

Resonance

(Wiegel, 1964, 1970)

Tsunami, having long periods of 100–2,000 seconds, can also be excited or amplified in height within harbours and bays if their period approximates some harmonic of the natural frequency of the basin. The term *tsunami* in Japanese literally means "harbour wave" because of this phenomenon. Here tsunami can oscillate back and forth for twenty-four hours or more. The oscillations are termed *seiches*, a German word used to describe long, atmospherically induced waves in Swiss alpine lakes. Seiches are independent of the forcing mechanism and are related simply to the three-dimensional form of the bay or harbour as follows:

Closed basin: $T_s = 2L_b(gd)^{-0.5}$ 2.16

Open basin: $T_s = 4L_b(gd)^{-0.5}$ 2.17

where L_b = length of a basin or harbour (m)
               T_s = wave period of seiching in a bay, basin, or harbour(s)

Equation 2.16 is appropriate for enclosed basins and is known as Merian's Formula. In this case, the forcing mechanism need have no link to the open ocean. As an example, an Olympic-sized swimming pool measuring 50 m long and 2 m deep would have a natural resonance period of 22 seconds. Any vibration with a periodicity of 5.5, 11, and 22 seconds could induce water motion back and forth along the length of the pool. If sustained, the oscillations or seiching would increase in amplitude and water could spill out of the pool. Seismic waves from earthquakes can provide the energy for seiching in swimming pools, and the Northridge earthquake of 17 January 1994 was very effective at emptying pools in Los Angeles. Seiching was also induced in bays in Texas and the Great Lakes of North America about 30 minutes after the Great Alaskan Earthquake of 1964. Volcano-induced, atmospheric pressure waves can generate seiching as well. The eruption of Krakatau in 1883 produced a 0.5-m high seiche in Lake Taupo in the middle of the North Island of New Zealand via this process. Whether or not either of these phenomena technically are tsunami is a moot point.

Resonance can also occur in any semienclosed body of water with the forcing mechanism being a sudden change in barometric pressure, semi- or diurnal tides, and tsunami. In these cases the wave period of the forcing mechanism determines whether or not the semienclosed body of water will undergo excitation. The effects can be quite dramatic. For example the Bay of Fundy, on the East Coast of Canada, has a resonance period that is within 6 minutes of the diurnal tidal period of 12.42 hours. One-metre high tides in the open ocean are amplified to 14 m within the bay, while atmospheric pressure disturbances during storms moving up the coast can generate storm surges equal to another 16 m. Tsunami have the same potential. The predominant wave period of the tsunami that hit Hawaii on 1 April 1946 was 15 minutes. The tsunami was most devastating around Hilo Bay, which has a critical resonant length of about 30 minutes. While most tsunami usually approach a coastline parallel to shore, those in Hilo Bay often run obliquely alongshore because of resonance and edge-wave formation. Damage in Hilo due to tsunami has always been a combination of the tsunami and a tsunami-generated seiche. The above treatment of resonance is cursory. Harbour widths can also affect seiching and it is possible to generate subharmonics of the main resonant period that can complicate tsunami behaviour in any harbour or bay. These aspects are beyond the scope of this text.

Other Wave Theories
(Mader, 1974, 1988; Murty, 1977, 1984; Titov and González, 1997; Titov and Synolakis, 1997)

The preceding theories model either small amplitude or long waves. They cannot do both at same time. Tsunami behave as small amplitude, long waves. Their height-to-length ratio may be higher than 1 : 100,000. If tsunami are modelled simply as long waves, they become too steep as they shoal towards shore and break too early. This is called the long-wave paradox. About 75% of tsunami do not break during run-up. Because their relative height is so low, tsunami are also very shallow waves. Under these conditions, tsunami characteristics can be modelled more realistically by using Navier–Stokes incompressible shallow-water long-wave equations having the following form:

$$h_t + (uh)_x + (vh)_y \quad = 0 \qquad\qquad 2.18$$

$$u_t + uu_x + vu_y + gh_x = gd_x \qquad\qquad 2.19$$

$$v_t + uv_x + vv_y + gh_y = gd_y \qquad\qquad 2.20$$

where h $= \eta\ (x,y,t) + d(x,y,t)$
       $\eta(x, y, t)$ = the amplitude of the wave above mean sea level
       $d(x, y, t)$ = the water depth below mean sea level (m)
       $u(x, y, t)$ = depth averaged velocity in the onshore x-direction (m s^{-1})
       $v(x, y, t)$ = depth averaged velocity in the longshore y-direction (m s^{-1})
       t = time (s)

In these equations, water pressure at any point is hydrostatic and a linear function of depth, which is small compared to wavelength. In the simplest form, vertical velocity components under the wave can be assumed negligible. If latitude and longitude are used on a spherical globe that rotates, Equations 2.18–2.20 must be converted to spherical coordinates and include Coriolis force. Coriolis force accounts for the apparent deflection of particle motion on a rotating sphere – the Earth. Because tsunami can take hours to cross a body of water, this apparent deflection can be significant.

Shallow-water, long-wave equations are solved using nonlinear, finite-difference techniques. A grid is placed over a study area, and height and velocity values are calculated over time across the grid using depths interpolated either at the centre or corners of the grid. A simple grid is shown in Figure 2.10. Problems of grid interpolation immediately become obvious. The ocean's depths are not measured on a regular grid, and depths must be interpolated from survey lines that crisscross an ocean. The density of these survey lines is not constant, but increases shoreward. Error in interpolation always exists using these survey values. For example in Figure 2.10 on the cell marked B, the elevation in the bottom right-hand corner is clearly greater than 10 m. However, in the top right-hand corner, the value is less obvious. Interpolation at this latter point can be performed objectively using spline or polynomial fitting techniques, but a degree of error will always exist. In deep water, depths vary little and the effect of changing bathymetry upon a tsunami is minimal. However in shallower water, increases in topographic variability impact more on behaviour. Once a long wave crosses the shelf, its behaviour is influenced by both longshore and onshore variations in topography. As a rule of thumb, there should be thirty grid points covering the wavelength of a tsunami wave. For a tsunami with a wave period of 5 minutes, this criterion requires grid sizes of 50 m, 100 m, and 500 m where the depth

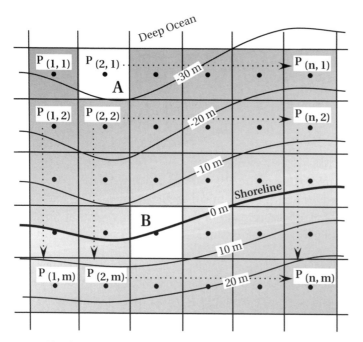

2.10 Simple representation of gridded bathymetry. Interpolation of depth values occurs at the centre of grid cells – as in this figure – or at the grid intersects. Computer simulations of actual tsunami use grids of much higher resolution than presented here.

of water exceeds 2.5 m, 10 m, and 250 m respectively. In recent years, with the demise of the Cold War, this type of detailed bathymetry has become available in the open ocean; however, inshore bathymetry is not as accessible because it usually comes under the domain of individual countries. More advanced modelling techniques can accommodate for variable grid sizes. Accurate depth information appears to be more important than the width of the grid in modelling tsunami behaviour precisely in the nearshore zone. Calculations can also be performing using triangular grid cells. The size of the triangles decreases towards shore. For example, run-up heights, current velocities and inundation limits, were modelled accurately for the 12 July 1993 Hokkaido Nansei–Oki tsunami, around Okushiri Island, using a finite element grid with triangular elements. The simulations involved a grid having 20,307 nodes and 39,668 triangles ranging in size from 7,683 m² to 311.8 km².

At present, the most exact and commonly used computer simulation to solve shallow-water long-wave equations is the SWAN code, developed by Charles Mader in the mid-1970s. In this procedure, a tsunami wave is not traced iteratively into shore; rather, changes in water level are calculated simultaneously across the whole depth grid iteratively over time. The technique is ideal for determining the amount of reflection of tsunami wave energy from a slope or the behaviour of a tsunami wave in

2.11 Computer simulation of the tsunami wave train generated by the 10 June 1996 Andreanov earthquake on the tip of the Alaskan Peninsula. The simulation uses shallow-water long-wave equations (Titov and González, 1997). For presentation purposes, the complex reflected wave crests behind the main wave front have been removed. Wave height nowhere exceeds 50 cm.

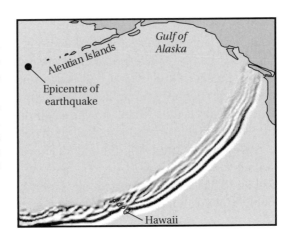

an enclosed bay or harbour. It is also possible to add any wave generated by aftershocks to the calculations. Finally, the method mimics the withdrawal and subsequent flooding of a coast over time as a tsunami wave approaches shore. The stability of solutions is a crucial function of the time increment used in the calculations. The modelling becomes unstable at time increments of less than 4 seconds. For a 10 minute period tsunami wave, crossing a shelf 100 km wide, the shallow-water long-wave equations can only be solved using super computers. An example of shallow-water, long-wave techniques using Equations 2.18–2.20 and including Coriolis force is shown in Figure 2.11 for the 10 June 1996 Andreanov earthquake originating from the Aleutian Islands in the North Pacific Ocean. Only the first three or four waves in the train are shown, and nowhere does wave height exceed 50 cm. An extensive array of tide gauges and buoys showed that heights and arrival times were well simulated in this model. Accurate solutions for tsunami wave trains can now be performed within hours of the occurrence of any major earthquake, using similar simulations.

The shallow-water long-wave equations work well in the open ocean, on continent slopes, around islands, and in harbours. On a steep continental slope greater than 4°, the techniques show that a tsunami wave will be amplified by a factor of three to four times. Because they incorporate both flooding and frictional dissipation, the equations overcome problems with linear theory where the wave breaks too far from shore. They also show that, because of reflection, the second and third waves in a tsunami wave train can be amplified as the first wave in the train interacts with shelf topography. The shallow-water long-wave equations do have limitations. If they do not include vertical velocity components, they cannot describe wave motion resulting from the formation of cavities in the ocean surface (meteorite impacts); replicate wave profiles generated by sea floor displacement, underwater landslides, or tsunami travelling over submerged barriers; or simulate the behaviour of short-wavelength tsunami. They still suffer in some cases from the long-wave paradox and break too early. In these cases, a full solution for incompressible waves must be used that includes the vertical components of motion in the Navier–Stokes equations. This increases the total computation time; however, more complicated situations can be modelled more accurately. For example, the effect of tsunami crossing reefs or underwater barriers can be simulated. Effectively, an underwater barrier does not become significant in attenuating the tsunami wave height until the barrier height is more than 50% the water depth. Even where the height of this barrier is 90% of the water depth, half of the tsunami wave height can be transmitted across it. Modelling using the full shallow-water long-wave equations shows that submerged offshore reefs do not necessarily protect a coast from the effects of tsunami. This is important because it indicates that a barrier such as the Great Barrier Reef of Australia may not protect the mainland coast from tsunami.

Freak Waves and Storm Surges
(Wiegel, 1964; Bryant, 1991; Rabinovich and Monserrat, 1996; Hamer, 1999)

One of the most unusual phenomena to explain is the occurrence of freak waves arriving at a coastline on fine days. These waves are probably solitary waves that have a peak rising above mean water level, but no associated trough. Solitary waves may only have a height of several centimetres in deep water, but when they enter shallow water, their height can increase dramatically. For example, very fast boats such as catamaran ferries can produce a wake that behaves as a solitary wave. In shallow water, the wakes have reached heights of 5 m, overturning fishing boats and swamping beaches under placid seas.

Meteorological phenomena can generate long period waves in the tsunami window. These waves have been referred to as meteorological tsunami. They take on various local names: *rissaga* in the Balearic Islands in the Eastern Mediterranean, *abiki* or *yota* in bays in Japan, *marubbio* along the coast of Sicily, *stigazzi* in the Gulf of Fiume, and *Seebär* in the Baltic Sea. They also occur in the Adriatic Sea, the South Kuril Islands, Korea, China, the Great Lakes of North America, and numerous other lakes that can come under the influence of atmospheric activity. Meteorological tsunami can be significant recurrent phenomena. For example, the south end of Lake Michigan near Chicago has experienced many atmospheric events, with one of the largest generating a 3-m wave in 1954. In Nagasaki Bay, Japan, eighteen *abiki* events have occurred between 1961 and 1979. The event of 31 March 1979 produced 35-minute oscillations having amplitudes of 2.8–4.8 m. In Longkou Harbour China, thirteen seiches have occurred between 1957 and 1980 having a maximum amplitude of 2.9 m. Finally, in the Mediterranean Sea, meteorological tsunami with heights up to 3 m have been recorded at numerous locations.

Meteorological tsunami are distinct from storm surges, although in some cases both consist of a single wave. For example, a meteorological tsunami was probably the cause of the single wave that swept Daytona Beach, Florida, late at night on 3 July 1992. The wave swamped hundreds of parked cars and injured seventy-five people. However, isolated occurrences and single waves are rare because meteorological tsunami tend to recur at specific locations and travel in wave trains. The periodicities of meteorological tsunami appear constant at many locations – a fact indicating that resonance controls the phenomenon due to the geometry and topography of a specific section of coastline. It is often noted that the phenomenon only affects a particular inlet or bay along a coast. Certainly meteorological tsunami appear restricted to harbours and bays rather than being prevalent along open coastlines. Friction and nonlinear processes weaken the formation or propagation of meteorological tsunami so that they disappear in narrow or shallow inlets. This phenomenon is associated with the passage of typhoons, fronts, atmospheric pressure jumps, or atmospheric gravity waves; however, not all of the latter produce meteorological tsunami even at favourable locations. Other forcing mechanisms may be involved. For example, tide-generated internal waves play an essential role in the formation of seiches in the Philippines and Puerto Rico, while wind waves can generate seiching in many harbours.

Not all isolated waves in the tsunami window have a meteorological origin. They can be caused by the occurrence of a small, localised, submarine landslide – in some cases without an attendant earthquake. Such freak waves are usually treated as a novelty and consequently have not received much attention in the scientific literature. Unlike meteorological tsunami that occur repetitively at some spots, freak waves are a sporadic phenomenon. In the Bahamas, isolated sets of large waves occurring on fair-weather days are referred to as *rages;* however, distant storms cannot be ruled out as a source. In Hawaii, the threat of freak waves is certainly recognised and they are attributed to tsunami of an uncertain origin. One of the more unusual events took place on the island of Majuro in the Marshall Islands in 1979. On a clear calm day, a single 6-m-high wave appeared from the northeast at low tide, crossed the reefs protecting the shoreline, and crashed through the residential and business districts in the town of Rita, washing away 144 homes. The next day at high tide the same thing happened again. After this second wave hit, the island was declared a natural disaster area by the U.S. government, which was administering the islands. Six days later, another series of waves up to 8 m high again swept the east coast of the island, destroying the hospital, communications centre, and more houses. The waves cost $20 million and affected the livelihood of two thirds of the island's 12,000 people. As recently as 16 September 1999, a 5- to 6-m wave struck the coastline of Omoa Bay on the Island of Fatu Hiva, south of the Marquesas in the central South Pacific Ocean, in the early afternoon on a quiet sunny day. Fortunately, the wave was preceded by a drop in sea level that was recognised as the imminent approach of a tsunami. While evacuations took place, the wave still hit a school, leaving some fleeing students hanging onto floating objects. Buildings close to the shore were destroyed, but no one was killed.

In Australia, two areas – Wollongong and Venus Bay – have reported freak waves.

At Wollongong, 40 km south of Sydney on the New South Wales South Coast, there have been two incidences of freak waves, both occurring under calm conditions. In the 1930s, water suddenly withdrew from a bathing beach and was followed less than a minute later by a single wave that washed above the high-tide line into the backing dunes. In January 1994, my son witnessed a lone wave that removed para-phernalia and sunbathers from a popular swimming beach, on a calm sunny day. At Venus Bay, which lies 60 km east of Melbourne, Victoria, single large waves often strike the coast on calm seas. The waves have been described as walls of water. One such wave took out a flock of sheep that were grazing near the shore, while another almost dunked a small fishing boat, which only survived because its owner cut the anchor and rode out the wave. Several fishing parties have disappeared along this coast on days when the sea was calm. None of these events can be linked to any tsunamigenic earthquake offshore in the Tasman Sea, Southern Ocean, or Pacific Ocean.

Finally, some of the evidence for tsunami has been interpreted as storm surge. Storm surges are generated by meteorological conditions, mainly tropical cyclones. Storm surges are not a problem along coastlines with narrow shelves, with steep offshore topography, or where the coastal alignment bends away from storm tracks. For example, along the East Coast of the United States, where the shelf is wide and the coast bends into the track of hurricanes, 7-m-high storm surges are possible. In contrast, storm surges rarely exceed 1 m along the South-east Coast of Australia, where the shelf is only 10–15 km wide and orientated northeast–southwest, and where tropical cyclones and east coast lows tend to move northwest-to-southeast. Dynamically, a storm surge is similar to a tsunami except that a storm surge has only one crest whereas a tsunami has several wave crests in a well-defined wave train. Storm surges can flood long distances inland, the same as tsunami. Tsunami waves however normally travel faster than storm surges. There are exceptions. For example, during the Long Island Hurricane of 1938, people described the storm surge as a 13-m high wall of water approaching the coast at breakneck speed. The description is not too different from descrip-tions of 1- to 15-m-high tsunami approaching coastlines.

RUN-UP AND INUNDATION

Run-up

(Wiegel, 1964, 1970; Synolakis, 1987, 1991; Yeh, 1991; Tadepalli and Synolakis, 1994; Camfield, 1994; Yeh et al., 1994; Briggs et al., 1995)

Tsunami are known for their dramatic run-up heights, which commonly are greater than the height of the tsunami approaching shore by a factor of two or more times. For example, in the Pacific Ocean region, forty-one tsunami have gen-erated wave run-up heights in excess of 6 m since 1900, while five events since

2.12 The remains of the Scotch Cap lighthouse, Unimak Island, Alaska, following the 1 April 1946 tsunami. A Coast Guard station, situated at the top of the cliff 32 m above sea level, was also destroyed. Five men in the lighthouse at the time perished. Source: United States Department of Commerce, National Geophysical Data Center.

1600 have produced run-up heights between 51 and 115 m. The 26 March 1947 earthquake offshore from Gisborne, New Zealand, generated a 10-m-high run-up. While not high by any standard, this wave height was maintained along a 13-km stretch of coast. The Alaskan tsunami of 1 April 1946 overtopped cliffs on Unimak Island and wiped out a radio mast standing 35 m above sea level (Figure 2.12). The eruption of Krakatoa in 1883 generated a wave that reached elevations up to 40 m high along the surrounding coastline. By far the largest run-up height recorded was that produced on 9 July 1958 by an earthquake-triggered landslide in Lituya Bay, Alaska. Water swept 524 m above sea level up the shoreline on the opposite side of the bay, and a 30- to 50-m-high tsunami propagated down the bay. In Japan, run-up heights as high as 38.2 m have been measured. The 1896 earthquake offshore from the Sanriku Coast sent a wave this high crashing into the towns of Yoshihama and Kamaishi.

Tsunami differ from wind-generated waves in that significant water motion occurs throughout the whole water column under the former wave. While this may not be important on the shelf, it causes the tsunami to take on the shape of a solitary wave in shallow water. A solitary wave maintains its form in shallow water, and, because the kinetic energy of the tsunami is evenly distributed throughout the water column, little energy is dissipated, especially on steep coasts. The maximum run-up height of a solitary wave can be approximated using the following formula:

$$H_{rmax} = 2.83 (\cot \beta)^{0.5} H_s^{1.25} \qquad\qquad 2.21$$

where H_{rmax} = maximum run-up height of a tsunami above sea level (m)
         H_s = wave height at shore or the toe of a beach (m)

The run-ups derived from Equation 2.21 are higher than those predicted using sinusoidal waves. If a leading trough precedes the tsunami, then its form is best characterised by an *N*-wave (Figure 2.4). These waves are more likely to be generated close to shore because the critical distance over which a tsunami wave develops is not long enough relative to the tsunami's wavelength to generate a wave with a leading crest. This critical distance may be as great as 100 km from shore – a value that encompasses many near-coastal tsunamigenic earthquakes. *N*-waves, as shown in Figure 2.4 can take on two forms: simple and double. The double wave is preceded by a smaller wave. Run-ups for these two types of waves can be approximated by the following formulae:

Simple *N*-wave $H_{rmax} = 3.86 (\cot \beta)^{0.5} H_s^{1.25}$ 2.22

Double *N*-wave $H_{rmax} = 4.55 (\cot \beta)^{0.5} H_s^{1.25}$ 2.23

The two equations are similar in form to Equation 2.21 for solitary waves. However, they result in run-ups that are 36% and 62% higher. In some cases, *N*-waves may account for the large run-ups produced by small earthquakes. For example, a very small earthquake (M_s magnitude of 7.3) preceded the tsunami that struck the island of Pentecost, Vanuatu, on 26 November 1999. The tsunami's arrival at shore was preceded by a leading depression and its run-up reached 5 m above sea level.

The run-up height of a tsunami also depends upon the configuration of the shore, diffraction, standing wave resonance, the generation of edge waves that run at right angles to the shoreline, the trapping of incident wave energy by refraction of reflected waves from the coast, and the formation of Mach–Stem waves. Mach–Stem waves are not a well-recognised feature in coastal dynamics. They have their origin in the study of flow dynamics along the edge of airplane wings, where energy tends to accumulate at the boundary between the wing and air flowing past it. In the coastal zone, Mach–Stem waves develop wherever the angle between the wave crest and a cliff face is greater than 70°. The portion of the wave nearest the cliff continues to grow in amplitude even if the cliff line

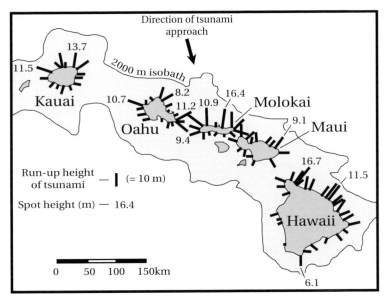

2.13 Run-up heights around the Hawaiian Islands for the Alaskan tsunami of 1 April 1946 (based on Shepard, 1977).

curves back from the ocean. The Mach–Stem wave process is insensitive to irregularities in the cliff face and can increase ocean swell by a factor of four times. The process often accounts for fishermen being swept off rock platforms during rough seas. The process explains how cliffs 30 m or more in height can be overtopped by a shoaling tsunami wave that is only 10 m high. Mach–Stem waves play a significant role in the generation of high-speed vortices responsible for bedrock sculpturing by large tsunami – a process that will be described in the following chapter.

All these processes, except Mach–Stem waves, are sensitive to changes in shoreline geometry. This variability accounts for the wide variation in tsunami wave heights over shore distances. Within some embayments, it takes several waves to build up peak tsunami wave heights. Figure 2.13 maps the run-up heights around Hawaii for the Alaskan tsunami of 1 April 1946. The northern coastline facing the tsunami received the highest run-up. However, there was also a tendency for waves to wrap around the islands and reach higher run-ups at supposedly protected sites, especially on the islands of Kauai and Hawaii. Because of refraction effects, almost every promontory also experienced large run-ups, often more than 5 m high. Steep coastlines were hardest hit because the tsunami waves could approach shore with minimal energy dissipation. For all of these reasons, run-up heights were spatially very variable. In some places, for example on the north shore of Molokai, heights exceeded 10 m, while several kilometres away they did not exceed 2.5 m.

2.14 The American warship *Wateree* in the foreground and the Peruvian warship *America* in the background. Both ships were carried inland 3 km by a 21-m-high tsunami wave during the Arica, Peru (now Chile) event of 13 August 1868. Retreat of the sea from the coast preceded the wave, bottoming both boats. The *Wateree*, being flat hulled, bottomed upright and then surfed the crest of the tsunami wave. The *America*, being keel-shaped, was rolled repeatedly by the tsunami. Photograph courtesy of the United States Geological Survey. Source: Catalogue of Disasters #A68H08-002.

Tsunami interaction with inshore topography also explains why larger waves often appear later in a tsunami wave train. For example during the 1868 tsunami off Arica, Peru (now Chile), the *USS Wateree* and the Peruvian ship *America* escaped the first two waves, but were picked up by a third wave 21 m high. The wave moved the two ships 5 km up the coast and 3 km inland, overtopping sand dunes (Figure 2.14). The ships came to rest at the foot of the coastal range, where run-up had surged to a height 14 m above sea level. Similarly, during the 1 April 1946 tsunami that devastated Hilo, Hawaii (the same tsunami that destroyed the Scotch Cap lighthouse shown in Figures 2.1 and 2.12), many people were killed by the third wave, which was much higher than the preceding two.

Shallow-water long-wave equations can accurately simulate run-up. Figure 2.15 presents the results for a tsunami originally 3 m high with a period of 900 seconds travelling across a shelf of 12 m depth onto a beach of 1% slope. Under these conditions, linear theory would have the wave breaking several kilometres from shore. However, the shallow-water long-wave equations indicate that the wave surges onto the beach with a wave front that is 3.5 m high. This is similar to many descriptions of tsunami approaching steep coasts. While flooding can occur long distances inland, the velocity of the wave front can slow dramatically. During the Oaxaca, Mexico, tsunami of 9 October 1995, people were able to out-

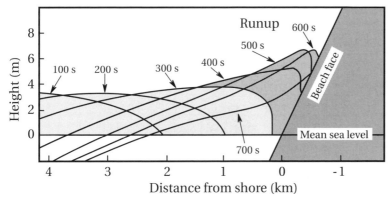

2.15 Run-up of a tsunami wave onto a beach modelled using shallow-water long-wave equations. The model used a grid spacing of 10 m and 0.5-second time increments. The original sinusoidal wave had a height of 3 m and a period of 900 seconds. Run-up peaked at 6 m above mean sea level and penetrated 600 m inland on a 1% slope (based on Mader, 1990).

run the wave as it progressed inland. A tsunami's backwash can be just as fast as, if not faster than, its run-up. The modelled wave shown in Figure 2.15 took 300 seconds to reach its most shoreward point, but just over 100 seconds to retreat from the coast. Tsunami backwash is potentially just as dangerous as run-up. Unfortunately, little work has been done on tsunami backwash.

The sheltered locations on the lee side of islands appear particularly vulnerable to tsunami run-up. Solitary waves propagate easily along steep shores, forming a trapped edge wave. Laboratory models show that the maximum run-up height of this trapped wave is greatest towards the rear of an island. More important, the run-up velocity here can be up to three times faster than at the front. For example, the 12 December 1992 tsunami along the North Coast of Flores Island, Indonesia, devastated two villages in the lee of Babi, a small coastal island lying 5 km offshore. Run-up having maximum heights of 5.6–7.1 m completely destroyed two villages and killed 2,200 people. Similarly, during the 12 July 1993 tsunami in the Sea of Japan, the town of Hamatsumae, lying behind the Island of Okusihir, was totally destroyed by a 30-m-high tsunami that killed 330 people.

Finally, tsunami run-up can also take on complex forms. Video images of tsunami waves approaching shore show that most decay into one or more bores. A bore is a special waveform in which the mass of water propagates shoreward with the wave. The leading edge of the wave is often turbulent. Waves in very shallow water can also break down into multiple bores or solitons. Soliton formation can be witnessed on many beaches where wind-generated waves cross a shallow shoal, particularly at low tide. Such waves are paradoxical because bores should dissipate their energy rapidly through turbulence and frictional attenuation, especially on dry land. However, tsunami bores are particularly damaging as they cross a shoreline. Detailed analysis indicates that the bore pushes a small wedge-

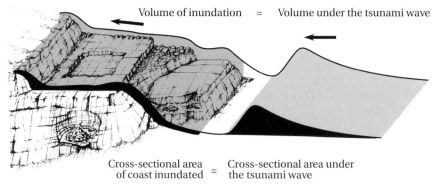

Volume of inundation = Volume under the tsunami wave

$$\frac{\text{Cross-sectional area}}{\text{of coast inundated}} = \frac{\text{Cross-sectional area under}}{\text{the tsunami wave}}$$

2.16 Schematic diagram showing that the cross-sectional area of coastline flooded and volume of inundation by a tsunami are equal to the cross-sectional area and volume of water under the tsunami wave crest. The landscape represented in this diagram will be described in Chapter 4.

shaped body of water shoreward as it approaches the shoreline. This transfers momentum to the wedge, increasing water velocity and turbulence by a factor of two. While there is a rapid decrease in velocity inland, material in the zone of turbulence can be subject to impact forces greater than those produced by ordinary waves. Often objects can travel so fast that they become water-borne missiles. This process can also transport a large amount of beach sediment inland. Tsunami that degenerate into bores are thus particularly effective in sweeping debris from the coast.

Inland Penetration
(Hills and Mader, 1997)

As a rough rule of thumb, the cross-sectional area of coastline flooded by a tsunami is equal to the cross-sectional area of water under the wave crest close to shore. This effect is illustrated by Figure 2.16. The bigger the tsunami, or the longer its wave period, the greater the volume of water carried onshore and the greater the extent of flooding. The maximum distance that run-up can penetrate inland can be calculated using the following formula:

$$x_{\text{max}} = (H_s)^{1.33} \, n^{-2} \, k \qquad\qquad 2.24$$

where x_{max} = limit of landward incursion (m)
       k = a constant

Very smooth terrain such as mud flats or pastures has a Manning's n of 0.015. Areas covered in buildings have a value of 0.03, and densely treed landscapes have a value of 0.07. The constant in Equation 2.24 has been evaluated for many

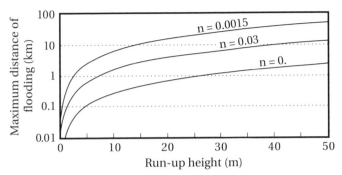

2.17   Tsunami run-up heights versus landward limit of flooding on a flat coastal plain of varying roughness. Roughness is represented by Manning's n, where n equals 0.015 for very smooth topography, 0.03 for developed land, and 0.07 for a densely treed landscape (based on Hills and Mader, 1997).

tsunami and has a value of 0.06. Using this value, the maximum distance that tsunami can flood inland is plotted in Figure 2.17 for different run-up heights, for the three values of Manning's n mentioned. For developed land on flat coastal plains, a tsunami 10 m high can penetrate 1.4 km inland. Exceptional tsunami 40–50 m high can race 9–12 km inland. Only large earthquakes, submarine landslides, and meteorite impacts with the ocean can generate these latter wave heights. For crops or pasture, the same waves could theoretically rush inland four times further – distances of 5.8 km for the 10-m-high wave and 36–49 km for the 40- to 50-m-high tsunami. Equation 2.24 also indicates that the effect of tsunami can be minimised on flat coastal plains by planting dense stands of trees. For example, the 10-m-high tsunami can only penetrate 260 m inland across a forested coastal plain, while extreme waves 40–50 m in height could not travel more than 2.3 km inland across the same terrain.

Equation 2.24 can be simplified further as follows:

$$x_{\mathrm{max}} = 1000(0.5H_{\mathrm{o}})^{1.33}$$ 2.25

This equation assumes that run-up height is equal to ten times the open ocean wave height and that Manning's n is equal to 0.03. Most earthquake-generated tsunami do not have a wave crest that exceeds 1 m above mean sea level in the open Pacific Ocean. Such waves would not flood more than a kilometre inland through a city. The validity of Equation 2.25 has been substantiated by observations at Hilo, Hawaii, following the Alaskan and Chilean earthquakes of 1946 and 1960 respectively. Tsunami generated by these events did not flood more than 750 m inland in this city. With the advent of real-time measurement of tsunami wave heights in the Pacific Ocean using pressure transducers located on the seabed, Equation 2.25 can provide a first approximation of the limit of likely flooding hours in advance of the arrival of any tsunami.

Depth and Velocity at Shore
(Blong, 1984; Camfield, 1994; Yeh et al., 1994; Nott, 1997)

Equation 2.2 indicates that the velocity of a tsunami wave is solely a function of water depth. Once a tsunami wave reaches dry land, then wave height equates with water depth and the following equations apply:

$$H_s = d \tag{2.26}$$

$$v_r = 2(gH_s)^{0.5} \tag{2.27}$$

where v_r = velocity of run-up (m s^{-1})

This equation yields velocities of 8–9 m s^{-1} for a two-metre high tsunami wave at shore. Slope and bed roughness can be incorporated into the calculation as follows:

$$v_r = H_s^{0.7} [\tan(\beta_w)]^{0.5} n^{-1} \tag{2.28}$$
$$\beta_w = \text{slope of the water surface (degrees)}$$

While the inclination of the water surface can be a difficult parameter to estimate, it can be determined after an event by measuring water lines on buildings and trees, and debris left stranded in vegetation. Generally, the inclination of the water surface ranges between 0.001 and 0.0025, increasing with slope. In Japan, it has been shown that for a Manning's n of 0.023, velocities under a tsunami wave can range between 1.3 and 5.6 m s^{-1} using Equation 2.28. In Hilo, Hawaii, the 8-m-high tsunami of 1946 mentioned earlier could obtain velocities between 5.9 and 9.3 m s^{-1} – far faster than people could run. Where tsunami behave as solitary waves and encircle steep islands, velocities in the lee of the island have been found to be three times higher than those calculated using this equation.

The velocities defined by Equations 2.27 and 2.28 have the potential to move sediment and erode bedrock, producing geomorphic features in the coastal land-scape that uniquely define the present of both present-day and past tsunami events. These signatures will be described in detail in the following chapter.

PART TWO

Tsunami-Formed Landscapes

CHAPTER THREE

Signatures of Tsunami in the Coastal Landscape

INTRODUCTION

Tsunami are high-magnitude phenomena that can achieve flow velocities at shore of 15 m s^{-1} or more. These flows have the potential to leave many depositional and erosional signatures near the coastline (Figure 3.1). The impact of palaeo-tsunami can be identified where these geomorphic signatures have been preserved, either singly or in combination. The depositional signatures of tsunami can be further subdivided into sedimentary deposits and geomorphic forms. The most commonly recognised depositional signature is the occurrence of anomalous sand sheets or lamina sandwiched in peats or muds on coastal plains. The sedimentary deposits, except for imbricated boulders, are less dramatic because they do not form prominent features in the landscape. Without detailed examination, they also could be attributed to other processes. Many of the signatures, such as aligned stacks of boulders, also reveal the direction of approach of a tsunami to a coastline. In many cases, tsunami have approached at an angle to the coast – a feature not commonly associated with the dynamics of tsunami described in Chapter 2. Suites of signatures define unique tsunami-dominated coastal landscapes. The signatures of tsunami will be described in this chapter, while the formation of tsunami-generated landscapes will be discussed in Chapter 4.

The signatures of tsunami were formulated from field evidence linked to earthquake-generated tsunami around the Pacific Rim and to the presence of tsunami identified along a 400-km stretch of the South Coast of New South Wales, Australia (Figure 3.2A). Many of the features summarised in Figure 3.1 are dramatic and allude to tsunami events an order of magnitude larger than those normally associated with earthquake-generated tsunami. These latter types of signatures have since been used to identify the presence of palaeo-tsunami along other sections of the Australian coastline, particularly in Northeastern Queensland, Northwest Aus-

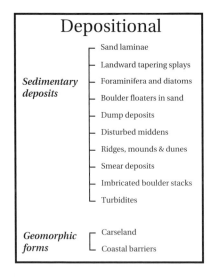

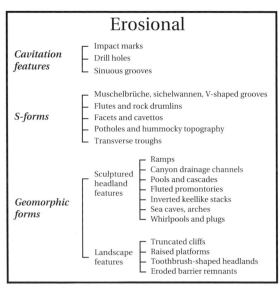

3.1   Depositional and erosional signatures of tsunami.

tralia, and on Lord Howe Island in the Tasman Sea, and in New Zealand, and along the East Coast of Scotland. It is not intended here to debate the merits of this evidence, as this has been done in many peer-reviewed scientific papers. The signatures of large tsunami are not related to storms. Submarine landslides, meteorite impacts with the ocean, and the largest earthquakes produce mega-tsunami features. This evidence will be presented in the second half of this book.

DEPOSITIONAL SIGNATURES OF TSUNAMI

Buried Sand or Anomalous Sediment Layers
(Atwater 1987; Dawson et al., 1988, 1991; Darienzo and Peterson, 1990; Minoura and Nakaya, 1991; Clague and Bobrowsky, 1994; Dawson, 1994; Minoura et al., 1994; Pinegina et al., 1996; Goff and Chagué-Goff, 1999)

The commonest signature of tsunami is the deposition of landward tapering sandy units up to 50-cm-thick sandwiched between finer material and peats on flat coastal plains. While similar lenses can be deposited by individual surging waves during tropical cyclones, such units are rarely longer than 10–20 m and do not form continuous deposits behind modern beaches. Tsunami sand units form part of a coherent landward thinning splay of fining sediment extending up to 10 km or more inland. The thickness of laminae decreases landward while that of units decreases upwards, implying waning energy conditions. All these characteristics match transport of sediment-rich flows by tsunami across marsh surfaces.

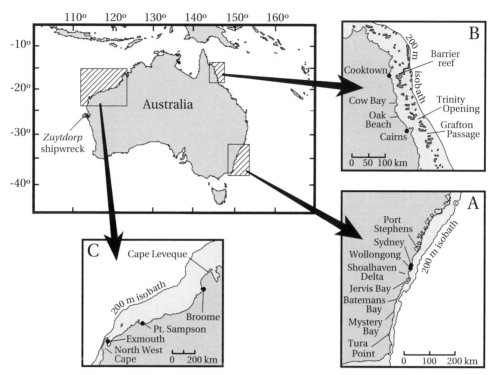

3.2 Locations of coastline around Australia showing the most prominent signatures of palaeo-tsunami: **A.** South Coast of New South Wales; **B.** Cairns Coast, Northeast Queensland; and **C.** Northwest West Australia.

Tsunami can also deposit discontinuous pencil-thin lenses or wavy wisps of silt in a matrix of much finer material. Thicker units are characterised by a series of fining-upward sequences stacked one upon each other. Each unit appears to be indicative of a single wave in the tsunami wave train, although these are indistinguishable from each other without detailed sedimentological analysis. Anomalous sand layers can have an erosional basal contact and incorporate rip-up clasts of muddier sediment. Where they are deposited over water-saturated sediment, the layers may press into the underlying surface, producing loading structures that can be preserved in the stratigraphic record. The layers can also contain shell- and land-based plant material such as twigs and leaves. This material is deposited towards the top of the unit, where it eventually decays into a humic-rich layer.

The coarsest sediment transported in suspension by a wave is deposited first as the wave's velocity slackens inland. This is followed by deposition of ever-decreasing grain sizes. The upper surface of the deposit may be truncated by backwash. However, because backwash is more likely to be channelised, truncation probably occurs as the next wave sweeps across a previously deposited sand unit. The coarsest unit in a deposit usually represents the biggest wave in the wave train.

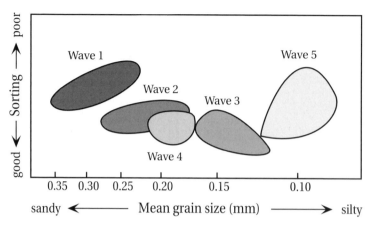

3.3 Grain size and sorting relationships throughout a 52-cm-thick sand unit at Ardmore Scotland deposited by a tsunami associated with the Storegga submarine slide.

Thus, the relative size of sediment in each unit gives an indication of the relative magnitude of each wave. The number of units is often characteristic of the causative mechanism of the wave. For example, tsunami generated by meteorite impacts or submarine landslides generally produce wave trains with two to five large waves. Each wave is capable of moving sediment inland. Multiple sand layers cannot be produced by storm surge because it occurs as a single wave event. In contrast, earthquake-generated wave trains, while consisting of tens of waves, tend to produce only a single wave that is large enough to transport sediment inland. Without additional evidence, it may be difficult to separate an earthquake-generated tsunami deposit from that produced by a storm surge.

Figure 3.3 shows the relative grain size and sorting in a 52-cm-thick sand unit at Ardmore, Scotland, deposited by large tsunami from the Storegga slide off the West Coast of Norway 7,950 years ago. This tsunami deposited fine sand and silt as much as 80 km inland across numerous estuarine flats – termed carseland – that are now raised along the East Coast of Scotland. The tsunami and its wider impact will be described in more detail in Chapter 6. Each wave in Figure 3.3 is numbered sequentially. The first wave, as expected, had the greatest wave energy and moved a greater range of grain sizes. Subsequent waves moved finer sediment, probably because of decreasing wave height or availability of coarser sediment. While the last wave was smaller, it transported a wider range of grain sizes than the previous three waves. This may reflect the recycling of coarse sediment seaward by channelised backwash.

Sediment layers have been deposited by numerous historical tsunami. The 1 November 1755 Lisbon tsunami, generated by one of the largest earthquakes ever, deposited sand layers along the Portuguese Coast and as far as the Isles of Scilly off the West Coast of England. In the twentieth century, the Grand Banks tsunami of 18 November 1929, which was produced by a submarine landslide, laid down sand layers on the Burin Peninsula in Newfoundland. Following the Chilean tsunami of 22 May 1960, sand layers were deposited in the Río Lingue estuary of south-central Chile, while the Alaskan tsunami of 27 March 1964 deposited sand units along the coasts of British Columbia and Kodiak Island. More recently, this signature was observed on the island of Flores, Indonesia, following the tsunami of 12 December 1992 and along the Aitape Coast, Papua New Guinea, as the result of the tsunami of 17 July 1998. In both of the latter cases, landward-tapering wedges of sand were deposited over 500 m inland from the shoreline. These examples will be described in more detail in the chapter on earthquake-generated tsunami.

Sand units also allude to palaeo-tsunami. For example, a sand unit was deposited by a tsunami within clayey estuarine sediment, up to 10 km inland, on the Shoalhaven delta in Southeastern New South Wales 4,700–5,100 years ago. The source of this event is most likely attributable to a meteorite impact in the Tasman Sea. Considerable attempts have been made to date anomalous sand layers and relate them to historical tsunami events. However, because tsunami events occur so infrequently along many coasts, the true value of such dating programs lies in the delineation of the recurrence interval of such a hazard, especially along coastlines that have only been settled in the last few centuries. For example, on the Southeastern Coast of the Kamchatka Peninsula of Eastern Russia, sand layers are preserved in peats or organic-rich alluvium. These sequences also contain volcanic ash layers or tephras that can be used to date the sequences. The tsunamigenic layers consist of coarse marine sand mixed with gravel and pebbles in landward tapering sheets 2–3 cm thick. The largest tsunami overrode terraces 15–30 m above sea level up to 10 km from the coast. Many of the layers show evidence of suspension transport of sediment, in some cases aided by the passage of tsunami over frozen or icy ground. Forty events can be identified over the past 2,000 years, revealing a recurrence interval of one major event every 50 years. Not all events, though, may have been preserved, because the recurrence interval since 1737 averages one tsunami event every 12.3 years.

In Northern Japan rhythmic sedimentary units, consisting of coarsening upward sequences lying unconformably on top of each other, have been found in lagoons and linked to historical tsunami over the last 250 years. In a back barrier lagoon at Lake Jusan on the northern end of Honshu Island, facing the Sea of Japan, thicker layers of medium sand 40 cm or more in thickness are sandwiched within organic ooze. Sedimentological analysis indicates that the sands originated from sand dunes or the open ocean beaches. These deposits have been linked to large tsunami events with wave heights in excess of six metres occurring at intervals of 250–400 years over the past 1,800 years. On the Pacific Coast of Honshu, on the Sendai Plain, as many as five sand layers are sandwiched in peats up to 4 km inland. Two of the upper layers can be correlated to killer tsunami that swept over the plain in A.D. 869 and A.D. 1611. Along the Sanriku Coast of Northeast Honshu Island – renowned for deadly tsunami – up to thirteen well-sorted sand layers can be found intercalated within black organic muds in swamps and ponds. The sand layers are spatially extensive but do not form erosional contacts with the underlying muds. This suggests that sediment settled from suspension in quiescent waters stranded in depressions after rapid drowning by tsunami that exceeded 1 m in height. The layers have been correlated to known tsunami events originating off the coast and from distant sources in the Pacific Ocean (Table 3.1). Seven of the tsunami formed locally; however, four of the events originated either from Chile or the Kuril Islands to the north. Surprisingly, the severe Chilean tsunami of 22 May 1960 did not flood any of the marshes. In addition, no event prior to 1710 is preserved in the sediments despite this coastline's having historical records of earlier events. The stratigraphic evidence designates this coastline as the most frequently threatened inhabited coastline in the world.

Along the West Coast of North America, the sand layers are trapped within peats and show sharp nonerosional upper contacts, but erosional bottom ones. The sand layers are 1–30 cm thick (Figure 3.4) and similar in grain size to beach and dune sands on the adjacent seaward coast. Both optical luminescence dating of quartz sand and radiocarbon dating of buried carbon indicate that a great earthquake occurred along 700 km of the Cascadia subduction zone 300 years ago.

TABLE 3.1   Correspondence between the Inferred Age of Anomalous Sand Layers and Dated Tsunami Events on the Sanriku Coast of Northeast Honshu Island, Japan

Inferred Age	Closest Corresponding Event	Source
1948	4 March 1952	Hokkaido
1930	3 March 1933	Sanriku
1905	5 June 1896	Sanriku
1887	9 May 1877	Chile
1861	13 August 1868	Chile
1853	23 July 1856	Sanriku
1843	20 July 1835	Sanriku
1805	7 January 1793	Sanriku
1787	29 June 1780	Kuril Islands
1772	16 December 1763	Sanriku
1769	15 March 1763	Sanriku
1742	25 May 1751	Chile
1710	8 July 1730	Chile

Source: From Minoura et al., 1994.

3.4 Sand layer, deposited by tsunami, sandwiched between peats at Cultus Bay, Washington State. The markings on the shovel are at 0.1-m intervals. The tsunami event was radiocarbon dated from *Triglochin* rhizomes in the upper peat at A.D. 1040–1150. Photograph courtesy of Prof. John Clague, Simon Fraser University.

The exact age has been inferred from the recording of a tsunami in Japan on 26 January 1700. The Kwenaitchechat North American Indian legend in Chapter 1 referred to this event. At least five other events have been identified, with three of the largest occurring somewhere between 600 and 900, 1,000 and 1,400, and 2,800 and 3,200 years ago. At Crescent City, California, up to twelve additional sand layers have been found in a peat bog. This record is similar to the Kamchatka Peninsula in frequency. Interestingly, the tsunami from the Great Alaskan Earthquake of 27 March 1964 only produced a thin sand layer about a centimetre thick here. The lack of cross-bedding in the units indicates deposition out of turbulent suspension, while alteration within the same unit between sand and silty clay suggests pulses of sediment entrainment, transport, and deposition. Storm waves or surges can be ruled out as mechanisms of emplacement because most of the sites have a high degree of protection from the open ocean.

An unusual type of layering has been found in New Zealand in the Abel Tasman National Park on Cook Strait. Here, estuaries are infilled with medium-coarse sand and pebbles derived from the local granite. In sheltered locations up to 2.5 km from the coast, tsunami signatures occur as a peak in clay-to-silt-sized material concomitantly with higher concentrations of sulphur and iron. The muddy sedi-

ment originates from the seabed of Tasman Sea, while the isotopes indicate catastrophic inundation of the wetlands by saltwater. Three tsunami events have been
dated at A.D. 1250, A.D. 1430, and A.D. 1855. These dates coincide with ruptures of
either the Wellington Fault on the north side of Cook Strait or the Alpine Fault
along the West Coast of the South Island. The A.D. 1430 date also overlaps with a
volcanic eruption in Tonga, north of New Zealand, in A.D. 1453 that generated a 30-
m-high tsunami. The rupture in the Wellington Fault in A.D. 1855 generated a 9- to
10-m-high tsunami in Cook Strait. As will be shown in Chapter 8, meteorite impact
with the Tasman Sea also cannot be ruled out as a cause.

Foraminifera and Diatoms
(Dawson et al., 1996; Dawson, 1999; Haslett et al., 2000)

Silty sand units deposited inland by tsunami can also contain a distinct signature
of marine diatoms or foraminifera. Foraminifera are small unicellular animals,
usually about the size of a grain of sand, that secrete a calcium carbonate shell. On
the other hand, diatoms are similarly sized, single-celled plants that secrete a shell
made of silica. Both organisms vary in size and live either suspended in the water
column (planktonic) or on the seabed (benthic). Under fair-weather conditions,
only the larger benthonic species are transported shoreward as bedload under wave
action and deposited on the beach. Smaller benthonic and the suspended planktonic species are moved offshore in backwash or transported alongshore in currents
to quiescent locations such as estuaries or the low-energy end of a beach. Storm
wave conditions tend to move sediment, including diatoms and foraminifera, offshore in backwash, undertow, or rips. However, winds can blow surface waters to
shore. A storm assemblage includes very small diatoms or foraminifera diluted with
larger, benthonic foraminifera reworked from prestorm beach sediments. Tsunami
assemblages are chaotic because a tsunami wave moves water from a number of
distinctive habitats that include marine planktonic and benthic, intertidal, and terrestrial environments. A high proportion of the forams and diatoms are broken, with
spherical-shaped species being overrepresented because of their greater resistance
to erosion. Storm waves are incapable of flinging debris beyond cliff tops, whereas
tsunami can override headlands more than 30 m high. In the latter case, the occurrence of coarse, inshore benthonic species in the debris indicates a marine provenance for the sediment. Where this material is mixed with gravels or other coarse
material, it forms one of the strongest depositional signatures of tsunami.

Foram and diatom assemblages have been studied for a number of historic and
palaeo-tsunami events. For example, the 1-m-thick sand layers deposited by the Flores tsunami of 12 December 1992 contain planktonic species such as *Coscinodiscus*
and *Cocconeis scutellum* as well as the freshwater species *Pinnularia*. The Burin
Peninsula tsunami of 18 November 1929 deposited a distinct diatom signature
throughout the 25-cm-thick sand units in peat swamps. While the peats contain

only freshwater assemblages, the tsunami sands contain benthic and intertidal mudflat species such as *Paralia sulcata* and *Cocconeis scutellum*. Freshwater species are incorporated in the lowermost part of the sands, indicating that material was ripped up from the surface of the bogs as the tsunami wave swept over them. Finally, tsunami deposits in Eastern Scotland attributable to the Storegga slide 7,950 years ago contain *Paralia sulcata*, which is ubiquitous in the silty, tidal flat habitats of Eastern Scotland. The freshwater species *Pinnularia* is also present, indicating that bog deposits were eroded in many locations by the passage of the tsunami wave.

Boulder Floaters in Sand
(Bryant et al., 1992, 1996)

Boulders are not usually transported along sand beaches under normal wave conditions. Their presence as isolated floaters within a sand matrix is therefore suggestive of rapid, isolated transport under high-energy conditions. In many cases, deposits containing boulders are less than 1.5 m thick and lie raised above present sea level along coastlines with stable sea level histories. Either storm waves superimposed upon a storm surge or tsunami are known to be responsible for such deposits. For example, Hurricane Iniki in Hawaii in 1992 swept a thin carpet of sand containing boulders inland beyond beaches, while the tsunami that hit Flores, Indonesia, on 12 December 1992 did likewise (Figure 3.5). While storms can exhume boulders lying at the base of a beach and are known to move boulders alongshore, they cannot deposit sand and boulders together on a beach unless overwash is involved.

The cause of boulder floaters in palaeo-deposits is difficult to determine unless such deposits lie above the limits of storms. For example, 0.2–0.4 m boulder floaters are common within sand deposits south of Sydney along the East Coast of Australia (Figure 3.2A). The coastline is tectonically stable, and sea levels have not been more than 1 m higher than present. Here, many deposits lie perched on slopes to elevations of 20 m above sea level, well above the maximum limit of storm surges. Bouldery sands also appear behind beaches where the nearest source for boulders lies up to a kilometre away. Unfortunately, boulder floaters as a signature of tsunami are not an easily recognised one in the field. Boulder floaters often constitute less than 0.1% by volume of a deposit and lie buried beneath the surface. These facts make them difficult to find unless the deposits are trenched or intensively cored.

Dump Deposits
(Bryant et al., 1992, 1996; Dominey-Howes, 1996; Branney and Zalasiewicz, 1999)

Chaotic sediment mixtures, or dump deposits, are emplaced in coherent piles or layers above the limits of storm waves, mainly on rocky coasts. They can be problematic. Such deposits can also be formed by solifluction, ice push, slope wash, glowing volcanic avalanches, and human disturbance. Dump deposits were

3.5 Photograph of the coastal landscape at Riang–Kroko on the island of Flores, Indonesia, following the tsunami of 12 December 1992. The greatest run-up height of 26.2 m above sea level was recorded here. Boulders and gravels have been mixed chaotically into the sand sheet. Note the isolated transport of individual boulders. Photo Credit: Harry Yeh, University of Washington. Source: National Geophysical Data Centre.

first identified along the South Coast of New South Wales, Australia (Figure 3.2A). Because ice, volcanic activity, and ground freezing are not present along this coast, catastrophic tsunami became a viable mechanism for the transportation and deposition of large volumes of sediment with minimal sorting in a very short period of time. Coarser dump deposits, containing an added component of cobble and boulders, often are plastered against the sides or on the tops of headlands along this coast (Figure 3.6). Many recent deposits may also contain mud lumps. It would be tempting to assign these deposits to storms but for three facts. First, storm waves tend to separate sand and boulder material. Storms comb sand from beaches and transport it into the nearshore zone in backwash and rips. Storm swash, however, moves cobbles and boulders landward and deposits them in storm berms. Certainly, storms do not deposit muds mixed with coarser sediment on steep slopes. On the other hand, tsunami, because of their low height relative to a long wavelength, form constructional waves along most shorelines, transporting all sediment sizes shoreward. Second, while it may be possible for exceptional storms to toss sediment of varying sizes onto cliff tops more than 15 m above sea level, tsunami dump deposits can be found not only much higher above sea level (Figure 3.6), but also in sheltered positions. Finally, there is substantial observation from Hawaii and elsewhere of tsunami laying down dump deposits (Figure 3.5).

3.6 A chaotically sorted dump deposit on Minnamurra Headland, New South Wales, Australia. The deposit, set in a mud matrix, lies on basalt, yet contains rounded metamorphic pebbles and shell bits. The site is 40 m above sea level on a coast where sea level has been no more than 1–2 m higher than the present over the past 7,000 years.

The internal characteristics, or fabric, of tsunami dump deposits allude hydrodynamically to their mode of transport and deposition. This fabric is identical to that formed by pyroclastic density currents or ash clouds emanating from volcanic eruptions. In pyroclastic flow, fine particles are suspended and transported by turbulent whirlwinds of gas. While appearing as choking clouds of ash, the particles are so widely spaced that they rarely collide. As these turbulent eddies pass over a spot, they can deposit alternating layers of coarse and fine sediment as the velocity of the current waxes and wanes in a similar fashion to gusts of wind. Where the cloud meets the ground, conditions are different. Sediment concentrations increase, and sediment particles ranging in size from silts to boulders undergo billions of collisions. The momentum of the flow is equalised between the particles and the flowing current of air. In some cases, the grains may flow independent of any fluid, a phenomenon known as granular flow. Granular flow tends to expel coarse particles to the surface; however, if fluid moves upwards through the flow, then a process called fluidisation may allow sediment particles in dense slurries to move as a fluid and remain unsorted. Different processes probably operate at different levels in pyroclastic flows. Turbulence lifts finer particles into the current higher up, while at ground level it enhances fluidisation. Similarly, the falling out of particles from slurries near the ground entraps smaller particles into a deposit while expelling water. The latter process also enhances fluidisation. From time to time, turbulent vortices penetrate to the bed, allowing the deposition of alternating layers of fine and coarse material. The resulting product is a disorganised deposit containing a wide range of particle sizes with evidence of layering.

All of these characteristics appear in tsunami dump deposits with water taking the place of gas. Alternating layers of fine and coarse sands can be found as small, eroded blocks embedded in chaotically sorted piles of bouldery sand. The layers were deposited from downward turbulent pulses of water that penetrated to the bed. Chipped gravels and shells form deposits that can also contain fragments that show no evidence of violent transport. The former are milled by the myriad of collisions occurring in granular flow towards the base of the current, while the latter have settled from the less-dense upper regions of flow. Preservation of fragile particles indicates that deposition must have been rapid. Mud clasts are evidence of the erosive power of turbulent vortices impinging upon the bed. Some of the mud clasts are caught in the granular flow and are disaggregated to form the mud matrix. Some clasts are suspended higher into the less dense part of the flow and are later mixed into the deposit unscathed. Not only do turbulent vortices create spatial variation in the internal fabric of the deposits, they also account for the rapid spatial variation in the degree of erosion of the landscape upstream. Hence, the eroded upstream sides of headlands that provide the material incorporated into *dump* deposits can still evince weathered soil profiles within metres of bedrock surfaces that have undergone the most intense erosion by vortices.

The rocky headlands of the New South Wales coast also show an unusual variation of dump deposits with a strong anthropogenic constituent. At sixteen locations, Aboriginal kitchen middens have been reworked by tsunami. Aboriginal middens are trash heaps containing discontinuous layers of edible shell species mixed with charcoal, bone fragments, and artifact stone chips, set in humic-rich sands. All of the disturbed sites incorporate, as an added component, marine shell grit, water-worn shells, rounded pebbles, or pumice. Humus is usually missing from the deposits. Chronology, based upon the suite of signatures present along this coast, suggests that the deposits are only 500 years old. In many cases, the reworked deposits are overlain by undisturbed midden devoid of pumice. Storm waves can be ruled out because many of the sites are situated in sheltered locations or at excessive heights. Storm waves also tend to erode shell and sand-sized sediment seaward and do not deposit bodies of mixed debris several metres thick at the point of run-up on steep backshores at these high elevations. The fact that many of these disturbed middens still contain a high proportion of midden material indicates that disturbance was rapid and incomplete with an input of contaminants from a nearby marine source. Other signatures of tsunami surround all sites. The importance of these disturbed middens will be discussed later in Chapter 8.

Dump deposits have also been described elsewhere, for example on the Island of Astypalaea, Greece, and the southern Ryukyu Islands of Japan. The dump deposit on Astypalaea, in the Aegean Sea, was deposited by a modern tsunami event that occurred on 9 July 1956 as the result of an earthquake-triggered submarine slide. The initial wave height was 30 m with run-up on Astypalaea Island reaching 4–20 m above sea level. At Livadia, a dump deposit consisting of clast-supported, rounded cobbles is located 1.5–2.0 m above sea level. The basal contact between the cobble bed and the underlying sediments is sharp and erosional. The cobbles are crudely aligned and are mixed with foraminifera from the inner shelf, which contains *Cibicides refulgens, Rosalina bradyi Cushman,* and *Eponides* spp. At Stavros, the gravel deposits cover several areas ranging in size between 2.5 and 250 m², deposited up to 40 m from shore. Some of these deposits are located on cliff tops 10 m above sea level or contain marine molluscs such as *Barbatia barbatia, Chamelea striatula, Mitra cornicula,* and *Monodonta turbinata.*

Mounds and Ridges
(Minoura and Nakaya 1991; Bryant et al., 1992, 1997)

Chaotic dump deposits can be moulded into isolated ridges and mounds. These deposits often rise to heights of 3 m or more above the limits of modern storm-wave activity. None of these deposits can be explained satisfactorily by ordinary wave processes. For example, at Bass Point 80 km south of Sydney, Australia, a ridge has been sculptured into mounds standing 2–3 m in elevation. The mounds consist of chaotically sorted shell hash mixed with rounded cobble and boulder-sized debris. The mounds extend 100 m alongshore and lie 5–10 m seaward of a scarp cut into sand dunes. As the shoreline becomes more sheltered, the mounds merge into a 10- to 20-m-wide bench consisting of the same material and rising steeply to a height of 4–5 m above the present storm wave limit. The mounds are not unlike the shell mounds formed by tsunami in Japan; however, they are an order of magnitude larger. Slopes on both sides of the mounds exceed 20° – a value that is much steeper than the 7–9° found on equilibrium profiles formed by storm waves in similarly sized material. A chaotic sheet of cobble and boulder material protruding through a shell hash matrix fronts the mounds. This sheet rests on a seaward-sloping bedrock platform. The fact that this debris is partially vegetated suggests that present storm wave run-up has not affected it. The mounds were deposited within vortices created in the lee of a 15-m-high headland that was overwashed by a large tsunami.

There are several examples of ridges formed by tsunami. On the island of Lanai in Hawaii, boulder deposits supposedly deposited by tsunami evince ripple forms 1 m high with a spacing of 100 m between crests. On the South Ryukyu Islands in Japan, ridges several metres high and 40 m wide have been linked to tsunami. However, among the most unusual coastal features are the chenier-like ridges formed in Batemans Bay on the New South Wales Coast of Australia (Figure 3.2A).

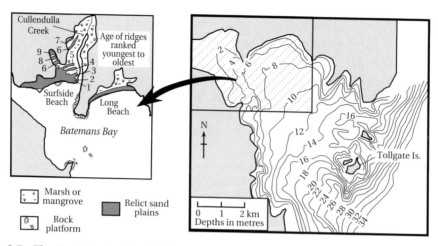

3.7 The chenier-like ridges in the Cullendulla Creek embayment at Batemans Bay, New South Wales, Australia (from Bryant et al., 1992).

Cheniers are ridges of coarser sediment usually deposited at the limit of storm waves in muddy environments. Batemans Bay is a 14.4-km-long, funnel-shaped, sand-dominated embayment, averaging 11 m in depth and semicompartmentalised by north–south structurally controlled headlands (Figure 3.7). This shape is conducive to resonance and enhancement of tsunami entering from the open ocean. The basin's resonance periods are 4.5, 13.4, 22.5, and 31.0 minutes – values that fall within the range of typical tsunami periods measured in harbours. The cheniers in the bay consist of a series of six ridges deposited in a sheltered embayment presently occupied by Cullendulla Creek (Figure 3.7). The ridges rise 1.0–1.5 m above the surrounding estuarine flats and increase in width and volume towards the bay, where they merge into a barrier complex consisting of at least eight ridges. The cheniers consist of shell-rich sand overlying estuarine muds. They are asymmetric in shape, rising steeply to a height of 0.5–0.6 m and then dropping landward over a 30- to 40-m distance to the estuarine surface. The ridges also extend as raised banks 1 km up a creek entering on the western side of the embayment. The banks are 30–40 m wide and stand 0.75–1.0 m above the high-tide limit. The innermost bank contains shell species derived from sheltered rocky shores, open ocean beaches, rock platforms, and the inner continental shelf.

The formation of these chenier ridges by storm waves is difficult to justify because wave refraction reduces wave heights by 80 to 90%. Waves would have to travel in the most convoluted pathway possible to deposit ridges up the reentrant within Cullendulla Creek. This includes passing through a gap 600 m wide at the most sheltered part of the embayment, bending 500 m behind a large rock promontory, and travelling up the shallowing reentrant for a distance of 600 m.

The simpler explanation is that tsunami have deposited chenier-like ridges and banks in this sheltered reentrant. There is chronological evidence for at least three tsunami events operating within the bay 300, 1,300, and 2,800 years ago. While Batemans Bay contains the best example of tsunami ridge development along this coast, similar ridges and banks exist in the Port Hacking, Middle Harbour, and Patonga Beach estuaries in the Sydney area 200 km to the north. Spits and ridges lying within sheltered estuarine environments should be examined closely for evidence of a tsunami origin.

Dunes
(Ota et al., 1985; Bryant et al., 1997)

Tsunami can also create dunes. These take two forms: those that back beaches and can easily be interpreted as wind-blown sand dunes, and those that form as bedforms under catastrophic unidirectional flow generated by large tsunami. The former are sandy, but contain angular clasts or mud layers that cannot be transported or deposited by wind. These types of ridges exist on Hateruma Island in the South Ryukyus of Japan, where they consist of sand, gravels, and large boulders up to 2 m in width. They were deposited by the great tsunami of 24 April 1771, which had a maximum run-up height in the region of 85.4 m. Another example occurs at the south end of Jervis Bay, New South Wales (Figure 3.2A). Here parabolic dunes, rising more than 130 m above sea level, contain gravels and mud clasts (Figure 3.8). Because these tsunami swash features are symmetrical, they form chevron-like dunes (named after the stripes on the arms of a military uniform). The chevron features are contiguous in form, from the present beach to their peak 130 m above present sea level. This feature is an exemplary signature of mega-tsunami because its limits are far beyond those of either storm swash or earthquake-generated tsunami reported in the literature.

Dune bedforms are created under catastrophic flow. Such features were first described in the scablands of Washington State. The scablands consist of a series of enormous dry canyons occupied by gigantic rippled sandbars. The catastrophic flow originated from the collapse of an ice dam holding back glacial Lake Missoula in the Northern Great Plains of the United States during the last Ice Age. Repetitive floods up to 30 m deep spread across sandy plains, creating enormous dune bedforms. Similar dune bedforms can be produced by large tsunami. Two examples have been identified. The first is located at Crocodile Head, north of the entrance to Jervis Bay, Australia. Here, sandy ridges containing pebbles and gravels appear as a series of well-defined, undulatory-to-lingoidal giant ripples spread over a distance of at least 1.5 km atop 80-m-high cliffs. These megaripples have a relief of 6.0–7.5 m, are asymmetric in shape, and are spaced 160 m apart. The megaripple field is restricted to a 0.5- to 0.7-km-wide zone along the cliffs, and is bounded landward by a linear ridge of sand several metres high paralleling the coastline.

3.8 Fabric of sand and gravel deposited in a chevron-shaped dune by palaeo-tsunami at Steamers Beach, Jervis Bay, Australia. Car keys for scale. The rounded ball to the left of the keys consists of humate eroded elsewhere from the B horizon of a podsolic soil profile. This deposit lies 30 m above sea level, while the dune itself crests 130 m above sea level.

This ridge is flanked by small depressions. Further inland, deposits grade rapidly into hummocky topography, and then a 1- to 2-m-thick sandsheet. The megaripples were produced by sediment-laden tsunami overwashing the cliffs, with subsequent deposition of sediment into bedforms along the cliff top. Flow then formed an overwash-splay as water drained downslope. The flow over the dunes is theorised to have been 7.5–12.0 m deep and to have obtained velocities of 6.9–8.1 m s⁻¹. The second example occurs at Sampson Point, Western Australia (Figure 3.2C). Here gravelly mega-ripples have infilled a valley with bedforms that have a wavelength approaching 1,000 m and an amplitude of about 5 m. Flow depth is theorised to have been as great as 20 m with velocities of over 13 m s⁻¹.

Smear Deposits

More enigmatic are deposits on headlands containing a mud matrix. These deposits are labelled smear deposits because they are often spread in a continuous layer less than 30 cm thick over the steep sides and flatter tops of headlands. Along the New South Wales Coast, these deposits have been found at elevations 40 m above sea level. These deposits can contain 5 to 20% quartz sand, shell, and gravel. Smear deposits are not the products of in situ weathering because many can be found on volcanic sandstone or basalt, which lacks quartz. These smear deposits form the traction carpet at the base of a sediment-rich tsunami-generated flow overwashing headlands. The mud allows sediment to be spread smoothly under substantial pressure over surfaces. When subsequently dried, the packed clay minimises erosion of the deposit by slope wash on steep faces. The deposit has only been identified on rocky coasts where muddy sediment lies on the seabed close to shore.

Large Boulders and Piles of Imbricated Boulders
(Baker, 1978; Moore & Moore, 1988; Kawana and Pirazzoli, 1990; Kawana and Nakata, 1994; Young et al., 1996a; Bryant et al., 1997; Nott, 1997)

Tsunami differ from storm waves in that tsunami dissipate their power at shore rather than within any surf zone. The clearest evidence of this is the movement of colossal boulders onshore. For example, the Flores tsunami of 12 December 1992 destroyed sections of fringing reef and moved large coral boulders shoreward (Figure 3.5), often beyond the zone where trees were ripped up by the force of the waves. The Sea of Japan tsunami of 26 May 1983 produced a tsunami over 14 m high. A large block of concrete weighing over 1,000 tonnes was moved 150 m from the beach over dunes 7 m high. Boulders transported by tsunami have also been found in palaeo-settings. For instance, on the reefs of Rangiroa, Tuamoto archipelago in the Southeast Pacific, individual coral blocks measuring up to 750 m^3 have been linked to tsunami rather than to storms. Boulders have also been scattered by tsunami across the reefs at Agari-Hen'na Cape on the eastern side of Miyako Island in the South Ryukyu Islands (Figure 3.9). On Hateruma and Ishigaki Islands in the same group, coralline blocks measuring 100 m^3 in volume have been emplaced up to 30 m above present sea level, 2.5 km from the nearest beach. These boulders have been dated and indicate that tsunami, with a local source, have washed over the islands seven times in the last 4,500 years. Two of the largest events occurred 2,000 years ago and during the great tsunami of 24 April 1771.

Along the East Coast of Australia, anomalous boulders are incompatible with the storm wave regime. For example, exposed coastal rock platforms along this coast display little movement of boulders up to 1–2 m in diameter, despite the presence of 7- to 10-m-high storm waves. At Boat Harbour, Port Stephens (Figure 3.2A), blocks measuring 4 m × 3 m × 3 m not only have been moved shoreward more than 100 m, but also have been lifted 10 to 12 m above existing sea level. At Jervis Bay, preparation zones for block entrainment can be found along cliff tops 15 m above sea level. The only documented boulder of any size to move during a storm was reported in 1912 at Bondi Beach, Sydney. Here a 200-tonne boulder measuring 6.1 m × 4.9 m × 3.0 m was shifted 50 m across a rock platform. This block has not shifted since, despite a storm in May 1974 that was judged the worst in a hundred years.

3.9 Scattered boulders transported by tsunami and deposited across the reef at Agari-Hen'na Cape on the eastern side of Miyako Island, Ryukyu Island, Japan. Photograph courtesy of Prof. Toshio Kawana, Laboratory of Geography, College of Education, University of the Ryukyus, Nishihara, Okinawa.

While much higher storm events can be, and have been, invoked for the movement of the boulders now piled prodigiously along this coast, tsunami offer a simpler mechanism for the entrainment of joint-controlled blocks, the sweeping of these blocks across platforms, and their deposition into imbricated piles.

In exceptional cases, boulders have been deposited as a single-grained swash line at the upper limit of tsunami run-up. Some of these tsunami swash lines lie up to 20 m above sea level and involved tractive forces greater than 100 kg m^{-2}. Just as unusual are angular boulders jammed into crevices at the back of platforms. For instance, at Haycock Point, north of the Victorian border, angular blocks up to 2 m in length and 0.5 m in width have been jammed tightly into a crevice, often in an interlocking series three or four blocks deep. What is more unusual about the deposit is the presence along the adjacent cliff face of isolated blocks 0.4–0.5 m in length perched in crevices 4–5 m up the cliff face. Boulders are also found in completely sheltered locations along the coast. At Bass Point, which extends 2 km seaward from the coast, a boulder beach faces the mainland coast rather than the open sea. Similarly at Haycock Point, rounded boulders, some with volumes of 30 m^3 and weighting 75 tonnes, have been piled into a jumbled mass at the base of a ramp that begins 7 m above a vertical rock face on the sheltered side of the headland (Figure 3.10). Chatter marks on the ramp surface indicate that many of the boulders have been bounced down the inclined surface.

3.10 Boulders transported by tsunami down a ramp in the lee of the headland at Haycock Point, New South Wales, Australia. This corner of the headland is protected from storm waves. The ramp descends from a height of 7 m above sea level. Its smooth undular nature is a product of bedrock sculpturing.

Perhaps the most dramatic deposits are those containing piles of imbricated boulders. These piles take many forms, but include boulders up to 106 m^3 in volume and weighing as much as 286 tonnes. The boulders lie *en echelon* one against the other like fallen dominoes, often in parallel lines. At Jervis Bay, New South Wales, blocks weighing almost 100 tonnes have clearly been moved in suspension and deposited in this fashion above the limits of storm waves on top of cliffs 33 m above present sea level (Figure 3.11). The longest train of imbricated boulders exists at Tuross Head where 2- to 3-m-diameter boulders stand as sentinels one against the other, over a distance of 200 m at an angle to the coast.

The velocity of water necessary to move these large boulders can be related to their widths as follows:

$$\bar{v} = 5.2\, b_\mathrm{I}^{0.487} \qquad\qquad 3.1$$

$$v_\mathrm{min} = 2.06\, b^{0.5} \qquad\qquad 3.2$$

where $\bar{v}$ = mean flow velocity (m s^{-1})
   v_min = minimum flow velocity (m s^{-1})
   b = the intermediate axis or width of a boulder (m)
   b_I = intermediate diameter of largest boulders (m)

3.11 Dumped and imbricated sandstone boulders deposited 33 m above sea level along the cliffs at Mermaids Inlet, Jervis Bay, Australia. Note the intensely eroded surface across which waves flowed from left to right.

Table 3.2 presents the theorised velocities of tsunami flow required to move the boulders found along the New South Wales Coast using these two equations. The minimum theoretical flow velocity ranges between 2.2–4.2 m s^{-1}. Mean flows appear to have exceeded 5 m s^{-1} with values of 7.8–10.3 m s^{-1} being obtained on exposed ramps or at the top of cliffs.

It is more useful to be able to determine flow depth because, as shown in the previous chapter, this equates to the height of the tsunami wave at shore (Equation 2.26). The velocity necessary initially to move any boulder is related to the forces acting on that boulder. A boulder will move when the combined drag and lift forces exceed the restraining force of the boulder. This relationship can be expressed as follows:

$$F_d + F_l \geq F_{res} \qquad\qquad 3.3$$

where F_d =   the drag force

          F_l =   the lift force

        F_{res} =   the restraining force

$$F_d = 0.125\ \rho_w\ C_d\ ac^2\ v^2 \qquad\qquad 3.4$$

$$F_l = 0.125\ \rho_w\ C_l\ b^2c\ v^2 \qquad\qquad 3.5$$

$$F_{res} = 0.5\ (\rho_s - \rho_w)\ (ab^2c)\ g \qquad\qquad 3.6$$

          a =   boulder length (m)

          c =   boulder thickness (m)

TABLE 3.2 Velocities and Wave Heights of Tsunami as Determined from Boulders Found on Headlands along the New South Wales Coast

Location	Size (m)	Mean Velocity (m s⁻¹)	Lowest Velocity (m s⁻¹)	Wave Height (m)
Jervis Bay				
Mermaids Inlet	2.3	7.8	3.1	1.4
Little Beecroft Head				
ramp	4.1	10.3	4.2	2.7
clifftop	1.2	5.6	2.2	0.7
Honeysuckle Point	2.8	8.6	3.4	1.8
Tuross Head	1.3	5.9	2.3	0.8
Bingie Bingie Point	2.8	8.6	3.4	1.8
O'Hara Headland	1.1	5.5	2.2	0.7

Note: Sediment size refers to the mean width of the five largest boulders:
Source: Based on Young et al., 1996a.

ρ_s = density of a boulder (usually 2.7 g cm⁻³)
$\rho\rho_w$ = density of sea water (usually 1.024 g cm⁻³)
C_d = the coefficient of drag (typically 1.2 on dry land)
C_l = the coefficient of lift, typically 0.178
v = flow velocity (m s⁻¹)

If the relationship between tsunami velocity and height at the toe of the beach is substituted into these equations and solved for wave height, H_s, then the following equation is obtained:

$$H_s \geq 0.5\, a\, [(\rho_s - \rho_w)\, \rho_w^{-1}]\, [C_d\, (ac\, b^{-2}) + C_l]^{-1} \qquad 3.7$$

Note that the tsunami height estimated by this equation is conservative because the equation uses the velocity of a tsunami wave in shallow water (Equation 2.2) rather than the higher velocity that is possible across dry land (Equation 2.27). Equation 3.7 can be simplified by substituting in the values listed under Equation 3.6. It can be further simplified if the boulders being transported are nearly spherical. These simplifications are expressed as follows:

Simplified: $\qquad H_s \geq 0.82\, a\, (1.2\, ac\, b^{-2} + 0.178)^{-1} \qquad\qquad 3.8$

For spheres: $\qquad H_s \geq 0.6\, b \qquad\qquad\qquad\qquad\qquad\qquad\qquad 3.9$

It should be noted that the crucial parameter in the movement of boulders is the drag force, and that this is very sensitive to the thickness of the boulder. The thinner the boulder, the greater the velocity of flow required to initiate movement. Thickness is even more important than mass or weight. This point is illustrated in Figure 3.12 for two boulders of equal length and width but different thicknesses.

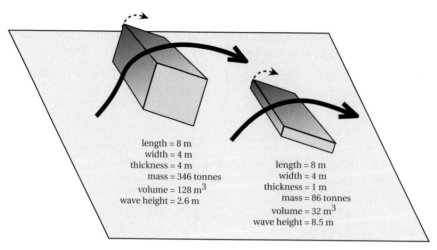

length = 8 m
width = 4 m
thickness = 4 m
mass = 346 tonnes
volume = 128 m^3
wave height = 2.6 m

length = 8 m
width = 4 m
thickness = 1 m
mass = 86 tonnes
volume = 32 m^3
wave height = 8.5 m

3.12   Illustration of the forces necessary to entrain two boulders having the same length and width, but different thicknesses.

Despite being four times larger in volume and weight, the rectangular boulder only requires a tsunami wave that is one third of the height of the wave needed to move the platy one. More realistically, a slab of concrete is more difficult to move than the office safe sitting on it. This effect is also illustrated in Figure 3.10. All of the boulders shown on the ramp required the same depth of water to be moved, despite the spherical ones being twice as large as the flatter ones.

Equations 3.7–3.9 have practical application for determining the height of a tsunami wave based upon the size of debris found afterwards. For instance, the largest boulder transported by the Flores tsunami shown in Figure 3.5 required a tsunami wave of about 2 m in height to move. Note that these flow depths are minimum values, because the tsunami wave would have been much higher at the shoreline than where the boulder was dumped. Calculations of the mean tsunami wave heights required to move boulders in the Jervis Bay region are included in Table 3.2. The calculations were based upon Equation 3.8, which is more exact. Boulders in the Jervis Bay region only required a tsunami wave 3 m high to be moved, even though waves higher than this must have been involved in order to transport boulders in suspension up cliffs. These examples illustrate how efficient tsunami waves are at initiating movement of bouldery material. The height of storm waves required to move the same size material is not nearly as small – a point that will be discussed in Chapter 4.

If tsunami wave height at shore equates with flow depth (Equation 2.26), the height of the tsunami can also be calculated using the spacing between boulder bedforms as follows:

$$H_s = 0.5\,L_s\,\pi^{-1}$$ 3.10

L_s = bedform wavelength (m)

However, boulder ripples or dunes are rare in nature. In the Jervis Bay region, boulder piles suggestive of megaripples exist at Honeysuckle Point and have a spacing of 60 m. The alignment of individual boulders in one pile, at an elevation of 16 m above sea level, shows foreset and topset bedding characteristic of a ripple. The minimum tsunami wave height for these features using Equation 3.10 is 9.5 m. This agrees with the flow depth required to construct the giant megaripples located nearby at Crocodile Head and described earlier.

The following scenario can account for the formation and transport of boulders along rocky coasts. On headlands or rock platforms where there is a cliff or ledge facing the sea, waves have to drown the cliff face before overtopping them. Tsunami do this by jetting across the top of the cliffs and developing a roller vortex in front of the cliff edge. Flow is thus thin (depths of 2.5–3.5 m) and violent (minimum velocities typically between 5 and 10 m s^{-1}). In jets, tsunami flow velocities may exceed 200 m s^{-1}. High-velocity flow first strips weathered bedrock surfaces of debris, exposing the underlying bedrock to large lift forces. Blocks, controlled in size by the thickness of bedding planes and the spacing of joints, may be too large to be entrained by this tsunami flow, but the lift forces can continually jar blocks until they eventually fracture into smaller pieces. This process occurs in preparation zones at the front of cliffs. When blocks are small enough, they then are transported across headlands or down platform and ramp surfaces. The lack of percussion marks or chipping on most boulders, some of which are highly fretted by chemical weathering, is suggestive of boulder suspension in sediment–starved flows without bed contact until the boulder is deposited. The minimum flow depth or tsunami wave height along the coast of New South Wales required in this scenario is 4 m, although larger waves must certainly have been present to move such material up cliffs 30 m high (Figure 3.11).

Turbidites
(Bouma and Brouwer, 1964; Masson et al., 1996)

There are few geomorphic features linked to tsunami described in the ocean. However, one of the most notable, turbidites, has received considerable attention in the literature. Submarine landslides generate both tsunami and turbidites. As a submarine landslide moves downslope under the influence of gravity, it disintegrates and mixes with water. The sediment in the flow tends to separate according to size and density, forming a sediment gravity flow called a turbidity current. The slurry in a turbidity current moves along the seabed at velocities between 20 and 75 km hr^{-1}, and can travel thousands of kilometres onto the abyssal plains of the deep ocean on slopes as low as 0.1°. As current velocity decreases, splays of sediment, known as turbidites, are deposited in submarine fans. Turbidite thickness depends upon the distance of travel and the amount of sediment involved in the original submarine landslide. In the Atlantic Ocean, individual turbidites have volumes of 100–200 km^3 – values that are sufficient to have generated tsunami of several metres amplitude at their source. Turbidity currents have not been directly observed; however, there is substantial indirect evidence for their existence. One of the best of these is the sequential breaking of telegraph cables laid across the seabed. The first noteworthy record occurred following the Grand Banks earthquake on 18 November 1929 off the coast of Newfoundland. Similar events have occurred off the Magdalena River delta (Colombia), the Congo delta, in the Mediterranean Sea north of Orléansville and south of the Straits of Messina, and in the Kandavu Passage, Fiji.

Turbidites generally are less than 1 m in thickness and form a distinct layered unit known as a Bouma sequence (Figure 3.13). The upward structure of a Bouma unit shows erosional marks in the underlying clays called sole marks, overlain by a massive graded unit (T_a), parallel lamination (T_b), rippled cross-lamination or convoluted lamination (T_c), and an upper unit of parallel lamination (T_d). The unit contains gravels and pebbles closer to the source, and fine sand and coarse silt out on the abyssal plain. The unit is overlain by pelagic ooze (T_{ep}) that has settled under quiet conditions between events. The basal contact below the coarse layer is sharp, while that above is gradational. The coarse layer is also well sorted and contains microfossils characteristic of shallow water. Interpretation of this sequence, supported by laboratory experiments, indicates deposition from a current that initially erodes the seabed, then deposits coarse sediment that fines as velocity gradually diminishes. More has been written about sediment density flows in sedimentology than on any other topic, and the deposits form one of the most common sedimentary sequences preserved in the geological record. Each turbidity deposit preserved in this record potentially could be a diagnostic signature of a tsunami event. Submarine landslides and their resultant tsunami have been very common features in the world's oceans throughout geological time.

EROSIONAL SIGNATURES OF TSUNAMI

Small-Scale Features
(Dahl, 1965, Baker, 1981; Kor et al., 1991; Young and Bryant, 1992; Bryant and Young, 1996; Aalto et al., 1999)

Tsunami can also sculpture bedrock in a fashion analogous to the s-forms produced by high-velocity catastrophic floods or surges from beneath icecaps in subglacial environments. S-forms include features such as muschelbrüche, sichelwannen, V-shaped grooves, cavettos, and flutes (Figures 3.1 and 3.14). They have been linked to palaeo-floods in Canada, the northwestern United States, Scandinavia, Britain, the Alps, and the Northern Territory, Australia. Tsunami flow over rocky headlands, at the velocities outlined in Chapter 2, has the hydrodynamic potential to generate cavitation or small vortices capable of producing sculptured

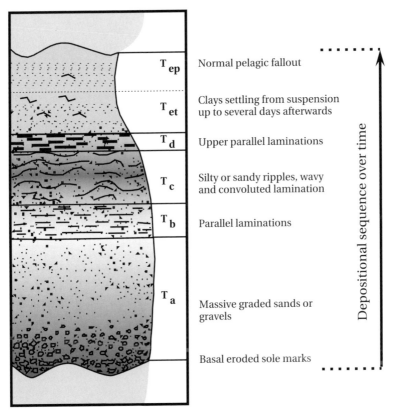

T_{ep}	Normal pelagic fallout	
T_{et}	Clays settling from suspension up to several days afterwards	
T_d	Upper parallel laminations	
T_c	Silty or sandy ripples, wavy and convoluted lamination	
T_b	Parallel laminations	
T_a	Massive graded sands or gravels	Depositional sequence over time
	Basal eroded sole marks	

3.13 A Bouma turbidite sequence deposited on the seabed following the passage of a turbidity current (based on Bouma and Brouwer, 1964).

forms. The spatial organisation of s-forms on headlands, often above the limits of storm waves, is a clear signature of tsunami in the absence of any other definable process. Individual s-forms and their hydrodynamics will be described in this chapter, while their spatial organisation into unique tsunami-generated land-scapes will be discussed in Chapter 4.

Cavitation is a product of high-velocity flow as great as 10 m s^{-1} in water depths as shallow as 2 m deep. At these velocities, small air bubbles appear in the flow. These bubbles are unstable and immediately collapse, generating impact forces up to thirty thousand times greater than normal atmospheric pressure. Cavitation bubble collapse is highly corrasive, and is the reason why propeller-driven ships cannot obtain higher speeds and dam spillways have a limited lifetime. In tsunami environments, cavitation produces small indents that develop instantaneously, parallel or at right angles to the flow, on vertical and horizontal bedrock surfaces. Cavitation features are widespread and consist of impact marks, drill holes, and sinuous grooves (Figures 3.1 and 3.14).

impact mark

drill holes and
comma marks

sinuous
grooves

muschelbrüche

sichelwanne

V-shaped
groove

flutes

pothole

hummocky
topography

trough

cavettos

3.14 Various cavitation features and s-forms produced by high-velocity tsunami flow over headlands. Note that the features are not scaled relative to each other.

Impact marks appear as pits or radiating star-shaped grooves on vertical faces facing the flow. It would be simple to suggest that such features represent the impact mark of a rock hurled at high velocity against a vertical rock face; however, such marks have also been found in sheltered positions or tucked into undercuts where such a process is unlikely (Figure 3.15). Drill holes are found over a range of locations on tsunami-swept headlands. Their distinguishing characteristic is a pit several centimetres in diameter bored into resistant bedrock such as tonalite or gabbroic diorite. Drill holes appear on vertical faces, either facing the flow or at right angles to it. Such marks also appear on the inner walls of large whirlpools. While it would be easy to attribute these features to marine borers, they often occur profusely above the limits of high tide.

3.15 A star-shaped impact mark on the face of the raised platform at Bass Point, New South Wales, Australia. Note as well the drill holes. The sheltered juxtaposition of these forms suggests that cavitation rather than the impact of rocks thrown against the rock face eroded them.

The most common type of drill mark appears at the end of a linear or sinuous groove and extends downwards at a slight angle for several centimetres into very resistant bedrock. In some cases, grooves also narrow with depth to form knifelike slashes a few centimetres deep. Sinuous drill marks are useful indicators of the direction of tsunami flow across bedrock surfaces. Sinuous grooves tend to extend no more than 2 m in length and have a width of 5–8 cm at most. Depth of cutting can vary from a few millimetres to several centimetres. In some cases, the sinuous grooves become highly fragmented longitudinally and form comma marks similar to those found in subglacial environments. Often they form *en echelon* in a chain-like fashion (Figure 3.16). They are not a product of storm waves or backwash because they evince internal drainage and do not join downslope. Sinuous grooves have been described for the Southeast Coast of New South Wales and for platforms near Crescent City, California, where tsunami appear to be a major process in coastal landscape evolution. It is tempting to credit their formation to chemical erosion along joints, microfractures, or igneous inclusions. Four facts suggest otherwise. First, while they may parallel joints, sinuous grooves diverge from such structures by up to 10°. Second, joints in bedrock are linear over the distances which grooves develop. The grooves described here are sinuous.

3.16 Sinuous grooves (by the hammer in the foreground) on a ramp at Tura Point, New South Wales, Australia. Flow is towards the camera. Note the flute forms in the background.

Third, sinuous grooves often appear as sets within the spacing of individual joint blocks. Finally, sinuous grooves occur only on polished and rounded surfaces characteristically produced by tsunami, and not on the highly weathered, untouched surfaces nearby.

S-forms also develop on surfaces that are smoothed and polished. This polishing appears to be the product of sediment abrasion. However, high water pressures impinging on bedrock surfaces can also polish rock surfaces. Flow vortices sculpture s-forms that can be categorised by the three-dimensional orientation of these eddies. The initial forms develop under small rollerlike vortices parallel to upslope surfaces. In this case muschelbrüche, sichelwanne, and V-shaped grooves are created (Figure 3.14). Muschelbrüche (literally mussel-shaped) are cavities scalloped out of bedrock, often as a myriad of overlapping features suggestive of continual or repetitive formation. While the features appear flat-bottomed, they have a slightly raised pedestal in the centre formed by unconstrained vortex impingement upslope onto the bedrock surface towards the apex of the scallop. They vary in amplitude from barely discernible forms to features having a relief greater than 15 cm. Their

3.17 Flutes developed on the crest of a ramp 14 m above sea level at Tura Point, New South Wales, Australia. The depressions on the sides of the flutes are cavettos. Flow is from left to right.

dimensions rarely exceed 1.0–1.5 m horizontally. Coastal muschelbrüche inevitably develop first on steeper slopes and appear to grade upslope into sichelwanne, V-shaped grooves, and flutes, as the vortices become more elongated and erosive. Sichelwanne have a more pronounced pedestal in the middle of the depression, while V-shaped grooves have a pointed rather than concave form downstream. V-shaped grooves have no subglacial equivalent and can span a large range of sizes. For instance, at Bass Point, New South Wales, V-shaped grooves approximately 10 m high and over 30 m wide have developed on slopes of more than 20°.

The term *flute* describes long linear forms that develop under unidirectional, high-velocity flow in the coastal environment. These are noticeable for their protrusion above, rather than their cutting below, bedrock surfaces (Figures 3.16 and 3.17). In a few instances, flutes taper downstream and are similar in shape to rock drumlins and rattails described for catastrophic flow in subglacial environments. In all cases, the steeper end faces the tsunami, while the spine is aligned parallel to the direction of tsunami flow. Flutes span a range of sizes, increasing in length to 30–50 m as slope decreases. However, their relief rarely exceeds 1–2 m. Larger features are called rock drumlins. The boulder trails at Tuross Head, mentioned above, are constrained by flutes. Fluted topography always appears on the seaward crest of rocky promontories where velocities are highest. Flutes often have faceting or cavettos superimposed on their flanks.

3.18 Small dissected potholes at the top of a 15-m-high headland at Atcheson Rock, 60 km south of Sydney, Australia. The potholes lie on the ocean side of the canyon structure shown in Figure 3.20.

(Faceting consists of chiselled depressions with sharp intervening ridges.) They represent either the impingement of vortices instantaneously upon a bedrock surface or hydraulic hammering of rock surfaces by high-velocity impacts. Cavettos are curvilinear grooves eroded into steep or vertical faces by erosive vortices. While cavettolike features can form due to chemical weathering in the coastal zone, especially in limestones, their presence on resistant bedrock at higher elevations above the zone of contemporary wave attack is one of the best indicators of high-velocity tsunami flow over a bedrock surface.

On flat surfaces, longitudinal vortices give way to vertical ones that can form hummocky topography and potholes (Figure 3.14). Potholes are one of the best features replicated at different scales by high-velocity tsunami flow. While large-scale forms can be up to 70 m in diameter, smaller features have dimensions of 4–5 metres. The smaller potholes also tend to be broader, with a relief of less than 1 m. The smaller forms can exhibit a central plug, but this is rare (Figure 3.18). Instead, the potholes tend to develop as flat-floored, steep-walled rectangular depressions, usually within the zone of greatest turbulence. While bedrock jointing may control this shape, the potholes' origin as bedrock-sculptured features is unmistakable because the inner walls are inevitably undercut or imprinted with cavettos. In places where vortices have eroded the connecting walls between potholes, a chaotic landscape of jutting bedrock with a relief of 1–2 m can be produced. This morphology – termed hummocky topography – forms where flow is unconstrained and turbulence is greatest. These areas occur where high-velocity water flow has changed direction suddenly, usually at the base of steep slopes or the seaward crest of headlands. The steep-sided, rounded, deep potholes found isolated on intertidal rock platforms, and attributed to mechanical abrasion under normal ocean wave action, could be catastrophically sculptured forms. Intriguingly, many of these latter features also evince undercutting and cavettos along their walls.

At the crests of headlands, flow can separate from the bedrock surface, forming a transverse roller vortex capable of eroding very smooth-sided, low, transverse troughs over 50 m in length and 10 m in width. Under optimum conditions, the bedrock surface is carved smooth and undular. In some cases, the troughs are difficult to discern because they have formed where flow was still highly turbulent after overwashing the crest of a headland. In these cases, the troughs are embedded in chaotic, hummocky topography. This is especially common on very low-angled slopes. Transverse troughs can also form on upflow slopes where the bed locally flattens or slopes downwards. Under these circumstances, troughs are usually short, rarely exceeding 5 m. The smoothest and largest features develop on the crests of broad undulating headlands.

Large-Scale Features
(Baker, 1978, 1981; Young and Bryant, 1992; Bryant and Young, 1996)

Large-scale features can usually be found sculptured or eroded on rock promontories, which protrude seaward onto the continental shelf. Such features require extreme run-up velocities that can only be produced by the higher or longer waves (mega-tsunami) generated by large submarine landslides or meteorite impacts in the ocean. One of the most common features of high-velocity overwashing is the stripping of joint blocks from the front of cliffs or platforms forming inclined surfaces or ramps (Figure 3.10). In many cases, this stripping is aided by the detachment of flow from surfaces, a process that generates enormous lift forces that can pluck joint-controlled rock slabs from the underlying bedrock. Where standing waves have formed, then bedrock plucking can remove two or three layers of bedrock from a restricted area, leaving a shallow, closed depression on the ramp surface devoid of rubble and unconnected to the open ocean (Figure 3.19). Ramps are obviously structurally controlled and have an unusual juxtaposition beginning in cliffs up to 30 m above sea level and, sloping down, flow often into a cliff. If these high velocities are channelised, erosion can produce linear canyon features 2–7 m deep and pool-and-cascade features incised into resistant bedrock on the lee side of steep headlands. These features are most prevalent on platforms raised 7–8 m above modern sea level. All features bear a resemblance to

3.19 The ramp at Bannisters Head on the New South Wales South Coast. The tsunami wave approached from the bottom right-hand corner of the photograph. Erosion increases up the ramp that rises 16 m above sea level. The blocky boulders at the top of the ramp are over 4 m in diameter.

3.20 Canyon feature cut through the 20-m-high headland at Atcheson Rock by a tsunami moving from bottom left to top right. Evidence exists in the cutting for subsequent downcutting of 2–4 m by a more recent tsunami. The latter event also draped a 0.5- to 2.0-m-thick dump deposit over the headland to the right. This deposit has been radiocarbon dated at around A.D. 1500. The small potholes in Figure 3.18 are located on the leftmost portion of the headland, while the large whirlpool shown in Figure 3.23 is located on the far side of the canyon.

3.21 Inverted keellike forms at Cathedral Rocks, 90 km south of Sydney, Australia. Flow came from the right. The stacks formed between horizontal, helical (eggbeater) vortices. A sea cave is bored into the cliff's downflow.

the larger canyon-and-cascade forms carved in the channelled scabland of the western United States. Wave-breaking may also leave a raised buttelike structure at the seaward edge of a headland, separated from the shoreline by an eroded depression. This feature looks similar to an inverted toothbrush (Figure 3.20). Large-scale fluting of a headland can also occur, and on smaller rock promontories, where the baseline for erosion terminates near mean sea level, the resulting form looks like the inverted keel of a sailboat. Often this keel has a cockscomb peak produced by the rapid, random erosion of multiple vortices (Figure 3.21). While technically a sculpturing feature, the cockscomb looks as if it has been hydraulically hammered. On narrow promontories, vortices can create arches. While arches have been treated offhandedly in the literature as products of chemical weathering or long term wave attack along structure weaknesses, close investigation shows that many are formed by vortices. It is also possible for these horizontal vortices to form in the lee of stacks and erode promontories from their landward side (Figure 3.22).

Perhaps the most impressive features are whirlpools formed in bedrock on the sides of headlands. Whirlpools and smaller potholes commonly formed under catastrophic flow in the channelled scabland of Washington State. In coastal environments, whirlpools often contain a central plug of rock and show evidence

3.22 An arch at Narooma on the South Coast of New South Wales, Australia. The tsunami wave swept towards the viewer along the platform in the background. A vortex on the seaward side eroded most of the arch. Joints in the rocks do not control the arch, nor does the rock face show evidence of major chemical weathering.

of smaller vortices around their rim. Whirlpools can reach 50–70 m in diameter with central plugs protruding 2–3 m vertically upwards from the floor of the pit at the quiescent centre of the vortex. The best coastal example occurs on the south side of Atcheson Rock south of Bass Point, New South Wales (Figure 3.23). Here a large vortex, spinning in a counterclockwise direction, produced smaller vortices rotating around its edge on the upflow side of a headland. The overall whirlpool is 10 m wide and 8–9 m high. The central plug stands 5 m high and is surrounded by four 3-m-diameter potholes, one of which bores another 3 m below the floor of the pit into resistant basalt. The counterclockwise rotation of the overall vortex, produces downward-eroded helical spirals that undercut the sides of the pit, forming spiral benches. Circular or sickle-shaped holes were drilled, by cavitation, horizontally into the sides of the pothole and into the wall of the plug. Under exceptional circumstances, the whirlpool can be completely eroded, leaving only the plug behind. The frontispiece at the beginning of the text shows an example of this cut into aplite (granite) at Cape Woolamai on the South Coast of Victoria Australia.

3.23 Whirlpool bored into bedrock on the south side of Atcheson Rock, New South Wales, Australia. The presence of drill holes and smaller potholes at the bottom indicates the existence of cavitation and multiple vortices respectively. Helical flow operated in a counterclockwise direction around the central plug

Flow Dynamics

(Alexander, 1932; Dahl, 1965; Wiegel, 1970; Fujita, 1971; Baker, 1978, 1981; Allen, 1984; Kor et al., 1991; Grazulis, 1993; Shaw, 1994; Bryant and Young, 1996)

Any model of the flow dynamics responsible for tsunami-sculptured bedrock terrain must be able to explain a range of features varying from sinuous cavitation marks several centimetres wide to whirlpools over 10 m in diameter. One of the controlling variables for the spatial distribution of these features is bed slope that can be higher than 10° at the front of promontories. Even a slight change in angle can initiate a change in sculptured form. For instance, sinuous cavitation marks can form quickly, simply by steepening slope by 1–2°. Similarly, flutes can develop with the same increase in bed slope. This suggests that new vortex formation or flow disturbance through vortex stretching is required to initiate an organised pattern of flow vortices able to sculpture bedrock. Because bedrock sculpturing features rarely appear close to the edge of a platform or headland, vortices did not exist in the flow before the leading edge of the tsunami wave struck the coastline.

Bedrock-sculptured features are created by six flow phenomena: Mach–Stem waves, jetting, flow reattachment, vortex impingement, horseshoe or hairpin vortices, and multiple-vortex formation. Mach–Stem wave formation was described in Chapter 2. It occurs when waves travel obliquely along a cliff line. The wave height can increase at the boundary by a factor of two to four times. The process is insensitive to irregularities in the cliff face and is one of the mechanisms allowing tsunami waves to overtop cliffs up to 80 m high.

Tsunami, because of their long wavelength, behave like surging waves as they approach normal to a shoreline. Jetting is caused by the sudden interruption of the forward progress of a surging breaker by a rocky promontory. The immediate effect is twofold. First, there is a sudden increase in flow velocity as momentum is conserved and vortex formation is initiated. Second, the sudden velocity increase is sufficient for cavitation and creates lift forces that can pluck blocks of bedrock from the bed at the front of a platform. Equation 2.27 can be used to calculate theoretical velocities on some of the ramp surfaces. If the height of the tsunami wave is equivalent to the cliff fronting ramps, then the calculated velocities can exceed 20 m s^{-1}. These velocities are well in excess of the 10 m s^{-1} threshold required for cavitation.

The third phenomenon occurs if flow separates from the bed at the crest of a rocky promontory. This occurs at breaks of slope greater than 4°. Rock platforms overwashed by tsunami have changes in slope much greater than this. Some distance downstream, depending upon the velocity of the jet, water must reattach to the bed. Where it does, flow is turbulent and impingement on the bed is highly erosive. Standing waves may develop in the flow, leading to large vertical lift forces under crests. The bedrock plucking at the front of platforms and in the lee of crests is a product of this process.

S-forms spatially change their shape depending upon the degree of flow impingement and vortex orientation (Figure 3.24). In coastal environments, the flow by tsunami overwashing bedrock surfaces is not confined. The high velocity and sudden impact of the vortex on the bedrock surface causes the vortex to ricochet upwards, and the unconfined nature of the flow permits the vortex to lose contact rapidly with the bed. This produces features that begin as shallow depressions, scour downflow, and then terminate suddenly, leaving a form that is gouged into the bedrock surface with the steep rim downstream. Narrow longitudinal vortices impinging upon the bed at a low angle produce muschelbrüche that become sichelwannen and V-shaped grooves at higher angles of attack. Obstacles in the flow boundary layer form horseshoe vortices. As flow impinges upon an obstacle, higher pressures are generated that cause flow deflection and separation from the bed. This generates opposing, rotating vortices that wrap around the obstacle like a horseshoe or hairpin. The vortices scour into the bedrock surface downflow. Because these vortices lie within the boundary layer, they are subject to intense shearing by the overlying flow. This shearing fixes the vortices in position and keeps them straight. Horseshoe or hairpin vortices produce flutes that are progressively eroded into the face of platforms and headlands as long as high-velocity overwashing is maintained (Figure 3.17). At the largest scale the keellike stacks shown in Figure 3.21 formed between horizontal, helical (eggbeater) vortices. These helical vortices, besides eroding vertical stacks, have the capacity to bore caves into cliffs and form arches (Figure 3.22).

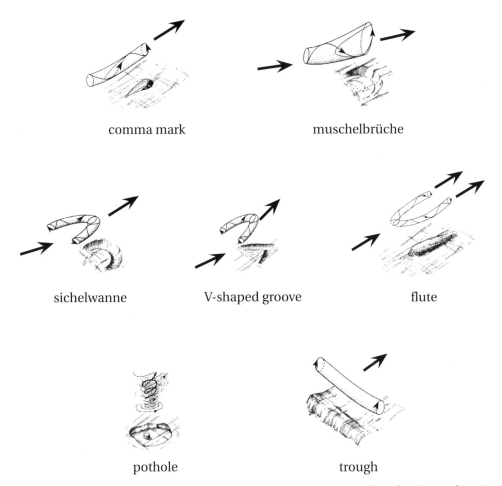

comma mark

muschelbrüche

sichelwanne

V-shaped groove

flute

pothole

trough

3.24 Types of vortices responsible for bedrock sculpturing by tsunami (based on Kor et al., 1991 and Shaw, 1994).

Multiple-vortex formation occurs at the largest scale and includes kolks and tornadic flow. Kolks are near-vertical vortices whose erosive power is aided by turbulent bursting. They are produced by intense energy dissipation in upward vortex action. The steep pressure gradients across the vortex produce enormous hydraulic lift forces. Kolks require a steep energy gradient and an irregular rough boundary to generate flow separation. These conditions are met when the tsunami waves first meet the steeper sides of headlands; however, the process does not account for the formation of whirlpools. Kolks may also be formed by macro turbulence whereby eddies grow within larger rotational flow. This latter concept has been invoked to account for the formation of tornadoes and incorporated into a model of multiple-vortex tornado formation that may be more appropriate in explaining the formation of whirlpools in tsunami-sculptured terrain.

Multiple-vortex tornadoes are produced by vortex breakdown within a tornado as air is pulled downwards from above into the low pressure of the tornado. Smaller secondary vortices rotating in the same direction develop around the circumference of the larger parent vortex. Tsunami-generated whirlpools are formed by the vertical vortices embedded in flow overwashing a headland. These vortices expand and increase in rotational velocity with time. When the vortex is wide enough, water is pulled down into the centre of the vortex and lifted upwards at the circumference (Figure 3.25). When velocity is high enough, the vortex bores into bedrock and locks into position. Velocities exceed that necessary for cavitation as shown by the presence of drill marks in the whirlpool at Atcheson Rock (Figure 3.23). Towards the centre of the vortex system, the directions of water movement in the minivortex and parent vortex are opposite and begin to cancel each other out. Here the resultant flow velocity is the rotational velocity of the parent vortex minus that of the mini-vortex. These lower rotational velocities aid the collapse of water into the centre of the vortex, but at velocities that are too low to erode bedrock for part of the time. This process leaves a plug of bedrock in the middle of the whirlpool. Once multiple vortices form, the system of flow becomes self-perpetuating as long as there is flow of water to maintain the parent vortex. The fact that the plug height is always lower than the pothole walls suggests that multiple vortices develop in the waning stages of tsunami overwash after the crest of the tsunami has swept past and established the parent vortex.

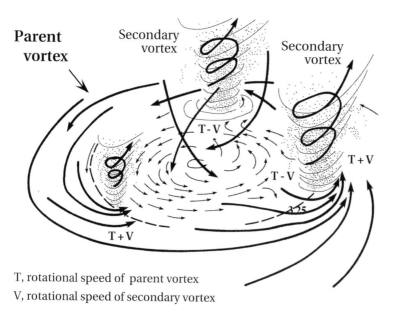

T, rotational speed of parent vortex

V, rotational speed of secondary vortex

3.25 Model for multiple-vortex formation in bedrock whirlpools (based on Fujita, 1971, and Grazulis, 1993).

Whirlpools form when flow velocities increase first through convergence of water over bedrock and then through funnelling at preferred points along the coast. Critical rotational velocities required to erode resistant bedrock also take time to build up; however, the flow under the crest of a tsunami wave is only sustained for a few minutes at most. Once a critical velocity is reached, bedrock erosion commences. The spin-up process causes vortex erosion to develop and terminate quickly, as is demonstrated by the fact that some potholes are only partially formed. Whirlpools, such as the one in Figure 3.23, are formed instantaneously in the space of minutes rather than by the cumulative effect of many wave events. Whereas minivortices in tornadoes can freely circumscribe paths around the wall of the tornado, those in whirlpools are constrained by the bedrock they are eroding.

This chapter has attempted to show that a myriad of distinct signatures produced by tsunami exist besides the descriptions of anomalous sediment layers that pervade the current literature. The most impressive of these additional signatures is produced by bedrock sculpturing. Most of the descriptions of individual signatures presented in this chapter indicate that they do not occur as isolated features but appear in combination with each other to form spatially organised suites that dominate some coastal landscape – such as rocky headlands. The description of these varied tsunami-generated landscapes constitutes the focus of Chapter 4.

CHAPTER FOUR

Coastal Landscape Evolution

INTRODUCTION

Large seismic disturbances on the sea floor, submarine slides, and meteorite impacts with the ocean can create tsunami waves that spread oceanwide with profound effects on coastal landscapes. They can generate run-up heights thirty times greater than their open ocean wave height and sweep several kilometres inland. This penetration inland can only be duplicated on flat coastlines by storms if they are accompanied by a significant storm surge. Tsunami are thus catastrophic events and can leave a permanent imprint on the landscape. There has been little appreciation in the literature that coastal landscapes may reflect tsunami processes rather than those induced by wind-generated waves and wind. Catastrophic events – termed catastrophism – are not well respected in modern science. This chapter will describe the role of catastrophism in the development of modern geological thinking, show in more detail how tsunami are different from storms, describe various models of tsunami-generated landscapes, and illustrate these models with examples from around the world.

CATASTROPHISM VERSUS UNIFORMITARIANISM
(Clifton 1988; Bryant, 1991; Alvarez, 1997)

A tsunami was involved in one of the pivotal debates of modern scientific development. On 1 November 1755, an earthquake, with a possible surface magnitude of 9.0 on the Richter scale, destroyed Lisbon, then a major centre of European civilisation. Shortly after the earthquake a tsunami swept into the city, and over the next few days fire consumed what was left of Lisbon. The event sent shock waves through the salons of Europe at the beginning of the Enlightenment. The earthquake struck on All Saints' Day, when many Christian believers were praying in church. John Wesley viewed the Lisbon earthquake as God's punishment for the licentious behaviour of believers in Lisbon, and retribution for the severity of the Portuguese Inquisition. Immanuel Kant and Jean-Jacques Rousseau viewed the disaster as a natural event and emphasised the need to avoid building in hazardous places. The Lisbon earthquake also gave birth to scientific study of geological events. In 1760, John Mitchell, geology professor at Cambridge University, documented the spatial effects of the earthquake on lake levels throughout Europe. He found seiching along the coastline of the North Sea and in Norwegian fjords, Scottish lochs, Swiss Alpine lakes, and rivers and canals in Western Germany and the Netherlands. He deduced that there must have been a progressive, wavelike tilting of the Earth outwards from the centre of the earthquake and that this was different from the type of wave produced by a volcanic explosion.

Mitchell's work in 1760 on the Lisbon earthquake effectively represented the separation of two completely different philosophies for viewing the physical behaviour of the natural world. Beforehand, the catastrophists – people who believed that the shape of the Earth's surface, the stratigraphic breaks evidenced in rock columns, and the large events that were associated with observable processes were cataclysmic – dominated geological methodology. More important, these catastrophic processes were Acts of God. The events had to be cataclysmic in order to fit the many observable sequences observed in the rock record into an age for the Earth of 4004 B.C., determined from Biblical genealogy. Charles Lyell, one of the fathers of geology, sought to replace this catastrophe theory with gradualism – the idea that geological and geomorphic features were the result of cumulative slow change by natural processes operating at relatively constant rates. This idea implied that processes that shape the Earth's surface followed laws of nature defined by physicists and mathematicians. Whewell, in a review of Lyell's work, coined the term *uniformitarianism,* and subsequently a protracted debate broke out on whether or not the slow processes we observe at present apply to past unobservable events. To add to the debate, the phrase "The present is the key to the past" was also coined.

In fact, the idea of uniformitarianism involves two concepts. The first implies that geological processes follow natural laws applicable to science. There are no Acts of God. This type of uniformitarianism was established to counter the arguments raised by the catastrophists. The second concept implies constancy of rates of change or material condition through time. This concept is nothing more than inductive reasoning: The type and rate of processes operating today characterise those that have operated over geological time. For example, waves break upon a beach today in the same manner as they would have a hundred million years ago, and prehistoric tsunami behave the same as modern ones described in our written records. If one wants to understand the sedimentary deposits of an ancient tidal estuary, one has to do no more than go to a modern estuary and study the processes at work. Included in this concept is the belief that physical landscapes such as modern floodplains and coastlines evolve slowly.

Few geomorphologists or geologists who study Earth surface processes and the evolution of modern landscapes would initially object to this concept. However, the concept does not withstand scrutiny. For example, there is no modern analogy to the nappe mountain building processes that formed the Alps, nor to the mass extinctions and sudden discontinuities that have dominated the geological record. Additionally, no one who has witnessed a fault line being upthrust during an earthquake or Mt. St. Helens wrenching itself apart in a cataclysmic eruption would agree that all landscapes develop slowly. As Thomas Huxley so aptly worded it, gradualists had saddled themselves with the tenet of *Natura non facit saltum* – Nature does not make sudden jumps. J Harlen Bretz from the University of Chicago challenged this tenet in the 1920s. Bretz attributed the formation of the scablands of Eastern Washington to catastrophic floods. He subsequently bore the ridicule and invective of the geological establishment for the next forty years for proposing this radical idea. Not until the 1960s was Bretz proved correct when Vic Baker of the University of Arizona interpreted space-probe images of enormous channels on Mars as features similar to the Washington scablands. At the age of eighty-three, Bretz finally received the recognition of his peers for his seminal work.

Convulsive events are important geological processes, and major tsunami can be defined as convulsive. More important, it will be shown in the remainder of this book that mega-tsunami – in many cases bigger than tsunami described in historic and scientific documents – have acted to shape coastlines. Some of these mega-tsunami events have occurred during the last millennium. In coastal geomorphology, existing scientific custom dictates that in the absence of convincing proof, the evidence for convulsive events must be explained by commoner events of lesser magnitude – such as storms. However, this restriction should not imply that storms can be ubiquitously invoked to account for all sediment deposits or coastal landscapes that, upon closer inspection, have anomalous attributes more correctly explained by a different and rarer convulsive process. The alternate phenomenon of tsunami certainly has the potential for moving sediment and moulding coastal landscapes to the same degree as, if not more efficiently than, storms. Tsunami can also operate further inland and at greater heights above sea level. Tsunami have for the most part been ignored in the geological and geomorphological literature as a major agent of coastal evolution. This neglect is unusual considering that tsunami are common, high-magnitude phenomena producing an on-surge with velocities up of 15 m s^{-1} or more.

TSUNAMI VERSUS STORMS
(Coleman, 1968; Bourgeois and Leithold, 1984; Morton, 1988; Bryant et al., 1996)

In the past, rapid coastal change, especially in sandy sediments, has been explained by invoking storms. Where sediments have previously incorporated boulders, the role of storms versus tsunami has become problematic. However much of the latter debate deals with isolated boulders or boulders chaotically mixed with sand deposited on low-lying coastal plains or atolls. The pattern of stacked and aligned boulders found along the New South Wales Coast has never been linked in the literature to storm waves or surges. More important, even a casual reconnaissance of the New South Wales coastline will show that storms inadequately account for the bedrock-sculptured features dominating the rocky coast. The uniform alignment of such forms, often not structurally controlled, also rules out chemical weathering. Along this coast, the largest storms measured this century in 1974 and 1978 only generated deep-water waves of 10.2 m, while the maximum probable wave for the coast is only a few metres higher. The effectiveness of these wave heights cannot be exacerbated at shore by storm surges because the narrowness of the shelf and the nature of storms limit surges here to less than 1.5 m. In addition, storm wave periods rarely exceed 15 seconds and then for only very short durations. Waves bigger than this, but more important, of several minutes duration, are required to account for the sustained high-velocity flows required to sculpture highly resistant bedrock into the features described in the previous chapter. Tsunami appear to be the only mechanism capable of providing these conditions along the coast.

The forms deposited by tsunami at the coastline can be easily interpreted as swell- or storm-wave–built features. However, the internal fabric of such deposits is different from that produced by storm waves. For example, ridges and mounds built above the high-tide line at Bass Point, New South Wales, contain a chaotic mixture of boulders, gravel, sand, and shell. The steep sides of these landforms are presently being combed down by storm waves. Features with such a wide range of grain sizes bear no relationship to anything described in the sedimentological literature as being storm- or swell-wave deposited. Their closest analogue is an ice-push ridge; however, the New South Wales Coast last came under the influence of sea ice during the Permian, 200 million years ago.

The effect of backwash in tsunami is rarely mentioned in the literature. Near the coast, such flow is generally channelised as the volume of overwash drains seaward through inlets and along defined drainage channels. More important, undertow whereby flow in the water column moves seaward along the seabed can occur out to the shelf edge in depths of 100–130 m of water. Continental shelf profiles may be a product of repetitive combing by tsunami-induced currents. If this undertow contains any sediment, then it behaves as a density current and powerful ebb currents of 2–3 m s^{-1} can sweep down the continental slope and along the abyssal plain under the passage of a tsunami wave. These currents can scour the seabed and deposit sand and gravels considerable distances from the shelf edge. Such flows may account for the presence of coarse, winnowed, lag deposits and 2- to 3-m-diameter boulders at the toe of the continental slope in water depths of 200–400 m. The above processes cannot be mimicked by storm waves.

Gravel and Cobble Beaches
(Bourgeois and Leithold, 1984; Shiki and Yamazaki, 1996; Dawson, 1999)

Coarse-grained beaches are invariably the product of shoreward movement of gravels, cobbles, and boulders by storm waves. Unless storm waves can overwash a beach, sand is generally moved shoreward only by fair-weather swell. Gravel and cobble beaches are characterised by shape and size sorting of particles. Larger disc-shaped particles tend to move to the top of the beach where they are deposited into a berm that may develop at the limit of storm-wave run-up. Smaller spherical particles tend to accumulate at the base of the foreshore. This difference is due to the greater potential for suspension transport of discs in swash and the greater rollability of spheres back down the beach face under backwash. A storm beach has four distinct zones. Towards the berm or high-tide mark, large disc-shaped cobbles are deposited. A zone of imbricated disc-shaped pebbles facing upslope fronts this deposit. The width of this zone increases as the amount of wave energy at the beach face increases. Lower down the beach, the spacing between cobbles is infilled by rod-shaped and spherical cobbles, while the low-tide mark is dominated by spherical cobbles. On bimodal beaches, sand tends to be trapped in the infill zone. The shape of particles is not created on the beach, but rather particles are shaped elsewhere and transported to the beach face. The internal fabric of coarse tsunami deposits is distinctly different from that deposited by storm waves. First, no such zonation exists in coarse-grained tsunami deposits because tsunami carry the full range of sediment sizes onto a beach. Second, the deposition of this sediment is usually chaotic, with the mixing of gravels, cobble, and boulders together. There also appears to be little sorting by shape. Third, tsunami deposits contain a significant fraction of freshly fractured, angular particles as well as broken, rounded ones. The rounded particles originate from a nearby beach overwashed by the wave, but the angular material is the product of vortex erosion of exposed bedrock surfaces. Finally, tsunami can deposit layers of sediment long distances from the coast, whereas storms appear to build up an asymmetric wedge of coarse-grained sediment that rarely extends 50–100 m from the shore.

Gravel and cobble storm-built beaches tend to be characteristic of eroding coasts, although there are exceptions. Thus, their preservation potential is poor. Coarse-grained tsunami beach deposits have a higher potential for preservation, if not in the longer geological record, then certainly at high sea-level stillstands over the last few million years on tectonically stable coasts. These sediments often are

deposited above the limits of storms, and unless eroded by subsequently larger
tsunami, will remain stranded above the active coastal zone on such coasts.

Movement of Boulders

(Bascom, 1959; Young et al., 1996a; Nott, 1997; Komar 1998; Solomon and
Forbes, 1999)

Tsunami and storm waves differ in the way that they transport bouldery material. The forces of storm waves and their ability to destroy stony breakwalls and move
large boulders on rock platforms are well documented. Under exceptional circumstances waves greater than 16 m have been recorded in the North Atlantic. There is
evidence of a single boulder being tossed 25 m above sea level on the island of Surt-
sey in Iceland by such waves. Waves with a force of 3 tonnes m^{-2} in the 1800s moved
blocks weighing 800 and 2,600 tonnes into the harbour at Wick, Scotland. Moreover,
in 1912, a boulder weighing 200 tonnes was moved 50 m on a rock platform at Sydney during a storm. Active pebble beaches sit 30 m above sea level on cliffs along the
West Coast of Ireland. In Hawaii, Hurricane Iniki in 1991 transported isolated boulders about 0.5 m in diameter along the coast of Maui. There are many stories of pebbles and even cobbles being hurled against lighthouse windows situated on cliff
tops. Probably one of the best documented incidences showing the ability of storms
to transport coarse material to the tops of cliffs occurred during Tropical Cyclone
Ofa on the South Coast of Niue in the Southwest Pacific on 5 February 1990. Niue is
a raised, relict coral atoll fringed by limestone cliffs rising up to 70 m above sea level.
A platform reef, up to 120 m wide, has developed at the base of the cliffs. Ofa generated winds of more than 170 km hr^{-1} and waves with a maximum significant height
of 8.1 m. As it approached Niue, it produced waves 18 m high along the coast. The
effect of these waves along the cliffs was dramatic. At Alofi, waves broke above the
roof of a hospital situated on an 18-m-high cliff. The lower floor of a hotel was
severely smashed by the impact of storm-tossed debris. Coarse gravel and boulders
2–3 m in diameter were flung inland over 100 m from the cliff line.

Isolated boulders tossed by storms, however, are different from the accumula-
tions of boulders deposited by tsunami. The analysis used in the previous chapter
to define the wave height of tsunami capable of moving boulders of different
shape and size can be applied to storm waves. The velocity of a wind-generated
storm wave breaking at the shoreline can be approximated as follows:

$$v = (gH_b)^{0.5} \hspace{4cm} 4.1$$

This is half the velocity of an equivalent tsunami wave travelling over dry land.
When Equation 4.1 is combined with Equation 3.3 and solved for wave height, the
following relationship is obtained:

$$H_b \geq 4 H_s \hspace{4cm} 4.2$$

TABLE 4.1   Comparison of Tsunami and Storm-Wave Heights Required to Transport the Boulders Mentioned to This Point in the Text

Location	Boulder Width (m)	Height of Tsunami at Shore (m)	Breaking Storm-Wave Height (m)
Jervis Bay			
Mermaids Inlet	2.3	1.4	5.6
Little Beecroft Head			
ramp	4.1	2.7	10.8
clifftop	1.1	0.7	2.8
Honeysuckle Point	2.8	1.8	7.2
Tuross Head	1.3	0.8	3.2
Bingie Bingie Point	2.8	1.8	7.2
O'Hara Headland	1.1	0.7	2.8
Sampson Pt., WA	1.0	0.6	2.4
Flores, Indonesia	1.5	2.0	8.0

Source: Based on Young et al., 1996a.

Table 4.1 presents a comparison of the wave heights of tsunami and storms necessary to transport the boulders described so far in this book. Clearly, tsunami waves are more efficient than storm waves at transporting boulders inland. This fact becomes more relevant knowing that storm waves break in water depths 1.28 times their wave height. Hence, the heights for storm waves shown in Table 4.1 require much larger waves offshore to overcome the effects of wave breaking. Storm waves lose little energy only along coasts where cliffs plunge into the ocean. Unless storm waves reach a platform before breaking, they do not have the capability to move boulders more than 2 m in diameter on most rocky coasts. While storm waves under ideal conditions can transport boulders, as shown by the effect of Cyclone Ofa on the island of Niue, they are very unlikely to transport boulders and deposit them in imbricated piles at the top of cliffs. Boulder imbrication in contrast to pebble imbrication is rarely referenced in the coastal literature. Along the New South Wales South Coast, imbrication is a dominant characteristic of boulder piles and a signature of tsunami. For example, at Tuross Heads, contact-imbricated boulders with an upstream dip of 30–50°, are piled en echelon in two single files over a distance of 150 m. This pattern is similar to that produced in erosive fluvial environments by high-magnitude, unidirectional flow. The size of the imbricated boulders not only matches that produced by high-magnitude flows in streams but also that produced by the catastrophic flows hypothesised for meltwaters in front of, or beneath, large glaciers. These imbricated piles do not require continuous flow, but rather a simple succession of three or more large waves – a feature that is commonly characteristic of a tsunami wave train.

TYPES OF COASTAL LANDSCAPES CREATED BY TSUNAMI

The modelling of spatial variation in modern sedimentary environments dominated by a specific process is termed a facies model. These models are well formulated for many processes such as tides, rivers, and waves. Facies models are used to interpret the geological rock record. Rarely do facies models consider bedrock erosion. For example, while much literature has been written on the formation of beaches with their attendant surf zones and the long-term development of sandy coastal barriers, little has been written on the erosion of rocky coasts by waves. What has been written is elementary and perfunctory. What is a sea cave, a coastal stack, or an arch? The literature frequently refers to such features but sheds little light as to their formation, especially in resistant and massively bedded bedrock. Even the formation of rock platforms, the most studied feature of rocky coasts, is ambiguous. Erosion of platforms across bedrock of differing lithologies or structures is attributed to long-term wave abrasion or chemical erosion – processes that have been measured at localised points but never broadly enough to establish them as the main processes. Catastrophic tsunami waves have the power to erode such surfaces in one instant. Many other aspects of cliffed and rocky coasts are treated in a similarly cursory fashion. If sea caves, stacks, arches, bedrock-ramped surfaces, sheared cliffs, imbricated boulder piles, and chevron ridges are the signature of tsunami, then how common are catastrophic tsunami in shaping the world's coastline?

Sandy Barrier Coasts
(Minoura and Nakaya, 1991; Andrade, 1992)

Large sections of the world's sandy coastline are characterised by barriers either welded to the coastline or separated from it by shallow lagoons. The origin of these barriers has been attributed to shoreward movement of sediment across the shelf by wind-generated waves, concomitantly with the Holocene rise of sea level. Lagoons form where the rate of migration of sand deposits lags the rate at which the rising sea drowns land. This theory of barrier formation ignores the role of tsunami as a possible mechanism, not only for shifting barriers landward, but also for building them up vertically. For example, at Bellambi along the New South Wales Coast of Australia, tsunami-deposited sands make up 20 to 90% of the vertical accretion of a barrier beach that stands 3–4 m above present sea level (Figure 4.1). This beach will be described in more detail subsequently.

Tsunami in shallow water are constructional waves with the potential to carry large amounts of sand and coarse-grained sediment shoreward. Large aeolian dunes that may develop on stable barriers do not necessary impede tsunami. Instead, tsunami can overwash such forms, reducing the height of the dunes, depositing sediment in dune hollows, and spreading sand as a thin sheet across backing lagoons (Figure 4.2). On the other hand, storm waves tend to surge through low-lying gaps in dune fields,

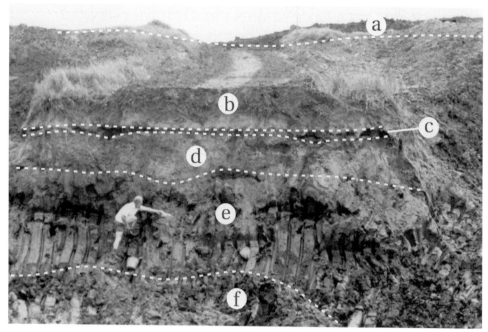

4.1 Section through the barrier beach at Bellambi, New South Wales, Australia: **a,** modern dune; **b,** Holocene barrier sand dated using thermoluminescence (TL) at 7,400 BP; **c,** clay unit; **d,** tsunami sand TL dated at 25,000 BP; **e,** Holocene estuarine clay radiocarbon dated at 5,100 BP; and **f,** Pleistocene estuarine clay TL dated at more than 45,000 years in age. The TL date of the tsunami sand indicates that the sands are anomalous. The Holocene ages overlap.

sporadically depositing lobate washover fans in lagoons. Rarely will these fans penetrate far into a lagoon or coalesce. The seaward part of the barrier can also be translated rapidly landward tens of metres by the passage of a tsunami. This occurred along the barrier fronting Sissano lagoon during the Papua New Guinea tsunami of 17 July 1998. The welding of a barrier to a coastline overtop lagoonal sediments may signify the presence of recent tsunami along a coast where historical evidence for such events is lacking. Preexisting tidal inlets are preferred conduits for tsunami. Sediment-laden tsunami may deposit large, coherent deltas at these locations, and these may be mistaken for flood tidal deltas. These features may be raised above present sea level or form shallow shoals inside inlets. Because the sediment was deposited rapidly, these deltas may end abruptly landward in lagoons and estuaries, forming sediment thresholds that are stable under present-day tidal flow regimes. Channels through the lagoon can also be scoured by tsunami, with sediment being deposited on the landward side of lagoons as splays.

If the volume of sand transported by a tsunami is large, then a raised backbarrier platform may form from coalescing overwash fans or smaller lagoons may be completely infilled. The height of either the backbarrier platform or infilled lagoon may lie several metres above existing sea level. These raised lagoons may

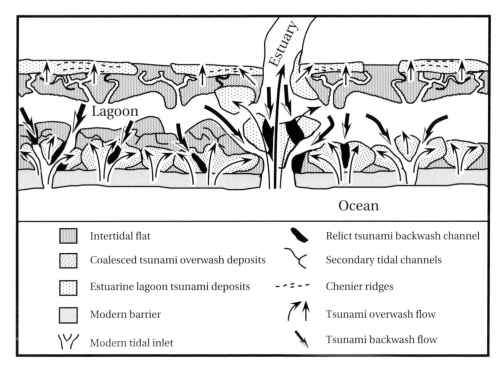

4.2 Model of the effect of tsunami upon a sandy barrier coastline (based on Minoura and Nakaya, 1991, and Andrade, 1992).

be misinterpreted as evidence for a higher sea level. If these surfaces are not covered by seawater or quickly vegetated, they may be subject to wind deflation, with the formation of small hummocky dunes. Under extreme conditions, tsunami waves may cross a lagoon, overwash the landward shoreline, and deposit marine sediment as chenier ridges. Such ridges ring the landward sides of lagoons in New South Wales. Finally, along ria coastlines, estuaries may be infilled with marine sediment for considerable distances up-river. High-velocity flood and backwash flows under large tsunami may form pool and riffle topography. This process may account for pools that are tens of metres deep and morphologically stable under present tidal flow regimes in coastal estuaries along the New South Wales Coast.

Water piled up behind barriers by tsunami overwash tends to drain seaward through existing channels. However, if the tsunami wave crest impinges at an angle to the coast, channels can be opened up, or widened, at the downdrift end of barriers. In some cases, on low barriers, water may simply drain back into the ocean as a sheet along the full length of the barrier. Because barriers are breached at so many locations, the resulting tidal inlets that form compete for the available tidal prism rushing into the lagoon under normal tides. As these inlets become less efficient in flushing out sediment, they lose their integrity and rapidly close. Hence, tsunami-swept barriers may show evidence of numerous relict tidal inlets without any obvious outlet to the sea. Reorganisation of tidal flow in the lagoon because of these openings may lead to the formation of a secondary, shallow, bifurcating distributary channel network. In contrast, under storm waves, new tidal inlets are usually located opposite, or close to, contemporary estuary mouths.

Deltas and Alluvial Plains
   (Young et al., 1995, 1996b)

   The pattern on deltas and alluvial plains is different. If these low-lying areas are cleared of vegetation, the low frictional coefficient permits the tsunami wave to penetrate far inland before its energy is dissipated. The limit of penetration is defined by Equation 2.24. In this instance, the wave can deposit silt or sand as a landward-tapering unit ranging from a few centimetres to over a metre in thickness. This feature is the most commonly identified signature of tsunami as described in Chapter 3. In some cases this sand unit can be deposited 10 km or more inland. In extreme cases, where sand is abundant, a swash bar or chevron ridge may be deposited at the landward limit of penetration. Very little attention has been paid to the resulting backwash, which according to hydrological principles must become concentrated into a network of interconnected channels that increase in size, but decrease in number, seaward. Only one description of such a network has appeared in the literature to date, and this is for the Shoalhaven Delta on the New South Wales South Coast (Figure 3.2A). Here a large tsunami event deposited a fine sand unit up to 10 km from the coast. The sand contains open marine shells, such as *Polinices didymus, Austrocochlea constricta,* and *Bankivia fasciata,* that are 4,730–5,050 years old.

   A network of meandering backflow channels drains off the delta to the southeast (Figure 4.3). Significantly, these smaller channels are elevated above the regional landscape and are bordered by broad swamps. These channels are distinct from the main channel of the Shoalhaven River in that they have developed within Holocene sediment, whereas the river is entrenched into a Pleistocene surface that is buried about 4 m below the surface of the delta. The backwash channels are not only an order of magnitude smaller than the main river, they also have a much lower carrying capacity. The channels increase in width from 40 m about 10 km upstream to 100 m at the mouth of the Crookhaven River, which exits to the sea at the sheltered southeast corner of the delta. The channels had a maximum discharge of 500 m^3 s^{-1} based upon the length of meanders. The modern Shoalhaven River has a bankfull discharge of 3,000 m^3 s^{-1} in flood. Many smaller streams were once tidal, but the channels have since undergone infilling with a reduction in their carrying capacity. The channels are 600 years old, coinciding with the regional age of a large palaeo-tsunami predating European settlement. This age was obtained from oyster shells found on a raised pile of imbricated boulders that

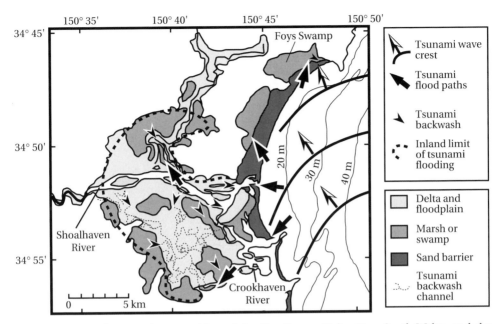

4.3 Hypothesised tsunami overwashing of the Shoalhaven Delta, New South Wales, and the meandering backwash channels draining water off the delta. A tsunami event 4,730–5,050 years ago – as determined from shell buried within a tapering sand layer within the delta – probably created these channels. The extent of this marine layer is also marked. A younger tsunami event around A.D. 1410 ± 60 years may also be implicated in the formation of the backwash channels (based on Young et al., 1996b).

were swept into the entrance of the Crookhaven River as the tsunami approached from the south. The wave then climbed onto the delta via this entrance and the main channel of the Shoalhaven River (Figure 4.3). The inferred direction of approach coincides with the alignment of nearby boulders deposited by tsunami on cliffs rising 16–33 m above sea level (Figure 3.11). The tsunami also swept over a coastal sand barrier and onto the northern part of the delta (Foys Swamp), depositing a layer of sand and cobble 1.8 km inland of the modern shore. The small meandering channels on the southern part of the delta were created by southeast drainage of backwash from the deltaic surface after the tsunami's passage northwards up the coast.

Rocky Coasts
(Bryant and Young, 1996)

There are two distinct models of sculptured landscape for rocky coasts: smooth, small-scale and irregular, large-scale. These are shown schematically in Figures 4.4 and 4.5 respectively. The two models can be differentiated from each other

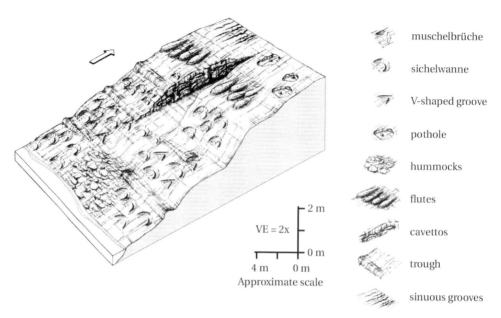

muschelbrüche

sichelwanne

V-shaped groove

pothole

hummocks

flutes

cavettos

trough

sinuous grooves

2 m

VE = 2x

0 m

4 m   0 m

Approximate scale

4.4 Model for smooth, small-scale, bedrock surfaces sculptured by tsunami. Model is for headlands within 7 m of sea level (Bryant and Young, 1996).

by their degree of dissection. Smooth, small-scale bedrock-sculptured landscapes
are restricted to headlands less than 7–8 m in height. Features consist of s-forms
and bedrock polishing, and rarely exceed a metre in relief. Dump deposits and
imbricated boulders usually are present nearby. The s-forms are directional, paralleling each other and the orientation of any imbricated boulders. The landscape is
dominated on the side of headlands facing the tsunami by fields of overlapping
muschelbrüche and sichelwannen grading into V-shaped grooves as slopes steepen.
Where vertical vortices develop, broad potholes may form, but rarely with preserved
central bedrock plugs. Crude transverse troughs develop wherever the slope levels
off. At the crest of headlands, muschelbrüche-like forms give way to elongated fluting. The flutes taper downflow into undulating surfaces on the lee slope. Sinuous
cavitation marks and drill holes develop on this gentler surface wherever flow accel-
erates because of steepening or flow impingement against the bed. On some surfaces, a zone of fluting and cavitation marks may reappear towards the bottom of
the lee slope because of flow acceleration. Cavettos or drill holes develop wherever
vertical faces are present. While cavettos are restricted to surfaces paralleling the
flow, cavitation marks are ubiquitous.

Irregular, large-scale, bedrock-sculptured landscapes are most likely created by
large tsunami generated by submarine landslides and meteorite impacts with the
ocean. The landscape typically forms on headlands rising above 7–8 m elevation in
exposed positions (Figure 3.20). Many facets of the small-scale model can be found
in this landscape. The large-scale model is characterised by ramps, whirlpools, and

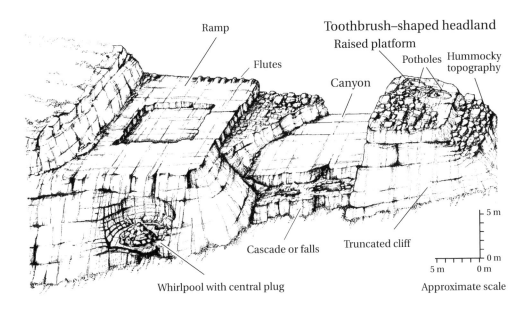

4.5   Model for irregular, large-scale, sculptured landscapes carved by tsunami. Model is for headlands 7–20 m above sea level (Bryant and Young, 1996).

canyons, forming toothbrush-shaped headlands (Figure 4.5). Ramps can extend from modern sea level to heights of 30 m and can evince zones of evacuated bedrock depressions (Figure 3.19), cascades, and canyons (Figure 3.20). Whirlpools up to 10–15 m deep are found primarily on the upflow side of the headlands, although they can also form on steep lee slopes. The base of whirlpools generally lies just above mean sea level, but some are drowned, with the central plugs forming stacks that are detached from the coastline. The base of whirlpools is controlled by the intensity of large-scale vortex formation rather than by the level of the sea at the time of formation. Smaller potholes are also found in these environments. Generally, canyon features are inclined downflow. However, where the effects of more than one event can be identified, earlier canyons provide conduits across the headland for subsequent, concentrated erosive flow. Irregular landscapes preserve a crude indication of the direction of tsunami approach, although the effects of wave refraction may complicate the pattern. Under extreme conditions, the complete coastal landscape can bear the imprint of catastrophic flow – headlands from 80 to 130 m high may be overwashed with sheets of water carving channels as they drained off the downflow side, headlands rising over 40 m above sea level may have their seaward ends truncated, talus and jagged bedrock may be stripped from cliff faces, platforms may be planed smooth to heights 20 m above sea level, and whole promontories may be sculptured into a fluted or drumlin-like shape. Finally, coastal tsunami flow is usually repetitive during a single erosive event, which consists of pulses of unidirectional, high-velocity flow as individual waves making up a tsunami wave train surge over bedrock promontories. In these instances, erosive vortices last for no more than a few minutes and the whole erosive event is over within a few hours.

Atolls

(Bourrouilh-Le Jan and Talandier, 1985, Talandier and Bourrouilh-Le Jan, 1988; Scoffin, 1993)

The islands of the South Pacific are exposed to both tropical cyclones and tsunami. One of the more notable features is the occurrence of alternating mounds and channel. Mounds called *motu,* consisting of sands and gravels derived from the beach and reef, separate the channels or *hoa* from each other (Figure 4.6). The *motu* are drumlin-shaped with tails of debris trailing off into the lagoon. In many respects, they could also be interpreted as moulded dump deposits. *Motu–hoa* topography appears to be restricted to atolls in the South Pacific. Many descriptions of this topography consider them features of storm waves. However, storm waves from tropical cyclones appear to modify prior *motu–hoa* topography rather than being responsible for it. Even under storm surge, bores generated by storms tend to flow through preexisting *hoa* into the lagoon. Despite the steepness of nearshore topography around atolls, storm waves dissipate much of their energy by breaking on coral aprons fringing islands. The

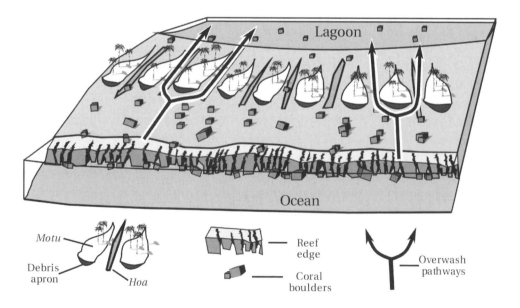

4.6 Model for the impact of tsunami upon coral atolls in the South Pacific Ocean (based on Bourrouilh-Le Jan and Talandier, 1985). Only a selection of overwash pathways around mounds (*motu*) and through channels (*hoa*) cutting across the atoll are marked.

effect of storms is primarily restricted to the accumulation of coarse debris aprons in front of *motu–hoa* topography. For example, Cyclone Bebe in 1972 built a ridge on Funafuti that was 19 km long, 30 km wide and 4 m high. The volume of sediment thrown into this ridge occupied 1.4×10^6 m^3 and weighed 2.8×10^6 tonnes. The ridge ran continuously in front of *motu–hoa* topography around the atoll. More often, changes in the landscape produced by storms are patchy, whereas *motu* and *hoa* form a regular pattern over considerable distances on many atolls. *Motu–hoa* topography can be formed by a single tsunami overwashing low-lying atolls. *Motu* or mounds form under helical flow where opposing vortices meet, while *hoa* or channels are excavated where these vortices diverge. The prominent nature of such topography suggests that mega-tsunami may be responsible.

Large boulders deposited on atolls are also difficult to link to storm waves. They have been ripped off the reef edge and deposited in trains or dumped in piles on the reef fronting *motu*. In some cases, isolated boulders have been carried through *hoa* and deposited in the backing lagoon (Figure 4.6). This process has not been observed during storms. While storm waves can erode boulders from the reef edge, they rarely transport them far. In fact, boulders wrenched from the reef edge by storms usually end up in a wedge-shaped apron of talus at the foot of the reef slope. More unusual is the sheer size of some of the boulders. One of the best examples occurs on the northwest corner of Rangiroa Island in the Tuamotu Archipelago. Here, one of the coral boulders measures 15 m × 10 m × 5 m and weighs over 1,400 tonnes. This boulder could have been moved easily by a tsunami wave 6 m high (Equation 3.7), but would have required a storm wave 24 m in height (Equation 4.2). Most boulders found scattered across atolls consist of coral that is more than 1,500 years old; however, the boulders rest on an atoll foundation that is as young as 300 years. This fact suggests that the boulders were deposited by a large tsunami at the beginning of the eighteenth century. Local legends in the South Pacific describe the occurrence of large catastrophic waves at this time concomitant with the abandonment of many islands throughout French Polynesia. Significantly, the legends have the sun shining at the time of the waves – a fact ruling out tropical cyclones.

EXAMPLES OF TSUNAMI-GENERATED LANDSCAPES: AUSTRALIA
(Young and Bryant, 1992; Bryant et al., 1992, 1996; Young et al., 1993, 1995, 1996a, b; Nott, 1997; Bryant and Nott, 2001)

South Coast of New South Wales

Many of the examples used as signatures of tsunami in Chapter 3 and in the construction of landscape models described in this chapter come from the South Coast of New South Wales (Figure 3.2A). Dating evidence indicates that tsunami here have been a repetitive feature over the past 7,000 years. These tsunami have acted profoundly on the landscape at two scales. At the smaller scale, beaches have been overwashed, chaotically sorted sediments that include gravel and boulders have been dumped onto headlands, imbricated boulders have been deposited in aligned piles, and bedrock surfaces have been sculptured. At the larger scale, complete barrier sequences and rocky headlands bear the unmistakable signature of high-velocity flow that only tsunami can explain. While the evidence for tsunami as a dominant element of the landscape is pervasive, only a few examples can be presented in the space available here.

Many of the deltas and low coastal plains along this coast have been swept by tsunami. The model of tsunami overwash for deltas shown in Figure 4.3 is found here. Buried sand layers have also been identified in many estuarine settings. These units contain cobbles and marine shell, are up to 1 m thick, and can be found 10 km inland of the modern beach in sheltered positions that all but rule out deposition by normal wind-generated waves. Similar shell-rich sands have been found along the New South Wales Coast trapped in sheltered embayments at Cullendulla Creek, Batemans Bay (Figure 3.7), and at Fingal Bay, Port Stephens. This coastline is also renowned for its sandy barriers entrapping large coastal lakes and enclosing lowlands. The evolution of these barriers has been explained in terms of high-frequency, low-magnitude marine and aeolian processes superimposed on the effects of changing sea level during the Late Quaternary. The sandy barriers were constructed at sea-level highstands during the Last Interglacial over 90,000 years ago, and during the Holocene between 3,000 and 7,000 years ago. Although sandy marine barrier deposits dating from the Last Interglacial are well preserved on the North Coast of New South Wales, they are paradoxically rare south of Sydney, where many of the signatures of tsunami have been identified. Many of the Holocene barrier deposits are in fact the product of tsunami overwash.

Two examples stand out – Bellambi Beach, Wollongong, and the sand barriers inside Jervis Bay about 100 km south of Sydney (Figure 3.2A). At Bellambi, a 1.0- to 1.2-m-thick humate-impregnated sand that contains isolated, rounded boulders lies sandwiched between Holocene estuarine clay and beach sand (Figure 4.1). The humate is anomalous because the sand dates much older than the Holocene, at an age of 22,000–25,600 years. At this time, the coastline was nowhere near its present location, but 130 m lower and 12 km offshore. Significantly, a 20- to 30-cm-thick pumice layer is trapped near the southern end of the barrier about 2 m above present sea level. The pumice is 25,000 years old and originated from volcanic eruptions north of New Zealand. Its composition does not match that of any modern pumice floating onto the coast. The pumice comes from relict beach deposits on the shelf. To maintain their older temporal signature, the humate sands were transported suddenly from the shelf to the present coastline during the building of the barrier. This process was carried out by a large tsunami about 4,500–5,000 years ago. The tsunami also picked up bouldery material from a headland 1 km away and mixed this with the sand. The pumice was not carried with any of this material. Being lighter than water, it floated to the surface of the ocean and drifted into the lagoon at the back of the barrier afterwards on sea breezes. If this scenario is correct, then the tsunami had to be large enough to generate bottom current velocities that could not only entrain sand but also erode humate-cemented sediment from water depths of 100 m at the edge of the continental shelf. For this to happen, distinct from the capability of storm waves, the tsunami had to be over 5 m high when it reached the shelf edge (Table 2.2).

Tsunami-deposited barriers in the Jervis Bay region consist of clean white sand that originated from the leached A2 horizon of podsolised dunes, formed at lower sea levels on the floor of the bay during the Last Glacial over 10,000 years ago. The barriers form raised platforms 1–2 km wide and 4–8 m above present sea level on a tectonically stable coast. The barriers supposedly were built up over the last 7,000 years; however, the sands yield an older age commensurate with their origin on the floor of the bay. Again, the sands must have been transported to the coast suddenly in order to maintain an older temporal signature. Transport occurred during one or more tsunami events.

The effect of tsunami on the rocky sections of this coast is even more dramatic. The scenic nature of many headlands is partially the consequence of intense tsunami erosion. Two headlands, Flagstaff Point and Kiama Headland, in the Wollongong area will be described here. Both headlands are similar in that they protrude seaward about 0.5 km beyond the trend of the coast. Both were affected by the same erosive tsunami travelling northwards along the coast. The raised dump deposit and keellike stacks shown in the previous chapter are located between the two headlands (Figures 3.6 and 3.21 respectively). Flagstaff Point consists of massively jointed, horizontally bedded volcanic sandstones that have been deeply weathered, while Kiama Headland consists of weakly weathered, resistant basalt. Overwashing by high-velocity tsunami has severely eroded the seaward facets of both headlands, but with subtle differences. On Flagstaff Point, the tsunami eroded the softer material, forming a reef at the seaward tip and a rounded cliff along the southern side (Figure 4.7). The wave planed the top of the headland smooth, spreading a smear deposit that consists of muds, angular gravel, and quartz sand across this surface. The wave also moved massive boulders into imbricated piles on the eroded platform surface on the south side. However, the most dramatic features were created by tornadic vortices on the northern side of the headland. These vortices were 8–20 m deep and rotated in a counterclockwise direction. The largest vortex was shed from the tip of the headland eroding a whirlpool with 20-m-high sides into the cliff. Fluted bedrock on the outer surface of erosion outlines the vortex and its helical flow structure. The whirlpool is incomplete, indicating that the vortex developed towards the end of the wave's passage over the headland. As the wave wrapped around the headland, it broke as an enormous plunging breaker, scouring out a canyon structure along the northern side. The canyon is separated from the

ocean by a buttress that rises 8 m above sea level at the most exposed corner, tapering downflow as the wave refracted around the headland. The tsunami then travelled across the sheltered bay behind the headland, carving giant cusps into a
bedrock cliff on the other side. Finally, it swept inland depositing a large pebble- and
cobble-laden sheet of sand up to 500 m inland.

At Kiama, the tsunami wave developed broad vortices as water was concentrated against the southern cliff face (Figure 4.8). While flow probably obtained
velocities similar to those at Flagstaff Point, the weakly weathered basalt was more
resistant to erosion. Vortices eroded into the headland along major joints forming
broad caves. At one location, a vortex penetrated 50–80 m along a joint, creating a
cave that blew out at its landward end to form the Kiama Blowhole. Where the
headland was lower in elevation, vortices began to scour out large muschelbrüche
about 50 m in length. The tip of the headland, instead of being eroded into a reef,
was heavily scoured into hummocky bedrock topography to form a surface 8–10 m
above sea level. Boulders were scattered across this surface. The wave also planed
the top off the headland and deposited a 5- to 10-cm-thick smear deposit that
thickens downflow. The wave then crossed a small embayment in the lee of the

4.7 Flagstaff Point, Wollongong, Australia. This headland, which is 20 m above sea level, was
overridden and severely eroded by a palaeo-tsunami that approached from the southeast (white
arrows). Main features (white lines) are as follows: **a,** boulder piles; **b,** incipient whirlpool with
fluted rim; **c,** plug; **d,** canyon; and **e,** smear deposit over headland. Sediment was transported
across the bay and deposited as a sand sheet in the distance (**f**).

4.8 Kiama Headland lying 40 km south of Flagstaff Point (Figure 4.7). A palaeo-tsunami also approached from the southeast (white arrows) and swept over the headland. Here caves have been bored into columnar basalts. Main features (white lines) are as follows: **a,** large muschel-brüche-like feature; **b,** incipient caves; **c,** cave leading to blowhole; **d,** hummocky topography on raised platform; **e,** smear deposit over headland; and **f,** sheared cliff face and planed platform. A sand sheet was again deposited across the bay (**g**).

headland, shearing the end off a cliff. Finally, the wave deposited quartz sand inland about a kilometre to the northwest.

Cairns Coast, Northeast Queensland

The signatures of tsunami are not restricted in Australia to the Southeast Coast. They also appear inside the Great Barrier Reef between Cairns and Cooktown along the Northern Queensland Coast (Figure 3.2B). This, at first, would appear to be unlikely. Even if tsunami have occurred in this part of the Coral Sea, the Great Barrier Reef should have protected the mainland coast. Evidence now suggests that tsunami have indeed reached the mainland coast with enough force, not only to transport boulders, but also to begin sculpturing bedrock. For example, at many locations, boulders weighing over a hundred tonnes can be found at elevations of 8–10 m above sea level. One of the largest boulders is found at Cow Bay. It has a volume of 106 m^3 and weighs 286 tonnes. At Oak Beach, imbricated boulders weighing up to 156 tonnes and measuring over 8 m in length can be found in piles (Figure 4.9). Using Equations 3.8 and 4.2, it can be shown that tsunami rather than storm

4.9 Boulders stacked by tsunami on the platform at the south end of Oak Beach, North Queensland. The largest boulder is 4.0 m in diameter. Note the smoothed bedrock surface with evidence of large overlapping muschelbrüche.

TABLE 4.2   Comparison of Tsunami and Storm-Wave Heights Required to Transport Boulders along the Queensland Coast North of Cairns

Location	Boulder Width (m)	Weight (tonnes)	Height of Tsunami at Shore (m)	Breaking Storm-Wave Height (m)
Cow Bay	6.3	247.0	11.2	44.8
Oak Beach	4.0	192.0	5.0	20.0
Taylor Point	4.2	90.0	9.1	36.4
Turtle Creek Beach	4.3	115.0	5.2	20.8
Cape Tribulation	4.1	86.0	5.8	23.2

Source: Based on Nott, 1997.

waves are the most feasible mechanism generating the flow velocities required to transport many of these boulders. These comparisons are presented in Table 4.2. All of the imbricated boulders can be transported by tsunami 5.0–11.2 m in height. The highest waves are required to move the boulders found at Cow Bay. The smallest storm wave must be 20 m high to move these boulders. This is virtually impossible because such a wave would break before reaching the coastline, even when superimposed on surges that can be up to 12 m high. The tsunami waves can generate theoretical flow velocities ranging from 3.7 to 10.4 m s^{-1}, with a mean value of 5.8 m s^{-1}. The highest of these velocities is more than sufficient to produce cavitation and sculpture bedrock. Indeed, many of the bedrock surfaces near the boulder piles evince bedrock-sculpturing signatures. For example, the boulders found at Oak Beach are emplaced on an undulatory, smooth surface characteristic of erosion by transverse roller vortices (Figure 4.9). S-forms are also evident on this bedrock surface. The northern headland of Oak Beach has also been dissected into a toothbrush shape that is covered with flutes, sinuous grooves, and cavitation drill holes (Figure 4.10).

The tsunami in the Cairns region have originated in the Coral Sea outside the Great Barrier Reef. It appears that probably palaeo-tsunami penetrated the reef through openings such as Trinity Opening and Grafton Passage, which are more than 10 km wide and between 60 and 70 m deep (Figure 3.2B). The alignment of boulders north of Cairns points directly towards Trinity Opening. At other locations, where the alignment of boulders is more alongshore, the tsunami waves appear to have been trapped through diffraction and refraction between the reef and the mainland, and by major headlands jutting from the coast at Cape Tribulation and Cairns.

4.10 The eroded headland at the north end of Oak Beach, North Queensland. The flutes protruding above the raised platform surface face towards the south-southeast, the same direction as the alignment of boulders shown in Figure 4.9.

Northwest West Australia

Signatures for both historical and palaeo-tsunami exist along the coast of West Australia. The 3 June 1994 tsunami that originated in Indonesia swept the coast of the North West Cape through gaps in the Ningaloo Reef, resulting in the inland deposition of marine fauna, sand, and isolated coral boulders up to 2 m in width. Coral boulders were also swept through gaps in the coastal dunes and deposited 1 km inland across a flat, elevated plain by the tsunami following the eruption of Krakatau in 1883. Tropical cyclone storm surges have never reached this far inland on this coast. While some of the boulders were organised into piles, none show any preferential alignment. These deposits can only be described as ephemeral dump deposits.

This contemporary evidence is dwarfed by the signature of palaeo-tsunami along 1,000 km of coastline between Cape Leveque and North West Cape (Figure 3.2C). The magnitude of some of this evidence is the largest yet found in Australia. At Cape Leveque, at least four waves from a palaeo-tsunami dumped gravelly sands and shell in sheets and mounds along the coast, overriding headlands 60 m above sea level in some places. The sandstone platform in front of Cape Leveque was sculptured smoothly with the clear signature of large muschelbrüche, fluting, and transverse troughs. The seaward edge of this platform shows evidence of a preparation zone for boulders. Large slabs of bedrock 1–2 m thick were lifted repetitively up and down along bedding planes until they broke into blocks 4–5 m wide. At this point the fractured pieces were transported by tsunami flow and dumped alongshore into imbricated piles (Figure 4.11). A similar process is present at Broome. Here sands and gravels were dumped 5 km inland at the back of a flat headland. The angle of approach of waves at both locations appears to have been from the southwest – a direction not matching the modelled approach of tsunami from Indonesia. In the Great Sandy Desert, aeolian ridges more than 30 km inland were truncated and remoulded into chevron ridges by a palaeo-tsunami. This distance is more than three times greater than the distance of penetration inland by any tsunami yet discovered for the South Coast of New South Wales. Marine shell and lateritic gravels were deposited in these chevron dunes; lateritic boulders were stacked in front. Seven hundred kilometres further south, at Exmouth on North West Cape, transverse dune ridges were overridden from the east by large waves. Coral, shell, and cobble were mixed with aeolian sand and spread more than 2 km inland across the upper surface of at least seven dune ridges paralleling the coast.

By far the most dramatic evidence of tsunami occurs at Point Sampson. Here, waves have impinged upon the coast from the Indian Ocean to the northeast. Shell deposits were deposited above the limits of storm surges (Figure 4.12) and on top of hills 15 m above sea level. In one extreme case, three layers of sand with a total thickness of 30 m were deposited in the lee of a hill more five hundred metres inland and over 60 m high. The sands contain boulder floaters, coral pieces, and shell. Each layer appears to represent an individual wave in a tsunami wave train. Dating of shell deposits in the region indicates that the tsunami occurred around A.D. 1080,

4.11 Platform at Cape Leveque, West Australia. A palaeo-tsunami swept over the island and cliffs in the background, and eroded the stack. Muschelbrüche outlined by the pools of standing water sculpture the platform surface. The bedrock surface was torn up along bedding planes. The boulders originated from the platform further seaward and were stacked in imbricated piles by flow travelling alongshore towards the camera.

4.12 Raised beds of cockles on a hill at Point Sampson, Northwest Western Australia. These shells extend 15 m above sea level. In this region, the palaeo-tsunami flowed up the valley and over the hills in the background. The minimum age for the event, based on radiocarbon dating of the shells, is A.D. 1080.

4.13 Chiselled boulders deposited in bedded gravels in a megaripple about 5 km inland of the coast at Point Sampson, Northwest Western Australia. The dip in bedding aligns with bedrock-sculptured features in the area (Figure 4.14) and shows flow from the northeast in the Indian Ocean.

before European discovery. In a valley leading back from the coast, large megaripples with a wavelength approaching 1,000 m and consisting of cross-bedded gravels, have been deposited up to 5 km inland. The spacing between the megaripples is an order of magnitude greater than that found at Jervis Bay, New South Wales. The flow depth is theorised to have been as great as 20 m with velocities of over 13 m s^{-1}. The dip in the bedding indicates flow transport from the Indian Ocean. The gravels also contain boulders over 1 m in diameter that have been chiselled into a spherical shape (Figure 4.13). The tsunami waves overrode hills 60 m high, another one kilometre inland, carving wind gaps 20 m deep through ridges. Sand and gravel were then deposited a further 2 km inland of these ridges. Finally, ridges 15 m high were sculptured into hull-shaped forms with bedrock plucked from the flanks and crest by helical flow that encompassed the whole height of the ridge (Figure 4.14). A cockscomb-like protur-

4.14 Streamlined inverted keel-shaped ridge with a cockscomb-like proturbance at Point Sampson, Northwest Western Australia. Note the resemblance of the cockscomb to that shown in Figure 3.21. The wave travelled from left to right. Intense vortices embedded in helical flow plucked blocks of bedrock from the sides of the ridge. The ridge is aligned with other bedrock features in the region. Scale is the person circled at the top.

bance similar to that shown on the keellike stack in Figure 3.21 was eroded towards the front of the ridge by intense vortices or hydraulic hammering.

OTHER EXAMPLES OF TSUNAMI-GENERATED LANDSCAPES

Grand Cayman
(Jones and Hunter, 1992)

Other examples exist in the world where the signatures of tsunami dominate in preference to those induced by storms. The first of these examples comes from the Cayman Islands in the middle of the Caribbean Sea (Figure 4.15A). The island exists in a region where tsunami have played a major role in the region historically. For example, in June 1692, a powerful tsunami devastated Port Royal on the nearby island of Jamaica. The region is also dominated by tropical cyclones. For example, in 1785 a tropical cyclone destroyed every house and tree except one on the Cayman Islands. In 1932, a hurricane with winds of 330 km hr^{-1} generated seas that carried huge rocks weighing several tonnes. The waves moved coral boulders 0.6–1.0 m in diameter shoreward from waters less than 15 m deep, constructing ramparts at shore.

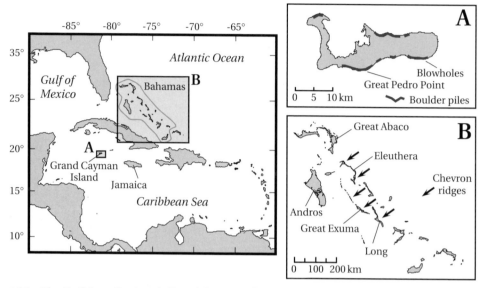

4.15   The Caribbean Region. **A.** Grand Cayman Island; and **B.** Barbados.

Bigger boulders exist than those transported by these storms. The boulders appear to be related to a high-energy, prehistoric event dating around A.D. 1662 that moved slabs as large as 5.5 m in length, depositing some in clusters up to 150 m inland. Some clusters contain imbricated stacks of boulders, aligned in a north–south direction, roughly at right angles to the shoreline. At Great Pedro Point, blocks weighing up to 10 tonnes were moved 18 m vertically and 50–60 m inland of the cliff line. The largest boulder measures 5.5 m × 2.8 m × 1.5 m. All of the boulders were emplaced at least 12 m above sea level. Many of the boulders originated as giant rip-up clasts torn from terraces formed in dolomite or from boulder deposits at the base of sea cliffs. The terraces had solutional weathering pinnacles and ridges with a relief of 2 m that were planed flat. According to Equations 3.8 and 4.2, the largest boulder required a storm wave 12.5 m high to move it. The equivalent tsunami wave was only 3.1 m high. The storm waves could only exist if water depths were more than 16 m deep at the base of cliffs, otherwise they would have broken. This condition occurs at only one of the locations where boulders are found. Even here, it is doubtful if storm waves could have maintained sufficient energy to transport boulders more than 100–150 m inland.

Bahamas

(Hearty, 1997; Hearty et al., 1998)

A second example comes from the Bahamas (Figure 4.15B), where the landscape has been created by either tsunami or storm waves. The Barbados offers a range of features suggestive of tsunami from the Last Interglacial. Most interesting are V-shaped ridges up to several kilometres in length that have penetrated inland from the exposed Atlantic Ocean side of the islands. The ridges are asymmetrical, averaging 3 km in length, with some exceeding 10 km. They are 20–100 m wide and stand 8–25 m high, increasing in elevation towards their tip. Some ridges indicate that waves must have run up to elevations of 40 m above sea level at the time of deposition. There are up to thirty ridges, some tucked into each other. Nowhere does this involve more than four ridges. The ridges have a consistent orientation to the west-southwest that varies by no more than 10° along 300 km of coastline, despite a 60° swing in the orientation of the shelf edge. Internally, the ridges contain low-angle cross-beds, scour and fill pockets, pebble layers, and bubbly textures characteristic of rapid deposition found at the swash limit of accreting sandy beaches. The steepest-dipping beds are found towards the landward margin of the ridges. Because of their shape and the fact that smaller ones lie nestled within large forms, they have been termed chevron ridges. It was similarities to these ridges, found at Jervis Bay, New South Wales, that led to this term being used to identify one of the prominent signatures of large tsunami.

The chevron ridges on Eleuthera Island are associated with huge boulders up to 970 m³ in size and weighing up to 1,850 tonnes (Table 4.3). These have been transported over the top of cliffs more than 20 m high. The boulders were emplaced at the end of the Last Interglacial sea level highstand. They were transported up to 500 m landward – a distance greater than present-day boulders have been transported. The modern boulders are much smaller, averaging 22 m³ in volume and weighing less than 175 tonnes (Table 4.3). However, some of these modern boulders are anomalous and suggestive of recent tsunami. For example, some have been transported up to 200 m landward and more than 10 m above sea level.

Again, Equations 3.7 and 4.2 can be used to resolve the difference between the capacity of theoretical tsunami and storm waves to transport these boulders. Note that in these calculations the density of 1.9 g cm⁻³ has been used for the coral boulders. The largest of the modern coral boulders requires a storm wave about 16 m in height to be transported, whereas the palaeo-boulders require maximum storm waves of about 24 m in height. Both of these sizes are difficult to obtain close to shore without breaking even under storm surges of 7–8 m that can be generated here by tropical cyclones. Local storms also do not account for the formation and consistent alignment of the chevron ridges. Tropical cyclones have winds that rotate around an eye that rarely exceeds 100 km in diameter. For the ridges to be produced by a cyclone, the storm would have had to maintain consistently strong winds and moved parallel to the islands over a much longer distance than is observed at present. The echelon nature of some chevron ridges, tucked one inside the other, also requires more than one storm – which appears unlikely.

Tsunami with a distant origin appear more feasible. The palaeo and modern

TABLE 4.3   Comparison of Tsunami and Storm-Wave Heights Required to Transport Boulders in the Bahamas

Location	Length (m)	Boulder Width (m)	Thickness (m)	Volume (m³)	Weight (tonnes)	Height of Tsunami at Shore (m)	Breaking Storm-Wave Height (m)
Palaeo-boulders	13.0	11.5	6.5	972	1846	5.9	23.7
	14.0	7.3	6.7	685	1301	2.6	10.5
	9.3	6.0	4.0	223	424	2.8	11.3
	8.1	5.7	5.5	254	482	1.9	7.7
	7.2	5.7	5.0	205	390	2.1	8.2
Modern	7.8	4.7	2.5	92	174	2.7	10.9
	4.9	4.5	1.2	26	50	4.0	16.1
	3.8	2.5	2.0	19	36	1.0	4.0
	3.8	3.2	1.5	18	35	1.9	7.8
	6.4	2.7	0.5	9	16	3.9	15.7

Source: Based on Hearty, 1997.

boulders require maximum tsunami wave heights of only 6 and 4 m respectively. Waves as small as 1.0–2.0 m in height could have moved some of the boulders. Tsunami generated by submarine landslides or meteorite impact with the ocean produce up to four waves in their wave train. This could easily account for up to four chevron ridges nestled one within another – all with the same orientation. While the Bahamas are subject to submarine landslides along the shelf margin, a local source can be ruled out because waves would have radiated outwards from this source and not been able to produce the consistent ridge alignment over such an extended distance. Either distant submarine landslides or a meteorite impact in the Atlantic Ocean accounts for the ridges and boulder deposits in the Bahamas. These possibilities will be discussed further in subsequent chapters.

Chilean Coast
(Lockridge, 1985; Paskoff, 1991; Ortlieb et al., 1995; Paskoff et al., 1995)

Of the world's entire coastline, the West Coast of South America is one of the most prone to recurrent large tsunami (Figure 4.16). In the twentieth century, twenty-three tsunamigenic earthquakes have occurred along the coasts of Chile and Peru. Of these, seven have had a surface magnitude of 8.0 or more on the Richter scale.

Seismicity is linked to subduction of the Nazca Plate beneath the South American Plate. Earthquake epicentres tend to cluster along coastlines or at the base of the Andes Mountains (Figure 4.16). A narrow, uplifting coastal plain is constrained by the Andes Mountains, and subject to some of the largest tsunamigenic earthquakes in the world. The coastal plain is also old, recording the signature of past marine planations in the form of raised terraces that date back to the Miocene, more than 12 million years ago. Lower terraces have been reoccupied repetitively by Interglacial highstands in sea level during the past 2 million years.

Both the older terraces and the modern coastal plain preserve the signature of palaeo-tsunami events. For example, at Herradurra Bay in the Coquimbo region, large tonalite boulders over 2 m in diameter are exposed within nearshore beach sands on a 200,000-year-old interglacial terrace that is now situated 35–40 m above sea level (Figure 4.17). The boulders originated 2 km away, on the Coquimbo Peninsula, which was an island at the time the terraces formed. The event that moved them was never repeated afterwards because boulders are not evident in younger deposits except where they have been eroded from the upper terrace. While boulders are uncommon in the present landscape, the signature of dump deposits is ubiquitous. One of the largest of these occurs at Michilla Bay in Northern Chile. This dump deposit is over 5 m thick, lies 7 m above present sea level, and consists of a massive bed of coarse sands interspersed with cobbles and large unbroken shells (Figure 4.18). Isolated boulders are scattered throughout the bottom of the deposit. Dating on the shell places the event about 7,000 years ago

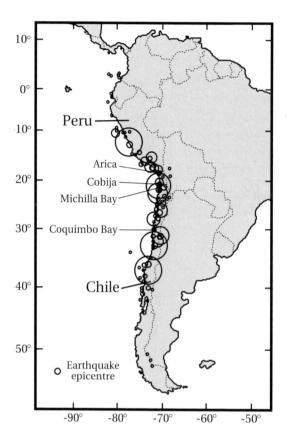

4.16 Location of historical tsunamigenic earthquakes since 1562 along
the West Coast of South America
(based on Intergovernmental Oceanographic Commission, 1999). Size of circle is proportional to the number of
events at the same site.

when sea level reached its present level following the Holocene marine transgression. Smaller Holocene events are also preserved as interstratified beds of pebbles
and coarse sand within prehistoric middens, particularly in the Cobija area. In
many respects, these deposits are similar to the disturbed Aboriginal middens of
Southeastern Australia described in the previous chapter.

Historically, tsunami have swept the total breadth of this plain and reached the
base of the Andes. The most tsunamigenic section of coastline occurs on the border
between Peru and Chile (Figure 4.16). Since historical records began in 1562, there
have been 230 tsunami generated by earthquakes. Five localities have had ten or more
tsunamigenic earthquakes over this period within a 110-km radius of each other.
Three events have had Pacific-wide impact: the events of 13 August 1868 and 10 May
1877, both near the town of Arica on the border between Peru and Chile, and the event
of 22 May 1960. This latter event will be discussed in more detail in Chapter 5. The
Arica events regionally had maximum run-ups of 21 and 24 m respectively along the
South American Coast. Prior to these events, Arica had been destroyed twice by
tsunami, in 1604 and 1705. The 1868 tsunami struck the town within half an hour of
the main shock. The sea rose initially 5 m and then withdrew, leaving a 2-km-wide

4.17 Boulders deposited by a catastrophic tsunami near Coquimbo Bay, 200,000 years ago. The boulders were transported more than 2 km and deposited in offshore waters. They have been tectonically uplifted since. Photograph courtesy Dr. Colin Murray-Wallace, School of Geosciences, University of Wollongong.

4.18 An elevated tsunami dump deposit at Michilla Bay, Northern Chile. The event occurred around 7,100 years ago, coincidentally with sea level reaching its present level following the Holocene marine transgression. The ocean is to the right. The top of the terrace stands 6–7 m above sea level and about 1 km inland. Photograph courtesy Dr. Colin Murray-Wallace, School of Geosciences, University of Wollongong.

strip of the seabed exposed. Several minutes later, the main wave came in and swept across the coastal plain (Figure 2.14).

Approximately 25,000 people lost their lives in this region alone. The tsunami then swept the Pacific Ocean, with damage being reported in New Zealand, Hawaii and Japan. In Antarctica, the wave broke up seaice. The 1877 event was just as large, if not more widespread. Its run-up was 20 m high at Arica and 24 m high at Tocopilla, 600 km south of the epicentre. The wave also swept the Pacific Ocean and had a particularly forceful impact on the coast of New Zealand, where run-up of 6 m was reported. In Eastern Australia, the wave was responsible for the largest tsunami, 1.07 m, recorded on the Sydney tide gauge. These historical tsunami have repetitively destroyed coastal towns and moved layers of coarse sand shoreward. For example, Figure 4.19 shows the ruins of Cobija in Northern Chile, wiped out by the Arica tsunami of 10 May 1877. The wave arrived 5 minutes after the earthquake and reached 11.9 m above mean sea level at this location, eroding into raised alluvial fans, which can be seen in the background. Backwash buried the ruins in gravels interspersed by layers of shelly sand brought ashore by the tsunami. Despite this precarious environment, rebuilding has always taken place even at the most

4.19 The ruins of Cobija in Northern Chile. The tsunami of 10 May 1877 destroyed the town, which was subsequently buried by alluvial gravels. The ruins contain layers of shelly sand brought ashore by the tsunami. Photograph courtesy Dr. Colin Murray-Wallace, School of Geosciences, University of Wollongong.

vulnerable sites. Today, in many places, evacuation from tsunami would be difficult even with adequate warning because sea cliffs back numerous towns.

Storms are significant events in the coastal environment. Nonetheless, tsunami
are an alternative mechanism for reworking coastal sediment and ultimately for
imprinting upon the landscape a signature of convulsive events that in some areas
has not been removed over thousands of years. The difficulty in invoking storms
for all coastal deposition and erosion lies not just in the insufficient magnitude of
observed events but also in the inapplicability of storm wave processes to account
for a suite of anomalous deposits. For instance, dump deposits, bouldery mounds,
and chevron ridges contain chaotic mixtures of sediment that cannot be explained
by storm waves because such waves erode sand from the shoreline, transport
coarse sediment shoreward as a body, and sort debris in a shore normal direction.
Storms also cannot account for bedrock sculpturing. Tsunami can move and
deposit highly bimodal sediment mixtures and create the suites of high-magnitude depositional and erosional signatures that dominate many landscapes world-
wide. This evidence cannot be ignored. The only question that remains is, "What
causes these tsunami"? To answer this question one must examine the type of
deposits and landscapes produced by earthquakes and submarine landslides that
are the main causative mechanisms of tsunami. In addition, meteorite impacts
with the ocean cannot be ignored despite having never occurred historically. As
will be shown, the flux of comets and meteorites has been substantially greater
over the past millennia than at present. This increased frequency has been
observed and reported in legends, but never in documents that can be interpreted
from a modern perspective, or that can withstand the scrutiny of contemporary
scientific methodology. Authentication of meteorite-generated tsunami as an
underrated and significant hazard will be presented in Chapter 8.

PART THREE

Causes of Tsunami

CHAPTER FIVE

Earthquake-Generated Tsunami

5.1 Artist's impression of the second and largest tsunami smashing into the Alaska Railway terminus at Seward in Prince William Sound, 27 March 1964. Locomotive 1828 was carried 50 m. Note that the locomotive has become a water-borne missile – a trait often generated by tsunami bores as described in Chapter 2. Drawing by Pierre Mion and appearing in December 1971 issue of *Popular Science*. Source: **http://wcatwc.gov/tpic4.htm.**

INTRODUCTION

The most common cause of tsunami is seismic activity. Over the past two millennia, earthquakes have produced 82.3% of all tsunami in the Pacific Ocean. Displacement of the Earth's crust by several metres during underwater earthquakes may cover tens of thousands of square kilometres and impart tremendous potential energy to the overlying water. These types of events are common; however, tsunamigenic earthquakes are rare. Between 1861 and 1948, over 15,000 earthquakes produced only 124 tsunami. Along the West Coast of South America, which is one of the most tsunami-prone coasts in the world, 1,098 offshore earthquakes have generated only 20 tsunami. This low frequency of occurrence may simply reflect the fact that most tsunami are small in amplitude and go unnoticed. Two thirds of damaging tsunami in the Pacific Ocean region have been associated with earthquakes with a surface magnitude of 7.5 or more. The majority of these earthquakes have been teleseismic events affecting distant coastlines as well as local ones. One out of every three of these teleseismic events has been generated in the twentieth century by earthquakes in Peru or Chile. This chapter discusses the mechanics of tsunamigenic earthquakes, and where possible, attempts to associate them with some of the signatures for tsunami presented in Chapter 3.

Seismic Waves
(Bryant, 1991; Geist, 1997b)

Earthquakes occurring mainly in the upper 100 km of the ocean's crust generate tsunami. However, earthquakes centred over adjacent landmass have also produced tsunami. Earthquakes are shock waves transmitted through the Earth from an epicentre that can lie as deep as 700 km beneath the Earth's crust. These seismic waves consist of four types: P, S, Rayleigh, and Love waves. P waves are primary waves that arrive first at a seismograph. The wave is compressional, consisting of alternating compression and dilation similar to waves produced by sound travelling through air. These waves can pass through gases, liquids, and solids. P waves can thus travel through the centre of the Earth; however, at the core–mantle boundary, they are refracted, producing two 3,000-km-wide shadow zones without any detectable P waves on the opposite side of the globe from an epicentre. To detect tsunami produced by earthquakes, seismic stations must be located outside these shadow zones. S, or shear, waves behave very much like the propagation of a wave down a skipping rope that has been shaken up and down. These waves travel 0.6 times slower than primary waves. The spatial distribution and time separation between the arrival of P and S waves at a seismograph station can be used to determine the location and intensity of an earthquake. Love and Rayleigh waves spread slowly outwards from the epicentre along the surface of the Earth's crust.

Love waves have horizontal motion and are responsible for much of the damage witnessed during earthquakes. Rayleigh waves have both longitudinal and vertical motions that produce an elliptical motion in the rock particles similar to that produced in water particles by the passage of an ocean wave. While it is logical to believe that earthquakes generate tsunami through the physical deformation of the seabed, they also obtain their energy from Rayleigh waves.

MAGNITUDE SCALES FOR EARTHQUAKES AND TSUNAMI

Earthquake Magnitude Scales
(Bryant, 1991; Okal et al., 1991; Schindelé et al., 1995)

Earthquakes are commonly measured using the Richter scale based upon the magnitude of surface seismic waves, M_s, at a period of 20 seconds. The Richter scale is so well recognised that it is commonly used to describe the size of tsunamigenic earthquakes. An M_s magnitude earthquake occurs about once per year, and only 10% of these occur under an ocean with movement along a fault that is favourable for the generation of a tsunami. Earthquake-generated tsunami are associated with seismic events registering more than 6.5 on the Richter scale. Around the coast of Japan any shallow submarine earthquake with a surface magnitude greater than 7.3 will generate a tsunami. The tsunami period is also proportion to the magnitude of uplift. Small earthquakes tend to produce short tsunami wavelengths. Most tsunami-generating earthquakes are shallow and occur at depths in the Earth's crust between 0 and 40 km.

Unfortunately, the M_s scale saturates around a magnitude of 8, precisely at the point where significant tsunami begin to form. A better measure of the size of an earthquake is its seismic moment, M_o, measured in Newton metres (N m). The seismic moment was developed and refined for both near- and far-field tsunami in 1987 at the French Polynesian Tsunami Warning Center (Centre Polynésien de Prévention des Tsunamis) in Papeete, Tahiti. This system uses an automated algorithm TREMORS (Tsunami Risk Evaluation through seismic MOment in a Real time System) to analyse, in real time, seismic data and detect P, S, Rayleigh, and Love waves for any earthquake in the Pacific Ocean. The epicentre is estimated from long period data using the difference between the arrival of P and S seismic waves, and the polarity of P waves in the horizontal plane. Rather than using surface waves to calculate earthquake magnitude, TREMORS uses the magnitude of seismic waves travelling through the mantle. This mantle moment, M_m, is calculated from Rayleigh or Love waves having periods between 30 and 300 seconds. These long wave periods are virtually independent of the focal geometry and depth of any earthquake. The seismic moment is then calculated from the mantle moment using the following equation:

$$M_m = \log_{10} M_o - 13 \qquad\qquad\qquad 5.1$$

where M_m = mantle moment scale (dimensionless)
       M_o = seismic moment measured (N m)

The seismic moment has a similar scaling to the better-known Richter scale, but has the advantage of growing in amplitude with earthquake size, rather than saturating when it reaches a value around 8.0.

Tsunami Earthquakes

(Kanamori and Kikuchi, 1993; Okal, 1988, 1993; Beroza, 1995; Geist, 1997b)

The preceding scales imply that the size of a tsunami should increase as the magnitude of the earthquake increases. This is true for most teleseismic tsunami in the Pacific Ocean; however, it is now known that many earthquakes with small and moderate seismic moments can produce large, devastating tsunami. The Great Meiji Sanriku earthquake of 1896 and the Alaskan earthquake of 1 April 1946 were of this type. The Sanriku earthquake was not felt widely along the adjacent coastline, yet the tsunami that arrived 30 minutes afterwards produced run-ups that exceeded 30 m in places and killed 22,000 people. These types of events are known as tsunami earthquakes. Submarine landslides are thought to be one of the reasons why some small earthquakes can generate large tsunami, but this explanation has not been proven conclusively. Submarine landslides as a cause of tsunami will be treated in more detail in Chapter 6. Presently, it is believed that slow rupturing along fault lines causes tsunami earthquakes. Only broadband seismometers, sensitive to low-frequency waves with wave periods greater than 100 seconds, can detect slow earthquakes that spawn "silent", killing tsunami. Figure 5.2 illustrates the difference between a tsunami earthquake and an ordinary one. The Hokkaido 1993 tsunami was an ordinary event. The earthquake that generated it lasted for about 80 seconds and consisted of five large and two minor shock waves. The earthquake was regionally felt

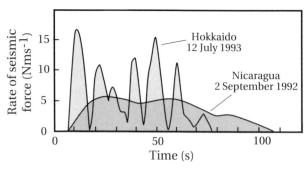

5.2 Comparison of the rate of seismic force between normal (Hokkaido) and slow (Nicaraguan) tsunamigenic earthquakes (based on Kikuchi and Kanamori, 1995).

along the Northwest Coast of Japan and produced a deadly tsunami. In contrast, the Nicaraguan tsunami of 1992 had no distinct peak in seismic wave activity. Rather, the movement along the faultline occurred as a moderate disturbance, for at least 80 seconds, tapering off over the next half minute. The earthquake was hardly felt along the nearby coast, yet it produced a killer tsunami. Both of these events will be described in detail later in this chapter.

The energy released by slow earthquakes cannot be measured accurately using surface-wave detection algorithms, because such waves do not produce a sharp enough change to trigger a noticeable difference in the recorded seismic amplitude. Instead, they are measured using the moment magnitude, M_w, determined from long period surface waves of more than 250 seconds. The following general relationship has been found between the seismic moment and this moment magnitude scale:

$$M_w = 0.67 \log_{10} M_o - 10.73 \qquad\qquad 5.2$$
where M_w = moment magnitude scale (dimensionless)

The moment magnitude can also be determined from seismic waves transferred through the Earth's mantle at a period of approximately 100 seconds. The TREMORS algorithm, described earlier, can detect these long waves. Technically, tsunami earthquakes are ones that occur in the ocean, where the difference in the M_s and M_w magnitudes is significantly large. Table 5.1 illustrates this difference for some modern tsunami. Besides the Sanriku and Alaskan events mentioned earlier, significant tsunami earthquakes occurred in the Kuril Islands on 20 October 1963, off Nicaragua on 2 September 1992, and off Java on 2 June 1994, with maximum run-ups of 15 m, 10.7 m, and 13.9 m respectively. Tsunami earthquakes happen under two conditions: where thick, accretional prisms develop at the junction of two crustal plates and wherever sediments are being subducted. Earthquakes under the former setting generated the 1896 Sanriku and 1946 Unimak Island tsunami. In contrast, the Peru, Kuril Islands, and Nicaragua tsunami listed in Table 5.1 were generated beneath subduction zones. The mechanisms of tsunami earthquake generation will be discussed in more detail later.

Tsunami Magnitude Scales
(Horikawa and Shuto, 1983; Abe, 1979, 1983; Hatori, 1986; Shuto, 1993)

Because of the high frequency of occurrence of tsunami around Japan, extensive research has been carried out there into predicting tsunami characteristics and magnitude. The following scale, known as the Imamura–Iida scale, was defined using approximately a hundred Japanese tsunami between 1700–1960:

$$m_{II} = \log_2 H_{rmax} \qquad\qquad 5.3$$

where m_{II} = Imamura–Iida's tsunami magnitude scale (dimensionless)
H_{rmax} = maximum tsunami run-up height (Equations 2.21–2.23)

TABLE 5.1   Disparity between the Seismic Magnitude, M_s, and the Moment Magnitude, M_w, of Recent Earthquakes Illustrating Tsunami Earthquakes

Event	Date	Seismic Magnitude (M_s)	Moment Magnitude (M_w)	Maximum Run-up (m)
Sanriku	15 June 1896	7.2	8.0	38.2
Unimak Island, Alaska	1 April 1946	7.4	8.2	35.0
Peru	20 November 1960	6.8	7.6	9.0
Kuril Islands	20 October 1963	6.9	7.8	15.0
Kuril Islands	10 June 1975	7.0	7.5	5.5
Nicaragua	2 September 1992	7.2	7.7	10.7
Java	2 June 1994	7.2	7.7	13.9

Source: Based on Geist, 1997b, and Intergovernmental Oceanographic Commission, 1999.

TABLE 5.2   Earthquake Magnitude, Tsunami Magnitude, and Tsunami Run-up Heights in Japan

Earthquake Magnitude Richter Scale	Tsunami Magnitude	Maximum Run-up (m)
6.0	−2	<0.3
6.5	−1	0.5–0.75
7.0	0	1.0–1.5
7.5	1	2.0–3.0
8.0	2	4.0–6.0
8.2	3	8.0–12.0
8.5	4	16.0–24.0
8.8	5	>32.0

Source: Based on Iida, 1963.

The small letter *m* usually represents this scale; however, it is subscribed here to differentiate it from other magnitude scales used in this book. On the Imamura–Iida scale, the biggest tsunami in Japan – the Meiji Great Sanriku tsunami of 1896, which had a run-up height of 38.2 m – had a magnitude of 4.0. The other great Sanriku tsunami of 1933 – which ranks as the second largest recorded tsunami in Japan – had a magnitude of 3.0. Japanese tsunami have between 1 and 10% of the total energy of the source earthquake. This can be related to a tsunami's magnitude using the Richter scale (Table 5.2). Only earthquakes of magnitude 7.0 or greater are responsible for significant tsunami waves in Japan with run-up heights in excess of 1 m. However, as an earthquake's magni-

TABLE 5.3   Soloviev's Scale of Tsunami Intensity

Tsunami Intensity	Mean Run-up Height (m)	Maximum Run-up Height (m)
−3.0	0.1	0.1
−2.0	0.2	0.2
−1.0	0.4	0.4
0.0	0.7	0.9
1.0	1.5	2.1
2.0	2.8	4.8
2.5	4.0	7.9
3.0	5.7	13.4
3.5	8.0	22.9
4.0	11.3	40.3
4.5	16.0	73.9

Source: From Horikawa and Shuto, 1983.

tude rises above 8.0, the run-up height and destructive energy of the wave dramatically increase. A magnitude 8.0 earthquake can produce a tsunami wave of between 4 and 6 m in height. The intensity needs to increase only to a value of 8.75 to generate the 38.2-m wave height of the Meiji Great Sanriku tsunami.

The Imamura–Iida magnitude scale has now acquired worldwide usage. However, because the maximum run-up height of a tsunami can be so variable along a coast, Soloviev proposed a more general scale as follows:

$$i_s = \log_2 (1.4 \, \bar{H}_r) \qquad\qquad 5.4$$

where i_s = Soloviev's tsunami intensity (dimensionless)
$\bar{H}_r$ = mean tsunami run-up height along a stretch of coast (m)

This scale and its relationship to both mean and maximum tsunami wave run-up heights are summarised in Table 5.3. Neither the Imamura–Iida nor the Soloviev scales relate transparently to earthquake magnitude. For example, both tsunami scales contain negative numbers and peak around a value of 4.0, whereas the seismic moment scale, M_o, for earthquakes tends to peak at values of 8.5–9.0. Most tsunami are also generated by earthquakes over a narrow range of magnitudes whereas the two tsunami scales span a broader range. Several attempts have been made to construct a more identifiable tsunami magnitude scale. Abe established one of the more widely used of these scales as follows:

$$M_t = \log_{10} H_r + 9.1 + \Delta C \qquad\qquad 5.5$$

where M_t = tsunami magnitude at a coast
ΔC = a small correction dependent on source region

Average ΔC corrections for Hilo, California, and Japan are –0.3, 0.2, and 0.0 respectively, irrespective of the source region of the tsunamigenic earthquake. There have been fifteen historical events in the Pacific Ocean with a tsunami magnitude greater than 8.5. The largest of these was the 22 May 1960 Chilean tsunami with an M_t value of 9.4. All Pacific-wide events have had a tsunami magnitude greater than 8.5. Ten of these events have occurred in the twentieth century, with the latest originating in the Aleutian Islands on 4 February 1965.

The tsunami magnitude scale has the advantage of being closely equated to the magnitude of earthquakes near their source because the average value of M_t for a coastline is set equal to the average M_w value of source earthquakes. Recently, emphasis has been placed upon near-field earthquakes and the M_t scale has been reformulated to include the exact distance between a coast and the epicentre of a tsunamigenic earthquake. For example, research on many tsunami on the East Coast of Japan shows the following relationship:

$$M_t = \log_{10}H_r + \log_{10}R_e + 5.80 \qquad\qquad 5.6$$

where M_t = tsunami magnitude (dimensionless)
        R_e = the shortest distance to the epicentre of a tsunamigenic
             earthquake (km)

The constant in this equation is dependent upon the source region. At present, few values have been calculated beyond Japanese waters, so there is no universal value that can be easily inserted into Equation 5.6. Tsunami waves clearly carry quantitative information about the details of earthquake-induced deformation of the seabed in the source region. Knowing the tsunami magnitude, M_t, it is possible to calculate the amount of seabed involved in its generation using the following formula:

$$M_t = \log_{10}S_t + 3.9 \qquad\qquad 5.7$$
where S_t = area of seabed generating a tsunami (m^2)

There is excellent agreement between the tsunami magnitude calculated using Equations 5.7 and 5.5. Finally, the Imamura–Iida scale has been converted to a form similar to Equation 5.6 as follows:

$$m_{II} = 2.7\,(\log_{10}H_r + \log_{10}R_e) - 4.3 \qqua\qquad 5.8$$

SEISMIC GAPS AND TSUNAMI OCCURRENCE
(Bryant, 1991; Satake, 1996; Bak, 1997)

The concept of earthquake cycles depends upon crustal movement occurring at constant rates over geological time and the buildup of frictional drag along fault-lines. Around the Pacific Rim, plates are moving at consistent rates. For instance, in the Alaskan region, the Pacific and North American Plates have generated continual earthquake activity over the past 150 years as stresses build up to crucial

limits and are periodically released at various points along the plate margin. However, stresses may not be released at some points. These appear in the historical
record as abnormally aseismic zones surrounded by seismically active regions. The
former locations are called seismic gaps and are believed to be prime sites for
future earthquake activity. The Alaskan earthquake of 1964 filled in one of these
gaps, and a major gap now exists in the Los Angeles area.

The seismic gap concept is flawed. Many earthquakes occur in swarms, with
the leading earthquake not necessarily being the largest one. Tsunami generation
also tends to occur in the area encompassing aftershocks. More significant, when
many tsunami events are examined, the arrival times of the first wave along different coastlines tend not to originate from a single point source. Finally, earthquakes
and the tsunami they generate are chaotic geophysical phenomena. They should
thus be generated by a spectrum of seismic waves with varying amplitudes and
periods. If this is the case, then the system of tsunamigenic earthquakes behaves
as white noise. One of the aspects of such systems is that earthquakes recur at the
same location rather than in areas that are more quiescent. This behaviour is characteristic of subduction zone earthquakes. For example, the 4 October 1994 Kuril
Islands tsunami occurred at the same location as a previous event in 1969. A
casual glance at the source location of tsunami over time in the Pacific will show
that tsunami originate repetitively within a 100-km radius of the same location in
many regions (Figure 1.2).

RELATIONSHIPS BETWEEN EARTHQUAKES AND TSUNAMI

How Earthquakes Generate Tsunami

(Wiegel, 1970; Ben-Menahem and Rosenman, 1972; Ward, 1980; Okal, 1988;
Satake, 1996; Geist, 1997b)

Rupturing along active fault lines where two sections of the Earth's crust are
moving opposite each other causes tsunamigenic earthquakes. Only three types
of faults can generate a tsunami: a strike-slip earthquake on a vertical fault, a
dip-slip earthquake on a vertical fault, and a thrusting earthquake on a dipping
plane (Figure 5.3). In each case rupturing can occur at any point along a faultline
deep in the Earth's crust. This location is known as the focal depth of the epicentre. The dip-slip and thrust faultline configurations are better at producing
tsunami than the strike-slip pattern. While the dip-slip mechanism seems to be a
logical one for tsunami generation because it abruptly displaces large sections of
the seafloor vertically, the area of uplift cancels out the area of subsidence, resulting in small or nonexistent tsunami. From a depth of 30 km below the seabed to
the surface, the thrust fault, which is characteristic of subduction zones (Figure
5.3), becomes the preferred fault mechanism for tsunami generation. The greater
the vertical displacement (or slip), the greater the amplitude of the tsunami.

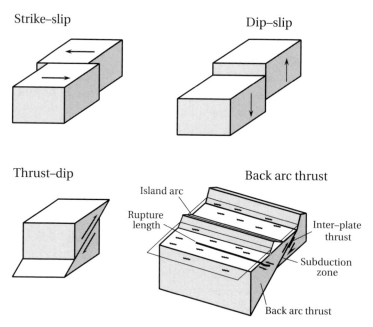

5.3   Types of faults giving rise to tsunami (based on Okal, 1988, and Geist, 1997b).

About 90% of earthquakes occur in subduction zones and these areas are the prime source for tsunami. Subduction zones typically have average dips of 25° ± 9°, while the largest tsunami run-up is associated with higher dip values of between 20° and 30°. As the dip angle decreases, the tsunami is more likely to have a leading trough. Faulting along subduction zones also results in subsidence at the coastline, a feature that compounds tsunami inundation for local earthquakes. Two tsunami actually develop in this case, one propagating shoreward and the other seaward. This tends to produce waveforms of different characteristics. The portion propagating seaward has a flatter crest and smaller amplitude than the wave moving landward.

Large tsunami are not restricted to subduction zones. For example, back-arc thrusting away from a plate boundary (Figure 5.3) produced the Flores, Indonesia, tsunami of 12 December 1992 and the Hokkaido Nansei–Oki tsunami of 12 July 1993. In addition, the 14 November 1994 Mindoro, Philippines, tsunami, which killed seventy-eight people, was generated along a strike-slip fault, while the 4 October 1994 Kuril Islands tsunami originated from the middle of a subducting slab. These anomalous events may also have involved a secondary mechanism such as a submarine landslide.

The greatest slip along faults tends to be concentrated towards the centre of a rupture, and this can result in a higher initial tsunami. This abnormal amplitude may not be detected if the degree of slip along a fault is averaged. Tsunami earthquakes tend to have much greater slip displacement than tsunamigenic earthquakes of comparable magnitude (Figure 5.4). For example, slip displacement for a tsunamigenic

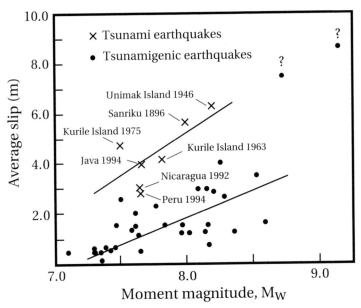

5.4 Relationship between moment magnitude, M_w, and the average slip distance of an earthquake (based on Geist, 1997b).

earthquake with a moment magnitude, M_w, of 8.0 is only 2 m. The equivalent tsunami earthquake has a value that is more than double this. The length and orientation of a rupture are also important in the generation of a tsunami. Length obviously correlates with the amount of seafloor displacement. Analyses show that the amplitude of a tsunami is proportional to the cube of the length of rupturing. Long ruptures such as those that occur along the coast of South America have the potential to produce the largest Pacific-wide tsunami. Rupturing over a 1,000-km length of faultline caused the exceptional 1960 Chilean tsunami. At the other extreme, if the tsunami wavelength that is generated by an earthquake is less than the rupture length, then there is a beaming effect. This certainly was the case with the Alaskan earthquake of 1964, whose tsunami concentrated along the Californian and Chilean Coasts. Beaming or directivity also depends upon the orientation of the rupture and can be calculated for any shoreline using the following formula:

$$D\,(\emptyset) \;=\; \sin\,\{0.5\;\Omega L_r\;C^{-1}[(C\upsilon_R^{-1}) - \cos\emptyset]\} \qquad\qquad 5.9$$

where $D\,(\emptyset)$ = direction of tsunami propagation relative to an observer
          (degrees)
    L_r = the length of fault rupture (km)
    υ_R = velocity of the rupture ($m\ s^{-1}$)
    $\emptyset$ = the azimuth of the observer relative to the direction of
          rupture (degrees)

The velocity of a rupture can be calculated from the nature of seismic waves arriving at various seismographs. There is an inverse correlation between the rupture velocity and directivity. Note that tsunami earthquakes have short rupture times (about 1.0 km s^{-1} compared to 2.5–3.5 km s^{-1} for normal subduction zone earthquakes). Tsunami earthquakes therefore have higher directivity and so can produce higher wave run-up along narrow sections of a coast. In the Pacific Ocean, rupturing occurs along major subduction zones that parallel coastlines. Tsunami should thus propagate at right angles to the rupture, or directly into the centre of the ocean. This is the reason why islands such as Hawaii and French Polynesia are so prone to teleseismic tsunami in the Pacific. An exception to this occurs along the Aleutian Island–Alaskan coastline. Here, faultlines are radial and project tsunami towards the Californian and Chilean Coasts. As one goes westwards along the Aleutian Island chain, the directivity of wave propagation sweeps towards Hawaii. Alaskan earthquakes do not affect French Polynesia because of topographic modification across the intervening ocean. For these reasons, once the epicentre of an earthquake has been located around the Pacific Rim, it is a simple task to plot the resulting tsunami's direction of travel and likely area of impact.

Finally, it is assumed that faulting occurs in massive rock units. Faulting is not this simple. In subduction zones, sedimentary rocks are often being buried. Accretional wedges can built up on the seabed as surface sediments are scraped off as one plate dips below another. This is especially prominent where low-angle trusting is occurring. Marine sediments are also water saturated. Close to continents these sediments can contain organic material that decomposes into methane, leading to gas-rich layers. Both of these types of sedimentary layers have low densities. Thrust rupturing into these types of sedimentary layers can increase the excitation of tsunami waves by a factor of one hundred. Only 10% of the force of the rupture needs to occur in the overlying sedimentary layer for this to happen. Tsunami earthquakes may also be generated by aftershocks associated with rupturing of an active fault through softer sediments near the seabed. In summary, the most prone area in the ocean for the generation of large tsunami is along a subduction zone where one plate moves upwards over another at a low angle, and where this movement propagates through less consolidated sediments near the seabed. In these circumstances, while the seismic moment, M_o, of the tsunamigenic earthquake may be an order of magnitude smaller than expected, the resulting tsunami can be very large.

Two theories can be used to model the formation of tsunami along faults. The first is the normal-mode theory and views tsunami as a long period, free oscillation of the Earth containing an outer layer of water characterised by an ocean of constant depth. This theory is a simpler version of the one used to describe Rayleigh waves moving through the surface of the Earth's crust. Normal-mode theory couples motion between the solid Earth and the ocean. This allows information about the source of the tsunami to be obtained from characteristics of the tsunami, even when it originates a long distance away. For example, information can be obtained about the focal depth of an earthquake. Generally, the amplitude of a tsunami is only reduced by 50% if the earthquake epicentre shifts from a depth of 30 km to 100 km. Only 4% of the energy in tsunami is stored elastically in the solid Earth, so only very large earthquakes, with a moment magnitude, M_m, greater than 8.6, can generate destructive teleseismic tsunami.

The second theory treats tsunami as linear gravity waves excited by the displacement of a large volume of water. Displacement of the ocean surface mimics the vertical displacement of the seabed if the length of rupture is at least three to four times the water depth. This displacement is calculated from theory on the dislocation of an elastic body, using as input the source parameters of the earthquake determined from seismographs. Increasing the focal depth of an earthquake decreases the vertical displacement of the seabed. However, the displacement is spread over a wider area. If the earthquake occurs on steep slopes, the horizontal component of displacement must also be considered. Once a tsunami wave is established, it travels across an ocean following the linear wave theory outlined in Chapter 2. It is now possible to propagate a tsunami wave forward from its source or trace it back from any coastline that it reaches. Linear wave theory is used in inversion schemes, where many of the characteristics of the earthquake that generated a tsunami can be determined simply from tsunami traces on tide gauge records or marigrams along a coastline – even a distant one. Sometimes just simple observations on a tsunami approaching a coast can be used. This was how Kenji Satake of the Geological Survey of Japan and his colleagues were able to link an obscure tsunami event along the East Coast of Japan on 26 January 1700 to an earthquake with a surface magnitude of 9.0 off the Northwest Coast of the United States. This earthquake formed the basis of the Kwenaitchechat Indian legend in Chapter 1.

Linking Tsunami Run-up to Earthquake Magnitude
(Iida, 1985)

For warning purposes, it is better to be able to predict the height of a tsunami along a coast given the magnitude of its source earthquake. Many tsunami approaching coasts tend to have a height that is consistent over long stretches of coastline. This certainly holds true along the East Coast of Japan and the West Coast of the United States. This fact can then be used to calculate the run-up height of a tsunami at various locations even if the slope varies (Equations 2.21–2.23). Figure 5.5 shows the relationships between the moment magnitude, M_w, of earthquakes and the amplitude of tsunami recorded on tide gauges for the East Coast of Japan and Papeete, Tahiti in the middle of the South Pacific Ocean. The data sets take into account both near-field and distant earthquakes, and have the following linear relationships respectively:

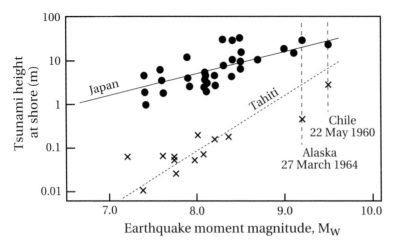

5.5   Relationship between the moment magnitude, M_w, and tsunami wave height for the East Coast of Japan and Tahiti (based on Okal, 1988, and Kajiura, 1983).

Japan: $\log_{10}\bar{H}_{tmax} = 0.5\,M_w - 3.3$ 5.10

Tahiti: $\log_{10}\bar{H}_{tmax} = 1.3\,M_w - 11.5$ 5.11
where $\bar{H}_{tmax}$ = mean maximum tsunami wave height along
                                                a coast (m)

The Japanese pattern characterises dispersive tsunami propagating from pointlike sources. Figure 5.5 indicates that tsunamigenic earthquakes generate tsunami that have less of an effect on Tahiti than they do on the East Coast of Japan. The reason for this has already been discussed in Chapter 2.

LARGE HISTORICAL TSUNAMIGENIC EARTHQUAKES

Lisbon, 1 November 1755
  (Reid, 1914; Myles, 1985; Foster et al., 1991; Andrade, 1992; Moreira, 1993;
  Dawson et al., 1995; Baptista et al., 1996; Hindson, et al., 1996)

   At 9:40 A.M. on 1 November 1755, All Saints' Day, one of the largest earthquakes ever documented devastated southern Portugal and Northwest Africa. Backward ray tracing simulations, using the type of Navier–Stokes incompressible shallow-water long-wave equations outlined in Chapter 2, position the epicentre on the continental shelf less than 100 km southwest of Lisbon (Figure 5.6). This location lies close to the boundary of the Azores–Gibraltar Plate, which historically has given rise to many tsunamigenic earthquakes in the region. The earthquake had an estimated surface magnitude, M_s, of 9.0 on the Richter scale and lasted for 10 minutes, during which time three severe jolts occurred. Heavy loss of life resulted

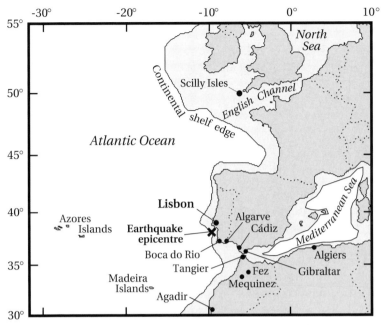

5.6 Location map for the Lisbon tsunami event of 1 November 1755.

in Lisbon and the Moroccan towns of Fez and Mequinez. Seismic waves were felt throughout Western Europe over an area of 2.5×10^6 km². Seiching occurred in ponds, canals, and lakes as far north as Scotland, Sweden and Finland.

Lisbon, a city of 275,000 inhabitants situated 13 kilomkmetres upstream on the Tagus River, was heavily damaged by the earthquake and consumed by fires. As the fires spread throughout the city, survivors moved down to the city's docks. Some even boarded boats moored in the Tagus River. Between 40 and 60 minutes after the earthquake, the water withdrew from the harbour, and a few minutes later one of the most devastating tsunami in history occurred as a 15-m-high wall of water swept up the river, over the docks, and into the city (Figure 5.7). Just as violently, the back-wash dragged bodies and debris back out into the estuary. Two other waves subsequently rolled in to the city an hour apart. The tsunami also caused widespread destruction along the coastline of Portugal, where it swept inland up to 2.5 km. At Porto Novo, north of Lisbon, run-up was 20 m high, while at Alvor and Sagres on the southwest tip of Portugal it reached 30 m above sea level. The wave had its greatest impact in southern Portugal. Coastal fortresses were destroyed, towns flooded, and in the city of Lagos, the city walls lying 11 m above sea level were overtopped.

The tsunami also caused widespread devastation in Southwest Spain and Western Morocco, as well as crossing the Atlantic Ocean and sweeping islands in the Caribbean 5,700 km away. In Southwest Spain, the tsunami caused damage to Cádiz and Huelva, and travelled up the Guadalquivir River as far as Seville.

5.7 Wood engraving by Justine of the tsunami sweeping up the North Tagus River, Lisbon, following the 1 November 1755 earthquake. Note that the earthquake, subsequent fires, and tsunami have been incorrectly drawn as occurring at the same time. Source: Mary Evans Picture Library Image No. 10047779/07.

At Cádiz, the wave had a run-up of 11–20 m. At Gibraltar, the sea rose suddenly by about 2 m; however, the wave's height rapidly decreased as it travelled into the Mediterranean Sea. The Moroccan Coast from Tangier to Agadir was severely affected, and in the latter city, the waves swept over the walls of the town, killing many people. The tsunami wave also swept up the West Coast of Europe into the North Sea, where it caused great disturbance to local shipping as boats in harbours were pulled from their moorings. Waves 3–4 m high moved through the English Channel on a high tide at about 2 P.M. The third and fourth waves were the largest. Oscillations in sea level with periods ranging from 10 to 20 minutes occurred over the next 5 hours in places. At Plymouth, the tsunami tore up muds and sandbanks at an alarming rate. The tsunami moved across the Atlantic, damaging the coastline of the Madeira and Azores Islands where wave heights of 15 m were reported. It reached the Caribbean Sea in the afternoon. Reports from Antigua, Martinique, and Barbados note that the sea first rose more than a metre about 3:30 P.M., followed by the arrival of large waves. Run-up heights of 7 m and 4.5 m were observed on the islands of Saba and St. Martin respectively. On the Leeward Island, Dutch colonists reported waves of 6.0–7.5 m in height. Run-ups of 3.0–4.0 m were typically observed elsewhere in the Eastern Caribbean. Significant oscillations continued at 5-minute intervals over the next three hours. In total, about 20,000 people may have been killed by the tsunami, although it is difficult to separate these deaths from those killed by the actual earthquake.

One of the clearest signatures of the Lisbon tsunami was the overwashing of barriers and transport of sands inland over peats. This evidence is clearest along the Algarve coastline of southern Portugal (Figure 5.6), where the tsunami reached its maximum height. The tsunami arrived on a low tide and was followed by up to eighteen secondary waves. At least four other tsunami subsequently affected the same coast between 1755 and 1769. Their impacts form the basis for the model of a tsunami effect on barrier coasts presented in the previous chapter (Figure 4.2). All five tsunami created extensive overwash deposits, infilled lagoons, and led to the creation of a backbarrier flat up to 800 m wide lying 4.0–4.5 m above high tide. This flat is now fronted seaward by a narrow foredune ridge. The flats today are poorly vegetated, with hummocky topography that traces out a labyrinth of second order drainage channels developed in response to tidal flooding through numerous inlets punched through the barrier by the tsunami. A lag of iron-stained gravels was left on the channel surfaces. The channels merge into first-order meandering ones that are incised progressively seaward into the backbarrier at the location of tidal inlets. The latter formed either major conduits for backwash or short-lived tidal inlets as the barriers recovered. Despite a tidal range of almost 4 m, most tidal inlets closed because the available tidal prism was insufficient to maintain strong enough currents to flush out sediment. Remnant channels are today truncated seaward by a foredune that has developed in the past 200 years. Some tidal inlets developed wide, flood-tidal deltas in the lagoon. These underwent extensive reworking as tidal currents moved the excessive amounts of sediment. These deltas now lie abandoned in the lagoon.

Along the South Portuguese Coast, the tsunami wave also ran up valleys. At Boca do Rio, the tsunami laid down a dump deposit and a tapering sand layer. The dump layer was deposited within 400 m of the coast by the first wave in the tsunami wave train. It consists of a chaotic mixture of muddy sand, cobbles, shell, sand-armoured mud balls, and the odd boulder up to 40 cm in diameter. The shells incorporate the littoral bivalve *Petricola lithophaga* and the subtidal sponge *Cliona* spp. The overlying sand layer was laid down by subsequent waves and consists of a 0.1- to 0.4-m-thick unit sandwiched between silty clays. The unit tapers landward over a distance of 1 km. The sand size within the layer coarsens upwards from a coarse, gritty sand to a silty or clayey, fine sand. The sequence is indicative of high-energy flow that decreased landward. The layer also contains clay 3–12 microns in size that originated from the weathered substratum that was eroded by the passage of the tsunami. Estuarine and intertidal shell species such as *Mytilus edulis, Scrobicularia plana,* and *Tellina tenius* are present in the lower part of the layer together with gravels and mud balls 0.5–5.0 cm in size. Foraminifera such as *Elphidium crispum* and *Quinqueloculina seminulum,* both of which are found in 20- to 30-m-depths of water, are present throughout the unit. Dating of the sands indicates that they were deposited by the Lisbon tsunami.

Sediment signatures of tsunami were also deposited on the Scilly Isles, 40 km southwest of Lands End, England. Here, shallow lagoons backing windswept dunefields and lying about a metre above the high-tide limit were inundated by tsunami swash. The first wave arrived at high tide and produced a 4.5-m-high run-up. The third and fourth waves were the largest. Coarse sandy layers 15–40 cm thick were deposited over sandy peats in three lagoons. Over the last 250 years, peat has subsequently covered these sands. Radiocarbon dating at the bottom of this peat indicates that the sands were deposited around the time of the Lisbon event.

Chile, 22 May 1960

(Wright and Mella, 1963; Myles, 1985; Lockridge, 1985, Lander and Lockridge, 1989; Pickering et al., 1991; Heinrich et al., 1996; Pararas-Carayannis, 1998a)

The most active area seismically producing tsunami is situated along the eastern edge of the Nazca crustal plate, along the coastline of Chile and Peru. This region has been inundated by destructive tsunami at roughly thirty-year intervals in recorded history in 1562, 1570, 1575, 1604, 1657, 1730, 1751, 1819, 1835, 1868, 1877, 1906, 1922, and 1960. The 22 May 1960 event was measured at 630 sites around the Pacific Ocean and is the most widespread tsunami to affect this basin. It was also responsible for the establishment of the modern Pacific Tsunami Warning System. The 22 May 1960 tsunami was generated by the last of over four dozen earthquakes occurring along 1,000 km of faultline parallel to the Chilean coastline. The first earthquake began at

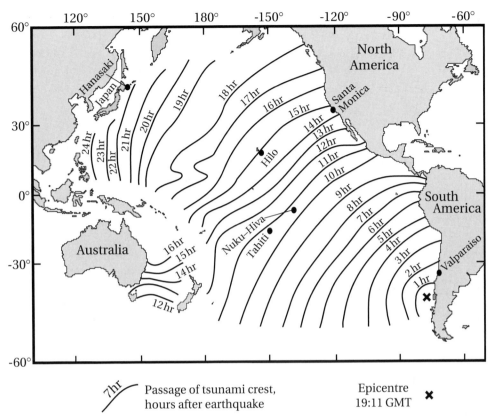

5.8 Passage of the tsunami wave crest across the Pacific Ocean following the 22 May 1960 Chilean earthquake (based upon Wiegel, 1964, and Pickering et al., 1991).

6:02 A.M. on Saturday, 21 May and destroyed the area around Concepción. Large after-shocks continued until, at 3:11 P.M. Sunday 22 May, the largest earthquake with M_s and M_w magnitudes of 8.9 and 9.5 respectively occurred with an epicentre at 39.5° S, 74.5° W and a focal depth of 33 km (Figure 5.8). Submarine uplift of 1 m and subsidence of 1.6 m ensued along a 300-km stretch of coast. Subsidence extended as far as 29 km inland with 13,000 km² of land sinking by 2–4 m. Many fisherman and their families quickly put out to sea to escape the flooding that was to come. Within 10 to 15 min-utes, the sea quickly rushed in as a smooth wave 4–5 m above normal tide level and just as quickly raced back out to sea taking with it boats and flotsam. This was only the harbinger of worse to come. Fifty minutes later, the sea returned as a thunderous 8-m wall of green water racing at 200 km hr⁻¹, drowning all those who had taken to the sea. An hour later, an even higher 11-metre wave came ashore at about half the speed of its predecessor. This was followed by a succession of waves that so obliterated coastal towns between Concepción and the south end of Isla Chiloe that the only evidence left of their existence were the remains of streets (Figure 5.9). Run-up along the

5.9 Aerial view of Isla Chiloe, Chile, showing damage produced by the 22 May 1960 tsunami. Two hundred deaths occurred here. Source: National Geophysical Data Center, http://www.ngdc.noaa.gov/seg/ image/geohazards_v3/images/648001/tif/64800112.tif.

Chilean Coast near the source area averaged 12.2 m above sea level and ranged between 8.5 and 25 m (Table 5.4). Dunes were eroded by overwashing, and sand was transported as a thin layer tapering inland over alluvial sediments. In the Valdivia region, 6–30 cm of beach sand was deposited inland over a distance of 500 m, while in the Rio Lingue Valley where the tsunami reached a height of 15 m, a thin layer of sand was deposited up to 6 km inland. The total loss of life in Chile is unknown but proba-bly lies between 5,000 and 10,000. The total property damage from the combined effects of the earthquake and tsunami in Chile was $417 million.

Over the next 24 hours, a series of tsunami wave crests spread across the Pacific, taking 2,231 lives and destroying property in such diverse places as Hawaii, Pitcairn Island, New Guinea, New Zealand, Japan, Okinawa, and the Philippines. The arrival time of the tsunami across the Pacific was accurately measured at many tide gauges (Figure 5.8). The initial wave travelled at a speed of 670–740 km hr^{-1}, depending upon the depth of the ocean. Individual waves in the tsunami wave train had wavelengths of 500–800 km and periods of 40–80 min-utes. In the open ocean, the wave height was only 40 cm high. Outside of South America, the tsunami reached landfall first along the Mexican, New Zealand, and Australian Coasts (Figure 5.8). The tsunami travelled quickest northwards along the coast of the Americas across the North Pacific Ocean to Japan. The wave crest

TABLE 5.4   Statistics on the Run-up Heights of the 22 May 1960 Chilean Tsunami around the Pacific Ocean

Region	Average Height (m)	Maximum Height (m)	Range (m)
Source area	12.2	25.0	8.5–25.0
Chile	2.7	5.0	0.4–5.0
Peru	2.0	3.9	1.0–3.9
Central America	0.5	1.4	0.2–1.4
U.S. West Coast	1.2	3.7	0.2–3.7
Canada	0.4	0.4	0.1–0.4
Alaska	1.3	3.3	0.4–3.3
Hawaii	3.1	10.5	1.5–10.5
Pacific Islands	4.3	12.2	0.5–12.2
Japan	2.7	6.4	0.2–6.4
New Zealand	0.6	0.9	0.4–0.9
Australia	0.5	1.8	0.2–1.8

underwent refraction around islands in the West Pacific, particularly those in the Izu-Bonin and Marianas island arcs. Detailed wave refraction analysis indicates that energy was focussed towards Japan in the Northwest Pacific, but dispersed elsewhere.

These effects are reflected in tidal gauge records around the Pacific Ocean. Tide gauges tend to record a tsunami's wave height near a coast and are a good indicator of its eventual run-up. Run-ups are summarised by region in Table 5.4, while marigrams are shown at select locations in Figure 5.10. In the Southwest Pacific, the tsunami entered the Tasman Sea from the south about 12 hours after the earthquake and caused rapid fluctuations in water levels in many harbours. Run-ups averaged 0.5–0.6 m respectively along the East Coasts of New Zealand and Australia, reaching maximum values of 1.8 m. Boats were torn from their moorings or beached by currents generated by tsunami-induced seiching. Along the coastline of the Americas, where the coast bends eastwards, the tsunami had minimal effect with run-ups averaging only 40 cm. On the exposed sections of the West Coast of the United States, wave run-up averaged 1.2 m, with values ranging between 0.2 and 3.7 m. The tsunami also had a variable impact on Pacific Ocean islands. On smaller islands fronted by steep offshore slopes and protecting fringing reefs, waves were only 1–2 m in height. However, where gradual bottom slopes existed and bays were present, wave heights were amplified by a factor of five and run-ups averaged 4.3 m. Hence, outside the source region, the Pacific Islands were affected the most by the tsunami. For example, on the Marquesas Islands, the wave height of the second wave exceeded 10 m at Nuku–Hiva and Hiva–Oa (Figure 5.10). Here, the wave ran up valleys 500 m inland. The largest run-up on any Pacific Island, 12.2 m, occurred on Pitcairn Island.

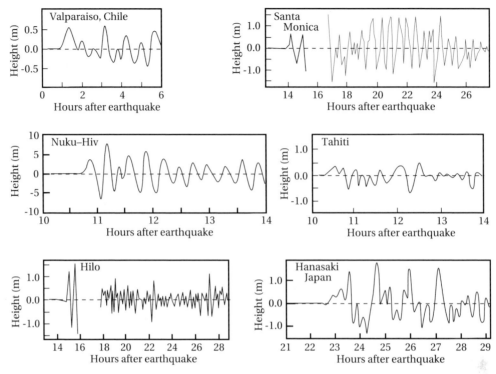

5.10 Tide records or marigrams of the 22 May 1960 Chilean tsunami around the Pacific Ocean. Records have had their daily tides removed (based upon Wilson et al., 1962, Tsuji, 1991, and Heinrich et al., 1996).

Hawaii was particularly affected by the tsunami that refracted around to the north side of islands. The greatest impact occurred at Hilo. Here, after travelling 10,000 km over a period of 14.8 hours, the tsunami's arrival time at 9:58 A.M. was predicted to within a minute. In spite of more than five hours' warning, only 33% of residents in the area affected in Hilo evacuated. Over 50% only evacuated after the first wave arrived, and 15% stayed behind even after the largest waves had beached. The first two waves did not do much damage, but the third wave was deadly. It swept inland 6 m above sea level, reaching a maximum run-up of 10.7 m. Sixty-one people, mainly sightseers, were killed as the wave swept ashore. Many of those killed were spectators who went back to see the action of a tsunami hitting the coast. The waterfront at Hilo was devastated as the waves swept inland over five city blocks (Figure 5.11). Ten-tonne vehicles were swept away and 20-tonne rocks were lifted off the harbour's breakwall and carried inland 180 m. In the area of maximum destruction, only buildings of reinforced concrete or structural steel remained standing. Wooden buildings were either destroyed or floated inland, and piled up at the limit of run-up. About 540 homes and businesses were destroyed or

5.11 Aftermath of the 22 May 1960 Chilean tsunami in the Waiakea area of Hilo, Hawaii, 10,000 km from the source area. The force of the debris-filled waves bent parking meters. Photograph Credit: U.S. Navy. Source: National Geophysical Data Center, **http://www.ngdc.noaa.gov/seg/ image/geohazards_v3/images/648001/tif/64800113.tif**

severely damaged. Damage in Hawaii was estimated at $24 million. Subsequently the flooded area was turned into parkland to prevent a recurrence of the disaster.

Although warnings were issued for Japan, the waves struck this country totally unexpectedly twenty-two hours after being generated. The tsunami rose from 40 to 70 cm height in 200 m depth of water despite losing 40% of its energy travelling across the Pacific. Dispersion dramatically reduced the height of the initial waves in the tsunami wave train. This effect is shown by the marigram for Hanasaki, Japan, in Figure 5.10. In addition, resonance effects tended to enhance isolated waves in the wave train such that the maximum run-up occurred hours after the arrival of the first waves. Run-up heights averaged 2.7 m along the east coast, with a maximum value of 6.4 m recorded at Rikuzen. At Shiogama and Ofunato on the Sanriku Coast of Northern Honshu – where local earthquakes, but not distant Pacific ones, had produced such devastating tsunami over the past century – fish-

ing boats were picked up and flung into business districts. Seiching at wave periods much lower than that of the main tsunami occurred in many harbours. Along
the shoreline of Hokkaido and Honshu, five thousand homes were washed away,
hundreds of ships were sunk, 251 bridges were destroyed, 190 people lost their
lives, and 854 were injured. Over 50,000 people were left homeless, with property
damage estimated at over $400 million.

Alaska, 27 March 1964

(Van Dorn, 1964; Hansen, 1965; Myles, 1985; Lander and Lockridge, 1989;
Pararas-Carayannis, 1998b; Johnson, 1999; Sokolowski, 1999a)

The Alaskan earthquake struck on Good Friday, 27 March 1964 at 5:36 P.M.
Alaska Standard Time (03:26 GMT, 28 March 1964) along a seismically active zone
paralleling the Aleutian Islands (Figure 5.12A). The area is noted for large tsunamigenic earthquakes that have had a continuing impact upon the Pacific Ocean (Figure 1.2B). Notable events that have affected Hawaii in the previous century
occurred on 20 January 1878, 1 April 1946, and 9 March 1957 (Table 1.3). The most
recent sequence of seismic activity began in 1938 with large earthquakes in 1938,
1946, 1957, 1964, and 1965. The last three events are amongst the ten largest earthquakes of the twentieth century. The southwards movement of Alaska over the
Pacific Plate at a shallow angle of 20° has generated these earthquakes, forming a
subduction zone known as the Aleutian–Alaska Megathrust Zone. Shallow dip
favours large trans-Pacific tsunami. The epicentre of the 1964 earthquake was
located in Northern Prince William Sound at 61.1° N, 147.5° W (Figure 5.12A). The
earthquake had a focal depth of 23 km and surface and moment magnitudes of 8.4
and 9.2 respectively – the largest ever measured in North America. The earthquake
rang the Earth like a bell and set up seiching in the Great Lakes of North America
and in Texas 5,000 km away. Water levels in wells oscillated in South Africa on the
other side of the globe. The ground motion was so severe around the epicentre
that the tops of trees were snapped off. More significant, ground displacement
occurred along 800 km of the Danali Fault system parallel to the Alaskan coastline.
The dislocations followed a dipole pattern of positive and negative displacements
on either side of a zero line running through the East Coast of Kodiak Island,
northeast to the Western side of Prince William Sound (Figure 5.12B). Maximum
subsidence of 3 m occurred west of this line, while as much as 11 m of uplift – the
greatest deformation from an earthquake yet measured – occurred to the east. At
some locations, individual fault scarps measured 6 m in relief. The earthquake was
felt over an area of 1.3×10^6 km^2 while land movement covered an area of 520,000
km^2. Most of Prince William Sound and the continental shelf were affected by the
latter deformation. This is the largest area known to be associated with a single
earthquake.

Approximately 215,000 km^2 of displacement contributed to the generation of the

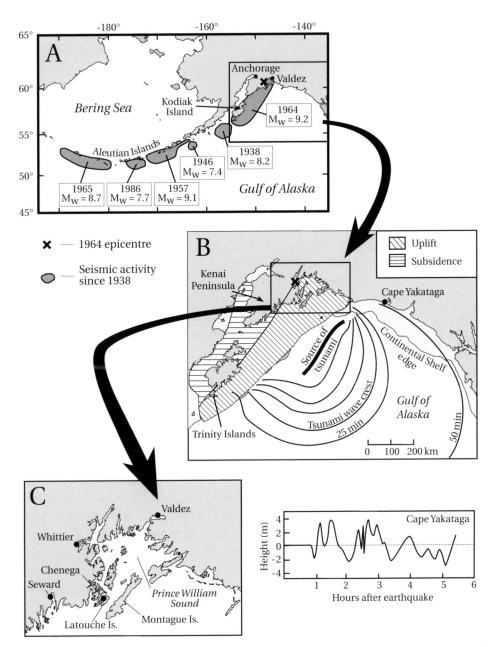

5.12 Earthquake and tsunami characteristics of the Great Alaskan Earthquake of 27 March 1964 (based on Van Dorn, 1965, Pararas-Carayannis, 1998b, and Johnson, 1999): **A.** Location of seismic activity since 1938. **B.** Gulf of Alaska land deformation caused by the earthquake and theorised open-Pacific tsunami wave front. **C.** Detail of Prince William Sound.

resulting tsunami over an area measuring 150 km × 700 km. The total volume of crust shifted amounted to 115–120 km³. More than 25,000 km³ of water were displaced. Tsunami generation was also aided by some of the fifty-two major aftershocks that occurred in the area of uplift. As a result, two main tsunami-generating areas can be distinguished, one along the continental shelf bordering the Gulf of Alaska and the other in Prince William Sound. As the tsunami moved away from the Gulf of Alaska, it was forecast by the Pacific Tsunami Warning System, which had been revamped following the 1960 Chilean tsunami. Within 46 minutes of the earthquake, a preliminary, Pacific-wide tsunami warning was issued by the Pacific Tsunami Warning Center in Honolulu. This was not sufficient to warn many Alaskan communities of impending tsunami. For example, the shores of Kenai Peninsula and Kodiak Island were struck by tsunami within 23 and 34 minutes respectively of the earthquake. Three major tsunami developed in Prince William Sound. One was related to the earthquake and had its origin near the West Coast of Montague Island, at the southern end of the Sound. The second was due to local landslides. The third developed much later in the Port of Valdez region, probably because of resonance within that part of the sound. Maximum positive crustal displacement in Prince William Sound occurred along the Northwest Coast of Montague Island and in the area immediately offshore. These earth movements caused a gradient in hydrostatic level and numerous large submarine slides in the area off Montague Island and at the north end of Latouche Island. Bathymetric surveys later showed that the combination of submarine slides and the tilting of the ocean floor due to uplift created the solitary wave observed in this region. The tsunami here did not escape the sound. At Chenega, a solitary wave reached 27.4 m above sea level within ten minutes of the earthquake.

Landslide-generated tsunami were confined to Prince William Sound where communities generally experienced wave run-ups of 12–21 m (Figure 5.12C). At the ports of Seward, Whittier, and Valdez, docks, railway track, and warehouses sank into the sea because of flow failures in marine sediments. The settlements were swamped by 7- to 10-m-high tsunami within an hour of the quake, but local run-ups were greater. None of these communities had any warning of the subsequent tsunami. At Seward, a swathe of the waterfront 100 m wide dropped into the ocean over a distance of 1 km. Twenty minutes later a 9-m-high wave rolled into the town picking up the rolling stock of the Alaska Railroad, tossing locomotives like Dinky toys (Figure 5.1), shearing off the pilings supporting the dock, destroying infrastructure and houses, and killing thirteen people. Near the harbour, the Texaco oil storage tanks burst and caught fire, spilling flaming oil into the receding sea. An hour later, a second, higher wave roared in. It could not be missed in the dark because it picked up the flaming oil and raced towards the town as a 10-m-high wall of flame. Eerily, the remains of the dock's pilings had caught fire and bobbed like large Roman candles in the waters of Resurrection Bay as successively smaller waves raced in.

At Whittier numerous slides occurred. One of the resulting tsunami reached a height of 32 m above sea level. At Valdez, a large submarine slide was generated at the

5.13 Limits of run-up of the tsunami that swept Valdez harbour following the 27 March 1964 Great Alaskan Earthquake. A submarine landslide triggered by the earthquake technically generated this tsunami. The height of run-up was 9 m. Photograph courtesy of the World Data Center A for Solid Earth Geophysics and United States National Geophysical Data Center. Source: Catalogue of Disasters #B64C28-205.

entrance to the port by collapse into the Bay of the terminal moraine at the end of Shoup Glacier. The resulting tsunami lifted driftwood to an elevation of 52 m above sea level and deposited silt and clay 15 m higher. In the town itself, which is situated on an outwash delta with a steep front, an area 180 m wide and 1.2 km long slid into the fjord, taking most of the waterfront with it. Within 2 or 3 minutes, a 9-m-high wave swept inland through the town (Figure 5.13). Thirty-two people lost their lives, many as the docks collapsed and were then swamped by water. Of the 106 deaths in Alaska due to tsunami related to the Good Friday earthquake, up to 82 were caused by these localised events. About five to six hours after the earthquake, further tsunami waves struck Valdez at high tide. The third wave came in at 11 P.M. March 27, and the fourth one at 1:45 A.M. March 28. This last wave took the form of a tidal bore and inundated the downtown section of Valdez. Apparently, these last tsunami were produced by resonance that had built up in the Bay over a five-hour period.

The main tsunami propagated southwards into the Pacific Ocean within 25 minutes of the earthquake (Figure 5.12B). Its wave period was exceptionally long, being an hour or more. This was caused by the long seiche period of the shallow shelf in the region where the tsunami originated. At many locations, the first wave arrived as a smooth, rapid rise in sea level rather than as a distinct wave. Seas then receded, to be followed by a bigger wave, often coincident with high tide. On Kodiak Island, the third and fourth waves were the highest and most destructive. Factors such as reflection, wave interaction, refraction, diffraction, and resonance had to be involved in the generation of this tsunami wave train. Kodiak Island sustained heavy damage with a maximum run-up of 10.6 m. The wave was focussed south down the West Coast of North America. Along the Canadian Coast, the wave's height registered about 1.4 m on tide gauges (Table 5.5). However, exposed locations such as Shield Bay recorded a maximum run-up of 9.8 m elevation. The tsunami became the largest historical tsunami disaster to sweep the West Coast of United States. Maximum wave heights reached 4.3–4.5 m in all of the three coastal states of Washington, Oregon, and California (Table 5.5). Many seaside towns received only 30 minutes' warning of the wave's arrival. Greatest damage occurred at Crescent City, California, where bottom topography concentrated the wave's energy. Crescent City has been affected by numerous tsunami geologically. Coring of marshes in the area shows evidence for at least thirteen previous events, one of which laid down a layer of sand 15 cm thick up to 500 m inland. The Alaskan tsunami only laid down a sand layer a centimetre thick in a marsh south of the city. Eleven people were killed in Crescent City by the Alaskan tsunami, most because they returned to low-lying areas before the advent of the fourth and largest wave. As with the Chilean tsunami, dispersion had greatly reduced the height of the first couple of waves in the tsunami wave train. The fourth wave was preceded by a general withdrawal of water that left the inner harbour dry. The wave then rapidly swept ashore, capsizing or beaching fishing boats, destroying piers, and flooding thirty city blocks backing the coast. Up to $8.8 million damage occurred in this one city out of the $12 million damage along the West Coast of the United States. Elsewhere the wave swept marinas, tore boats from moorings, and smashed piers. Fortunately, the waves diminished in height significantly before reaching San Francisco, where 10,000 people had flocked to the coast to witness the arrival of this rare event. Further south at San Diego, authorities were unable to clear sightseers from the beaches.

Following the arrival of the wave in California, warning sirens in Hawaii were sounded to alert the general population to evacuate threatened areas. The first wave arrived at Hilo 1.3 hours after striking Crescent City. Along the exposed shores of Hawaii, run-up averaged 2.3 m high, reaching a maximum of 4.3 m at Waimea on the island of Oahu. Hilo, which had been badly affected by the Chilean tsunami four years beforehand, experienced a run-up of only 3 m in height. Damage here was relatively minor because of the long warning time before the arrival of the waves and because much of the area affected by the Chilean tsunami had not been resettled. Over the next few hours, the wave reached Japan and the South American coasts. In Japan, the effect of the tsunami was minor. The wave averaged 0.4 m high on tide gauges reaching a maximum elevation of 1.6 m (Table 5.5). Little damage was reported. However, the tsunami continued to be a major event along the coast of

TABLE 5.5   Statistics on the Run-up Heights of the 27 March 1964 Alaskan Tsunami around the Pacific Ocean

Region	Average Height (m)	Maximum Height (m)	Range (m)
Source area	11.0	67.0	0.3–67.0
Canada	4.8	9.8	1.4–9.8
Washington State	1.8	4.5	0.1–4.5
Oregon	2.8	4.3	0.3–4.3
California	1.6	4.3	0.3–4.3
Central America	0.4	2.4	0.1–2.4
South America	1.2	4.0	0.1–4.0
Hawaii	2.3	4.9	1.0–4.9
Pacific Islands	0.2	0.6	0.1–0.6
Japan	0.4	1.6	0.1–1.6
Australia–New Zealand	0.2	0.2	—
Palmer Peninsula, Antarctica	0.4	—	—

South America. Here, the wave averaged 1.2 m in height, reaching a maximum value of 4 m at Coquimbo, Chile. Unlike the Chilean event four years previously, individual Pacific islands were not badly affected by the Alaskan tsunami. A maximum wave height of only 0.6 m was registered at the Galapagos Islands. Surprisingly, the Palmer Peninsula in the Antarctic recorded a wave height of 0.4 m – a value greater than that recorded on most Pacific Islands. South Pacific Islands appear to be immune from the effects of tsunami generated by Alaskan earthquakes.

EVENTS OF THE 1990S
(Satake and Imamura, 1995; Schindelé et al., 1995; González, 1999)

In the last decade of the twentieth century, there have been eighty-three tsunami events – a number much higher than the average historical rate of fifty-seven per decade. Like the majority of past events, these tsunami have involved earthquakes and have been concentrated in the Pacific Ocean region (Figure 5.14). The number of deaths during the 1990s has totalled more than four thousand (Table 5.6). Many of the events were unusual. For example, the 9 October 1995 tsunami in Mexico generated run-up heights of 1–5 m elevation. However, in some places the rapidity of inundation was so slow that people could outrun it at a trotting pace. A classic, slow tsunami earthquake caused the Nicaraguan event of 2 September 1992. People at the shore hardly noticed the vibrations, yet within an hour a deadly wave was racing overtop them. One of the most recent tsunami was the Aitape, Papua New Guinea, tsunami of 17 July 1998. It formed the background for one of the stories used in Chapter 1. Again, a tsunami earthquake occurred, with flow depths of over 15 m being recorded.

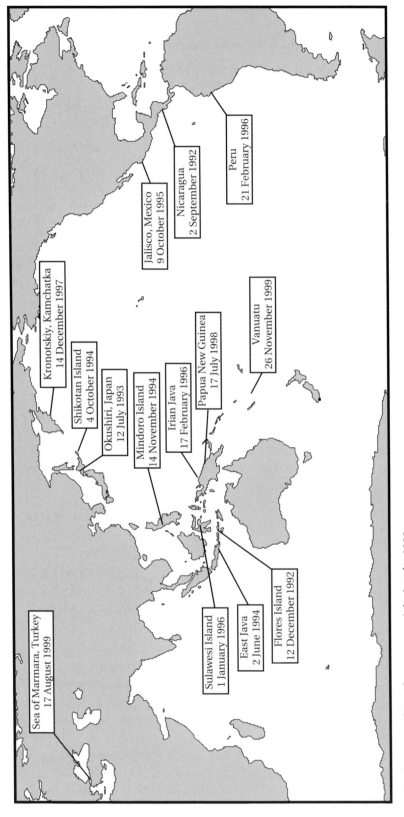

5.14 Location of significant tsunami during the 1990s.

The following labels appear on the map:

- Sea of Marmara, Turkey 17 August 1999
- Kronotskiy, Kamchatka 14 December 1997
- Shikotan Island 4 October 1994
- Okushiri, Japan 12 July 1993
- Mindoro Island 14 November 1994
- Irian Java 17 February 1996
- Papua New Guinea 17 July 1998
- Vanuatu 26 November 1999
- Jalisco, Mexico 9 October 1995
- Nicaragua 2 September 1992
- Peru 21 February 1996
- Sulawesi Island 1 January 1996
- East Java 2 June 1994
- Flores Island 12 December 1992

TABLE 5.6   Major Tsunami of the 1990s

Location	Date	Earthquake Magnitude	Maximum Height (m)	Death Toll
Nicaragua	2 September 1992	$M_s = 7.2$	10.7	170
Flores, Indonesia	12 December 1992	$M_s = 7.5$	26.2	1713
Okushiri Island, Sea of Japan	12 July 1993	$M_s = 7.6$	31.7	239
East Java	2 June 1994	$M_s = 7.2$	14.0	238
Shikotan, Kuril Islands	4 October 1994	$M_s = 8.1$	10.0	10
Mindoro, Philippines	14 November 1994	$M_s = 7.1$	7.0	71
Jalisco, Mexico	9 October 1995	$M_s = 7.9$	10.9	1
Sulawesi Island, Indonesia	1 January 1996	$M_s = 7.7$	3.4	24
Irian Jaya	17 February 1996	$M_s = 8.0$	7.7	108
Chimbote, Peru	21 February 1996	$M_s = 7.5$	5.0	2
Kronotskiy Cape, Kamchatka	14 December 1997	$M_s = 7.7$	8.0	—
Aitape, Papua New Guinea	17 July 1998	$M_s = 7.1$	15.0	2,202
Sea of Marmara, Turkey	17 August 1999	$M_s = 7.8$	2.5	?
Pentecost Island, Vanuatu	26 November 1999	$M_s = 7.3$	5.0	5

Source: Based on Satake and Imamura, 1995, and Internet sources.

The tsunami that struck the Sea of Marmara, Turkey, on 17 August 1999 was unusual for two reasons. First, it is a good example of a tsunami that was not restricted to an ocean, but that occurred in a small body of water. Second, it was caused by land subsidence because the earthquake occurred on a strike-slip rupture between the Eurasian and Turkish Plates – a feature not conducive to tsunami. Finally, the Vanuatu event of 26 November 1999 brought a slight glimmer of hope that we were getting it correct. The tsunami has not received the publicity of the Flores or PNG events, but on the island of Pentecost, it left a similar scene of destruction. The death toll was only five. People had fled to safety after they noticed a sudden drop in sea level. Though there was little memory of any previous tsunami, they had recently been shown an educational video based upon the PNG event and had reacted accordingly. Many of these events have already been referred to in this text, and not all will be described in detail here.

Slow Nicaraguan Tsunami Earthquake of 2 September 1992

(Abe et al., 1993; Kanamori and Kikuchi 1993; Satake et al., 1993; Geist, 1997b)

The Nicaraguan *tsunami earthquake* occurred at 7:16 P.M. (00:16 GMT) on 2 September 1992, 70 km offshore from Managua. The earthquake had a surface wave magnitude, M_s, of 7.0–7.2 and a seismic moment, M_o, of 3.7×10^{20} N m. It had a shallow focal depth of 45 km. Aftershocks occurred along the strike, parallel to the coast of Nicaragua in a band 100 km wide and 200 km long. The rupture occurred as the result of slow thrusting along the shallow dipping, subduction interface between the Cocos and Caribbean Plates near a previously identified seismic gap. The rupture propagated smoothly up-dip and alongshore at a velocity of 1.0–1.5 km s^{-1} over a period of 2 minutes (Figure 5.2). For these reasons, the earthquake was barely felt by residents along the coast. The earthquake's moment magnitude, M_w, was 7.7, a value at least half an order of magnitude greater than the surface wave magnitude, M_s, of 7.2 – a disparity that characterises tsunami earthquakes. The tsunami magnitude, M_t, was 7.9–8.0, much higher than should have been generated by an earthquake of this size.

The slow movement of the seabed generated a tsunami that reached the coastline 40–70 minutes later. Healthy adults, who were awake at the time, were able to outrun the tsunami. The tsunami killed 170 people, mostly children who were asleep and infirm people who could not flee. Run-up averaged 4 m along 2 km of coastline (Figure 5.15) and reached a maximum value of 10.7 m near El Transito (Figure 5.16). This is about ten times higher than the amplitude of a tsunami that should be generated by this size earthquake. Highest run-up appears to correlate with zones of greatest release of seismic force and maximum slip along the fault. Sand was eroded from beaches and transported, together with boulders and building rubble, tens of metres inland (Figure 5.16). The wave also propagated westwards into the Pacific Ocean. A 1-m wave was measured on tide gauges on Easter and Galapagos Islands, and a 10-cm wave was measured at Hilo, Hawaii, and Kesen'numa, Japan.

Flores, 12 December 1992

(Yeh et al., 1993; Shi et al., 1995; Tsuji et al., 1995; Minoura et al., 1997)

The Flores tsunami occurred at 1:29 P.M. (05:29 GMT) on 12 December 1992 along the North Coast of the island of Flores in the Indonesian Archipelago (Figure 5.17). An earthquake as the result of back-arc thrusting of the Indo-Australian Plate northward beneath the Eurasian Plate generated the tsunami. The earthquake had a surface wave magnitude, M_s, of 7.5. Damaging earthquakes and their tsunami have tended to increase in the Flores-Alor region since 1977, with at least nine earthquakes being reported beforehand. The 1992 earthquake had an epicentre at the coast and produced general subsidence of 0.5–1.0 m. Within 2 or 3 minutes a tsunami arrived ashore to the east, and within 5 minutes the tsunami reached most of the coastline where damage occurred. The tsunami wave train consisted of at least three waves, with the second one often being the largest. The first wave, which came in as a wall, was preceded by a general withdrawal of water. Submarine landslides triggered by the earthquake may explain many of the tsunami's features including the small number of waves in the wave train

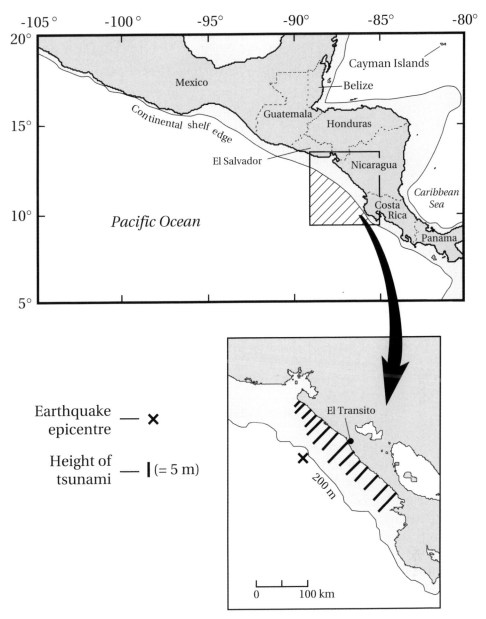

5.15 Location and height of Nicaragua tsunami of 2 September 1992. Height bars are scaled relative to each other.

and the larger run-up heights and shorter arrival times eastwards (Figure 5.17). Run-up heights varied from 2 to 5 m in the central part of the island, to as much as 26.2 m in the village of Riang–Kroko (Figure 3.5), on the northeast tip of Flores, where 137 of the 406 inhabitants were killed. This tsunami is only one of three

5.16   El Transito, Nicaragua, after the tsunami of 2 September 1992. While the second wave, which was 9 m high at this site, destroyed most of the houses, only sixteen out of one thousand people died because most fled following the arrival of the first wave, which was much smaller. Photo Credit: Harry Yeh, University of Washington. Source: NOAA National Geophysical Data Center.

that occurred in the twentieth century with run-up exceeding more than 20 m. Overall, two thousand deaths occurred as the waves destroyed entire villages. Damage was especially heavy on Babi Island, which lies 5 km offshore of the coast. Here the tsunami refracted around to the landward side of the island and ran up to heights of 7.2 m above sea level on the southwest corner. Of the 1,093 inhabitants on this island, 263 were drowned. The tsunami also affected the Southern Coast of Sulawesi Island where twenty-two people were killed. Two hours after the earthquake, smaller waves arrived at Ambon–Baguala Bay on Ambon Island.

Along much of the affected coastline, widespread erosion took place, denoted by coastal retreat, removal of weathered regolith and soils, and gullying. Small cliffs were often created by soil stripping, probably as the result of downward eroding vortices within turbulent flow. The tsunami spread a tapering wedge of sediment 0.01–0.5 m in thickness as much as 500 m from the coast. The wedge consisted of sand material swept from the beaches and shell and coral gravels torn up from the fringing reefs. Clay and silts appear to have been winnowed from the deposits by preceding tsunami backwash. Grain size tended to decrease, but sorting increased, towards the surface of these deposits. This pattern indi-

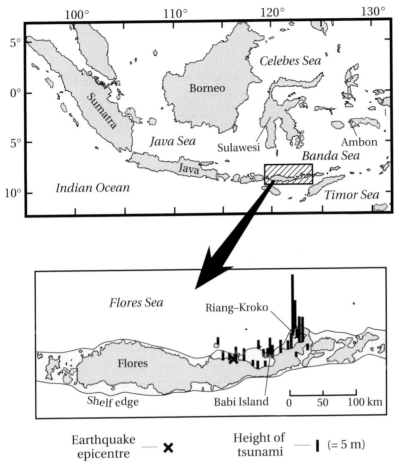

5.17 Location and height of the Flores, Indonesia, tsunami of 12 December 1992 (based on Tsuji et al., 1995). Height bars are scaled relative to each other.

cates an initial rapid rate of sedimentation. Grain size also tended to fine inland as the competence of the flow decreased concomitantly with a decrease in flow velocity. On Babi Island, there was a coarsening of grain size at the limit of run-up. Saw-tooth changes in grain size upwards through the deposits indicate that more than one wave was responsible for laying down the sand sheet. Sediment sequences on Babi Island imply that two different tsunami struck: the first from the direction of the earthquake and the second as the result of a trapped edge wave refracting around the island. This second wave was stronger than the first because it transported coarser material including large molluscs. Modelling results indicate that the first wave had a run-up velocity of 1 m s^{-1}, while the second one travelled faster at velocities of 2–3 m s^{-1}.

The Hokkaido Nansei-Oki Tsunami of 12 July 1993

(Yanev, 1993; Sato, 1995; Shimamoto et al., 1995; Shuto and Matsutomi, 1995;
Oh and Rabinovich, 1994)

In the late evening at 11:17 P.M. (13:17 GMT) on 12 July 1993 a strong earth-quake with a moment magnitude, M_w, of 7.8 was widely felt throughout Hokkaido, Northern Honshu, and adjacent islands (Figure 5.18). The earthquake occurred in the Sea of Japan, north of Okushiri Island. The Sea of Japan has his-torically experienced twenty tsunami, four of which have occurred in the twenti-eth century. The 1993 event occurred in a gap between the epicentres of the 1940 and 1983 earthquakes (Figure 5.18), and had a focal depth of about 34 km. This location coincides almost exactly with the epicentre of the Kampo earthquake of 29 August 1741, which produced a tsunami with a maximum run-up of 90 m along the adjacent Japanese Coast. Aftershocks from the 1993 event covered an ellipsoid 150 km long and 50 km wide close to Okushiri Island. About 150 km of faulting may have been involved in the event. The earthquake consisted of at least five intense jolts spaced about 10 seconds apart (Figure 5.2). Two to five minutes later a tsunami with an average run-up height of 5 m spread along the coast of Okushiri Island and killed 239 people – many of whom were still trying to flee the coastal area. On the southwestern corner of the island, run-up reached a maxi-mum elevation of 31.7 m in a narrow gully. This is the highest run-up of the cen-tury in Japan, surpassing that of the deadly Sanriku tsunami of 1923. Tsunami walls up to 4.5 m high protecting most of the populated areas were overtopped by the tsunami. Similar walls have been constructed in and around Tokyo and other metropolitan areas of Japan to protect urban areas from tsunami. They may be just as ineffective. In the town of Aonae, at the extreme southern tip of Okushiri Island, the first wave arrived from the west with a height of 7–10 m, overtopping the protective barriers and destroying the exposed southern section of the town (Figure 5.19). About 10 to 15 minutes later a second tsunami struck the sheltered, unprotected, eastern section of the town from the east, igniting fires that burnt most of the remaining buildings. The possibility that the second wave originated from aftershocks cannot be ruled out. The tsunami washed away half of the 690 houses in the town, although most were bolted to concrete foundations. The tsunami also severely damaged port facilities, power lines, and roads, stripping away pavement and depositing it inland. At Hamatsumae, which lies in the shel-tered southeast corner of the island, run-up measured 20 m above sea level. This high run-up was most likely due to refraction of a trapped soliton around the island – an effect similar to that produced around Babi Island during the Flores tsunami.

Within 5 minutes of the earthquake, the tsunami also struck the West Coast of Hokkaido with a maximum run-up of 7 m elevation. The simultaneous arrival

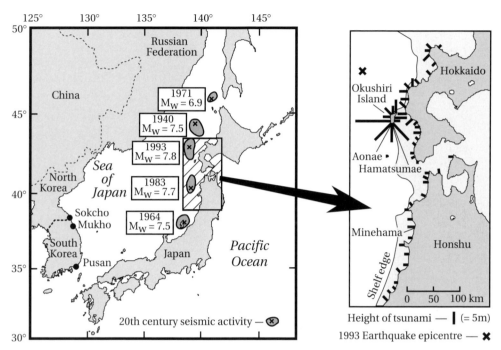

5.18 Areas and epicentres of twentieth century seismic activity in the Sea of Japan (based on Oh and Rabinovich, 1994). Detailed map shows location and height of the Hokkaido Nansei–Oki tsunami of 12 July 1993 (based on Shuto and Matsutomi, 1995). Height bars are scaled relative to each other.

times along Hokkaido and Okushiri Islands suggests that there may have been another tsunamigenic mechanism involved in generating the tsunami. Tsunami run-up decreased on the North Coast of Honshu; however, southwards, it reached a height of 3.5 m at Minehama. Fifty to seventy minutes after the earthquake, the tsunami reached the coastline on the opposite side of the Sea of Japan, striking the Russian Coast with an average run-up of 2–4 m elevation. Forty minutes after this, the wave reached the South Korean Coast at Sokcho and propagated southward to Pusan over the next ninety minutes. The tsunami wave height, as measured on tide gauges, was 0.2 m, 1.8 m, and 2.7 m respectively at Pusan, Sokcho, and Mukho. At Sokcho and Mukho, where the coastline and continental shelf edge are smooth and straight, waves were detectable for the next two days with periods averaging around 10 minutes. Along the Pusan Coast, dominated by bays and islands, wave periods were two to three times longer and decayed more slowly. It appears that tsunami amplification took place along the central Korean Peninsula, while seiching occurred along the South Coast in bays and harbours. There was also evidence of resonance effects in the Sea of Japan as a whole.

5.19   Damage at Aonae, Okushiri Island, due to the Hokkaido Nansei–Oki tsunami of 12 July 1993. The earthquake, and not the tsunami, damaged the leaning lighthouse. Concrete foundations in the foreground have been wiped clear of shops, houses, and kiosks. Note the gravel and boulder dump deposit, and the similarity of the unit to that deposited by the Flores tsunami in Figure 3.5. Photo Credit: Dennis J. Sigrist, International Tsunami Information Center at Honolulu, Hawaii. Source: National Geophysical Data Center

The tsunami's characteristics were heavily dependent upon the configuration of the coastline. At some locations sea level withdrew before the arrival of the tsunami crest, while at others the crest arrived first. Equal numbers of localities reported either the first or the second wave as the biggest. Run-up heights in many locations were two to three times greater than the initial height of the wave at shore. Calculations of the force required to remove houses bolted to concrete slabs indicate that flow velocities reached maximum values of between 10 and 18 m s^{-1}. The tsunami deposited 10-cm-thick sandy splays behind sand dunes. At Aonae, dump deposits several tens of centimetres thick were observed around obstacles and a sheet of poorly sorted sediment and debris was deposited throughout the village (Figure 5.19). This material rarely was transported inland more than 200 m. Internal, seaward-dipping bedding in some deposits indicates that upper flow regime antidunes formed. Where it first made landfall, the second tsunami cut parallel grooves up to 40 cm deep across the foreshore. Erosion also occurred with the formation of turbulent vortices around obstacles and in channelised backwash. Flow velocities interpreted from these sediment features agree with those interpreted from structural damage to buildings.

Papua New Guinea, 17 July 1998

(Gelfenbaum and Jaffe, 1998; McSaveney and Goff, 1998; Hovland, 1999;
Kawata et al., 1999; Tappin et al., 1999)

Historically, the Sissano Coast of Northwest Papua New Guinea (PNG) has been no more at risk from tsunami than any other South Pacific island in a zone of known seismic activity. Two previous earthquakes in 1907 and 1934 – neither of which appears to have generated a tsunami of any note – appear responsible for the formation of Sissano lagoon. Tsunami have occurred in the past, because a thin buried sand layer sandwiched within muds exists in the Arop area. At 6.49 P.M. (08:49 GMT) on July 1998, an earthquake with an M_s magnitude of 7.1 shook this coast. Twenty minutes later a moderate aftershock with a moment magnitude, M_w, of 5.75 jolted the coastline. Later analysis indicates that this second event was preceded by 30 seconds of slow ground disturbance. The location of the epicentre is still indeterminate; but the spread of aftershocks indicates that the earthquake was most likely centred offshore of Sissano lagoon, on the inner wall of the New Guinea trench that forms a convergent subduction zone where the Australian Plate is overriding the North Bismarck Sea. The Pacific Tsunami Warning Center detected the first earthquake and issued an innocuous tsunami information message about an hour later (Figure 5.20). In the meantime, a devastating tsunami with a tsunami magnitude, M_t, of 7.5 had already inundated the Sissano coastline shortly after the main aftershock. Tsunami flow depth averaged 10 m deep along 25 km of coastline (Figure 5.21), reaching a maximum 17.5-m elevation. The wave penetrated 4 km inland in low-lying areas. In places, the inundation of water was still 1–3 m deep 500 m inland. The wave also was measured at Wutung on the Indonesian border, where it reached a height of 2–3 metres. It then propagated northwards to Japan and Hawaii where 10- to 20-cm oscillations were observed on tide gauges about seven hours later. Over 2,200 people lost their lives.

The wave was unusual because it was associated with fire, bubbling water, foul-smelling air, and burning of bodies. Eyewitnesses reported that the crest of the tsunami was like a wall of fire with sparkles flying off it. In Chapter 1, this sparkling was attributed to bioluminescence, while the foul odour was linked to disturbance of methane-rich sediments in Sissano lagoon. The burnt bodies have been ascribed to friction as people were dragged hundreds of metres by the wave through debris and trees. These explanations may not be correct. Subduction zones incorporate organic material, which is converted to methane by anaerobic decomposition. The sudden withdrawal of 1–2 m depth of water can cause degassing of these sediments, leading to bubbling water. The tsunami crest approached the coastline at 100 km hr^{-1}. The atmospheric pressure pulse preceding this wave may have been sufficient to ignite this methane. Certainly, the pulse was strong enough to flatten people to the ground before the wave arrived. Those exposed to this flaming wall of water would have been severely burnt before being carried inland.

Subject: Tsunami Information Bulletin
Date: Fri, 17 Jul 1998 09:46:05 GMT
From: TWS Operations <ptwc@PTWC.NOAA.GOV>
To: TSUNAMI@ITIC.NOAA.GOVTSUNAMI

BULLETIN NO. 001
PACIFIC TSUNAMI WARNING CENTER/NOAA/NWS
ISSUED AT 0943Z 17 JUL 1998

THIS BULLETIN IS FOR ALL AREAS OF THE PACIFIC BASIN EXCEPT CALIFORNIA,
 OREGON, WASHINGTON, BRITISH COLUMBIA, AND ALASKA.

. . THIS IS A TSUNAMI INFORMATION MESSAGE, NO ACTION REQUIRED . .

AN EARTHQUAKE, PRELIMINARY MAGNITUDE 7.1 OCCURRED AT 0850 UTC
 17 JUL 1998, LOCATED NEAR   LATITUDE 2S   LONGITUDE 142E
IN THE VICINITY OF NORTH OF NEW GUINEA

EVALUATION: NO DESTRUCTIVE PACIFIC-WIDE TSUNAMI THREAT EXISTS.
          HOWEVER, SOME AREAS MAY EXPERIENCE SMALL SEA LEVEL
          CHANGES.

THIS WILL BE THE ONLY BULLETIN ISSUED UNLESS ADDITIONAL
INFORMATION BECOMES AVAILABLE.

. . . NO PACIFIC-WIDE TSUNAMI WARNING IS IN EFFECT . . .

5.20 Message issued by the Pacific Tsunami Warning Centre following the Papua New Guinea earthquake of 17 July 1998.

The wave also deposited one of the classic sedimentary signatures of tsunami by moving $1 \times 10^6 \text{m}^3$ of sand onshore and spreading it as a 5- to 15-cm-thick splay inland along the coast. At Arop, the sand deposit was up to 2 m thick and reached 60–675 m inland (Figure 5.22). The basal contact of the deposit was erosional. Mud rip-up clasts were evident in places, indicating high-velocity turbulent flow. Grain size decreased both upwards throughout the deposit and landward – facts reflecting the reduced capacity of the flow to carry sediment with time. Evidence of strong lee vortices in the form of 1- to 2-m-deep scour pits at the back of the barrier fronting Sissano lagoon, and the pattern of destruction to buildings, suggest that overland flow velocities attained values of 15–20 m s⁻¹.

Shock permeated the tsunami warning community following the PNG tsunami, and today the cause of the tsunami is still being hotly debated. Waves arriving in Japan indicated that the tsunami was generated by a steeply dipping reverse fault. However, modelling of tsunami waves locally indicates that a 7.1 magnitude earthquake could not have produced a tsunami higher than 2 m along the adjacent

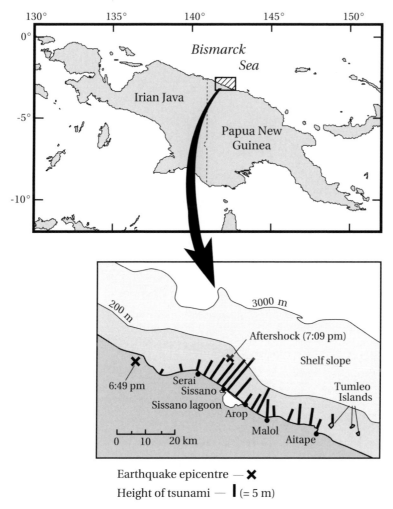

5.21 Location and height of Aitape, Papua New Guinea, tsunami of 17 July 1998 (based on Kawata et al., 1999, and Tappin et al., 1999). Height bars are scaled relative to each other.

coastline. While the event would appear to be a tsunami earthquake, its moment magnitude, M_w, was also too low, having a value of 7.1 – far less than that generated by tsunami earthquakes. Submarine landslides have been suggested as a cause for large tsunami following small tsunamigenic earthquakes. This now appears to be the main cause of this tragic event, and is supported by the fact that there were only three closely spaced waves in the tsunami wave train. Sea level withdrew from the coast, although there is evidence that about 0.4 m of submergence took place at the coastline. This pattern is similar to, though more rapid than, that of the 1960 Chilean tsunami along the Concepción Coast. Eastwards the wave did not approach shore-normal but travelled at an angle to the shore.

5.22 Overwash splay of sediment caused by the 17 July 1998 tsunami at Arop, Papua New Guinea. The splay is up to 2 m thick. Note the number of trees left standing despite the tsunami being over 15 m high and travelling at a velocity of 15–20 m s^{-1} near the shoreline. Source: Dr. Bruce Jaffe and Dr. Guy Gelfenbaum, U.S. Geological Survey.

This is unusual for seismically generated waves originating beyond the continental shelf but fits dispersion away from the point source of a submarine slump. Offshore mapping has now delineated slump scars, fissures, and amphitheatre structures associated with rotational slides. At present, no individual slump can be associated with the tsunami. Given the closeness of the epicentre to shore, the tsunami should have arrived within 10 minutes of the earthquake. It took 10 minutes longer. Slumping takes time after an earthquake to generate a tsunami, and this fact can account for the delay. If a submarine landslide generated the tsunami, theoretically 5 km^3 of material would have had to be involved. No evidence for such movement has been found. Alternatively, numerous slides could have coalesced instantaneously over an area of 1,000 km^2. Even if landslides were involved, topographic focussing must have occurred to produce the flow depths measured along the Aitape Coast.

The Papua New Guinea event raises a conundrum in ascribing tsunami to the occurrence of an antecedent earthquake no matter what its size. Secondary landslides were associated with many of the events described in this chapter. For example, many of the tsunami in Prince William Sound following the Alaskan earthquake were due to localised submarine landslides. At present, slides are perceived only as a minor contributor to tsunami. This view may be neglectful given the fact that many tsunamigenic earthquakes occur along the slopes of steep continental shelves prone to topographic instability. For example, the Tokyo earthquake of 1 September 1923 caused the seafloor to drop in elevation by 400 m in places. Under the circumstances, it is difficult to resolve what component of the PNG tsunami, which reached run-ups of over 15 m near Sissano Lagoon (Figure 5.21), was due to the earthquake or to the submarine landslide. This underrated aspect of tsunami will be discussed in depth in the next chapter.

CHAPTER SIX

Great Landslides

6.1 "All the sheep!" Artist's impression of a large tsunami breaking on a rocky coast at night. ©Kate Bryant, 1998.

INTRODUCTION
(Moore, 1978; Lockridge, 1990; Masson et al., 1996; Dawson, 1999)

The previous chapter on tsunamigenic earthquakes continually alluded to submarine landslides as being a contributing factor in the generation of anomalous tsunami. For example, the tsunami that swept Prince William Sound following the Great Alaskan Earthquake of 1964 and the one that swept the Pacific Ocean following the 1946 Alaskan earthquake are now recognised as being the products of submarine landslides. The latter event was large enough to affect Hawaii. Unfortunately, the evidence of sliding is difficult to obtain, especially in regions where detailed side-scan sonar surveys have not been performed beforehand.

About 70% of the Earth's surface consists of oceans containing tectonically and volcanically active areas near subduction zones. The oceans also consist of very steep topography along continental shelf margins, on the sides of ocean trenches, and on the myriad of oceanic volcanoes, seamounts, atolls, and guyots that blanket the ocean floor. Sediment moves under gravity down these slopes through a variety of processes that include slumps, slides, debris flows, grain flows, and turbidity currents. Substantial volumes of material are transported long distances, on slopes as low as 0.1°, across the deep ocean seabed by slides, flows, and turbidity currents. Slides consist of basal failure of topography that moves downslope in coherent blocks. As the slide progresses downslope it may disintegrate, producing a debris flow that mainly consists of disaggregated sediments. Debris flows can move without incorporating water; however, where material mixes with water, a dense turbid slurry of sediment can move as a current along the seabed under the effects of gravity. The latter are known as turbidity currents and form distinct deposits that were described in Chapter 3. Where large volumes of material are involved, these three processes can generate tsunami ranging from small events concentrated landward of the failure to mega-tsunami an order of magnitude larger than those generated historically by earthquakes. The most notable event to occur in the twentieth century has been the Grand Banks tsunami of 18 November 1929. This event, to be discussed in detail subsequently, resulted from a slide that had a volume of 185 km³. Geologically, larger slides with volumes over 5,000 km³ are known. For example, debris flows around the Hawaiian Islands have involved these volumes, while the Agulhas slide off the South Africa Coast contains a total of twenty thousand cubic kilometres of material (Table 6.1, Figure 6.2).

Landslides can take the form of indistinct slumping of rock and unconsolidated sediment, or rotation of material along planes of weakness in the rock. The latter often leaves a distinct scar or headwall eroded into the continental shelf slope or exposed on land as of amphitheatre forms in cliff lines. Rotational slides may also generate transverse cracks across the body of the slipped mass and tensional cracks above the head scarp, which gives rise to subsequent failure.

TABLE 6.1   Area and Volume of Large Submarine Slides and Their Associated Tsunami

Location	Area (km²)	Volume (km³)	Tsunami Features
Hawaiian Islands			
Nuuanu slide	23,000	5000	May be responsible for the
Alika 1 and 2	4,000	600	Lanai event, run-up >365 m
Storegga slides, Norway	112,500	5,580	
first	52,000	3880	
second	88,000	2470	
third	6,000	included in above	Maximum wave height of 5 m swept East Coast of Scotland
Agulhas, South Africa	—	20,000	
Sunda Arc, Burma	3,940	960	
Saharan Slide	48,000	600	
Canary Islands	40,000	400-1,000	May be the cause of tsunami that swept the Bahamas (see Chapter 4)
Grand Banks, 1929	160,000	760	3-m-high tsunami wave, Burin Peninsula, Newfoundland
Bulli, SE Australia	200	20?	May explain tsunami features in Sydney–Wollongong area

Source: Based on Moore, 1978, Harbitz, 1992, and Masson et al., 1996.

As shown in Chapter 5, earthquakes are the most likely triggering mechanism for landslides, especially submarine ones that have commonly been associated with tsunami. While terrestrial landslides falling into a body of water may be locally significant, this chapter will concentrate upon submarine landslides that have the potential to affect a much wider area of ocean.

CAUSES OF SUBMARINE LANDSLIDES
(Bryant, 1991; Carlson et al., 1991; Masson et al., 1996)

On dry land, slope failure can be modelled using the Mohr-Coulomb equation as follows:

$$\tau_s = c + (\sigma - \xi) \tan \phi \qquad\qquad 6.1$$

where τ_s =   the shear strength of the soil (kPa)
       c =   soil cohesion (kPa)
       σ =   the normal stress at right angles to the slope (kPa)
       ξ =   pore water pressure (kPa)
       ϕ =   the angle of internal friction or shearing resistance (degrees)

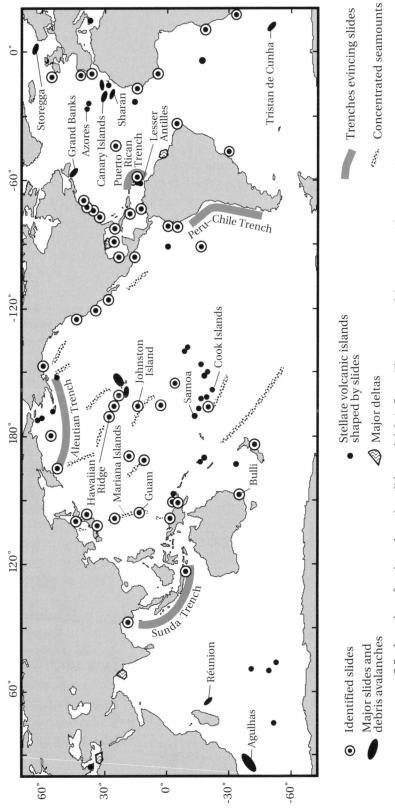

6.2 Location of major submarine slides and debris flows. Those parts of the oceans with topography susceptible to underwater landslides are also marked (based on Moore, 1978, Keating et al., 1987, and Holcomb and Searle, 1991).

181

In this equation, pore water pressure is a crucial determinant in failure. Sediments that are water-saturated are more prone to failure, while factors that temporarily increase pore water pressure such as the passage of seismic waves can reduce the term ($\sigma - \xi$) to zero. At this point, the strength of a soil becomes totally dependent upon the cohesion within loose sediment or within stratified rocks. Material on submerged continental shelves and island flanks are water saturated and devoid of vegetation. Their slopes were drowned and became prone to failure because of a Holocene rise in sea level amounting to 100–130 m in the last 3,000–6,000 years. This resulted in an additional weight on the seabed of 103–133×10^6 tonnes km^{-2}. Pore water pressure also increased internally within sediment lying at deeper depths with this rise in sea level. Slides in these environments usually occur close to the time of initial saturation. The Storegga slides off Norway may be the best example of such a process.

Changes in sea level do not have to be large to induce failure. Storm surges associated with the passage of a tropical cyclone can load and deload the shelf substantially. For example, along the East Coast of the United States, 7-m-high surges are common. The resulting increase in weight on the seabed can be 7.2×10^6 tonnes km^{-2}. If this is preceded by deloading due to the drop in air pressure as the cyclone approaches shore, than the total change in weight can amount to 10×10^6 tonnes km^{-2}. In areas where the Earth's crust is already under strain, this pressure change may be sufficient to trigger an earthquake. The classic example of a cyclone-induced earthquake occurred during the Great Tokyo Earthquake of 1923. A typhoon swept through the Tokyo area on 1 September and was followed by an earthquake, a submarine slide, and an 11-m-high tsunami that evening. In all, 143,000 people were killed. Measurements after the earthquake indicated that parts of the Sagami Bay had deepened by 100–200 m with a maximum displacement of 400 m. These changes are indicative of submarine landsliding. In Central America, the coincidence of earthquakes and tropical cyclones has a higher probability of occurrence than the joint probability of both events. Finally, failures on the submerged front of the Mississippi Delta have occurred primarily during tropical cyclones.

The internal pressure within submerged sediment can also be increased by the formation of natural gas through anaerobic decay of organic matter that has accumulated through the deposition of terrestrially derived material. This process is known as underconsolidation and occurs when fluid pressure within sediments exceeds the hydrostatic pressure. Methane is also locked into sediment on lower slopes, but because of the near-freezing water temperature and the extreme pressure, it is preserved as a solid gas hydrate. If this hydrate decomposes back to methane, then either the increased pressure due to the release of the gas into the sediment or the formation of voids can cause failure.

Slopes of ongoing sediment accumulation can become overloaded and oversteepened, leading to failure. This was a major process on continental slopes during glacials when sea level was lower. The increase in grade and exposure of continental shelves to subaerial weathering permitted increased volumes of sediment to be emplaced on the shelf slope. The process dominated the terminal end of major river deltas. Slides in these environments consist of rotational slumps. In solid rock, cohesion may be low because of shear planes that exist along the bedding planes between sedimentary layers or by joints running through the rock. Stratigraphically favourable conditions for landslides include massive beds overlying weaker ones, alternating permeable and impermeable beds, or clay layers. Structurally favourable conditions include steeply or moderately dipping foliations and cleavage, joint, fault, or bedding planes. Rock that is strongly fractured or jointed, or contains slickenside because of crushing, folding, faulting, earthquake shock, columnar cooling, or desiccation, is also likely to fail. Many rock units and sediment deposits on shelves thus have the potential for collapse.

Midocean volcanic islands are particularly prone to large landslides. On average, four such failures have occurred on these types of islands each century over the past 500 years. Shield volcanoes over hot spots accrete through successive lava flows that solidify quickly when they contact seawater. The islands consequently can rise 4,000 m or more above the seabed as steep pedestals emplaced on poorly consolidated, deep ocean clays. If weathering occurs between flow events, then subsequent flows are deposited higher in the edifice over weaker strata that may become zones for failure. If a landslide occurs early in the building phase of a volcano, then its seaward-dipping surface and its cover of blocky debris can become an unstable, low-friction foundation for ensuing lava flows. Eruptive events may inflate the volcano, steepening bedding planes and fracturing the domelike structure. Fracture lines may become zones for dyke intrusion that can overpressurise rock and increase its volume by more than 10%. For example, on the East Side of Oahu, Hawaii, dykes make up 57% of the horizontal width of the shield volcano. In addition, the weight of edifices standing over 4,000 m high can cause crustal subsidence that fractures the volcano further. Many volcanoes contain rift zones extending throughout the edifice along these zones of weakness. The flanks of these volcanoes are being pushed sideways, over the top of the sediment layer at their base under the force of gravity or by magma injection. Rates of spreading range between 1 and 10 cm per year. For example, the southern flank of Kilauea is presently moving at this higher rate away from a rift zone known as the Giant Crack. Giant landslides occur on the unbuttressed flanks of such volcanoes giving the volcano a tristar shape etched by steep headwall scars. This process is common on young volcanoes usually less than a million years in age, and has been a notable feature of the Hawaiian and Canary Islands.

HOW SUBMARINE LANDSLIDES GENERATE TSUNAMI

(Pelinovsky and Poplavsky, 1996; Watts, 1998)

The characteristics of tsunami generated by landslides are different from those simply generated by the displacement of the seabed by earthquakes. One of the more important differences is the fact that the direction of propagation of tsunami generated by landslides is more focussed. The slide moves in a downslope direction, and the wave propagates both upslope and parallel to the slide. While earthquake-generated waves are very symmetrical close to their source, landslide-generated ones have a shape that is best characterised by N-waves (Figure 2.4). The wave train is led by a very low-crested wave followed by a trough up to three times greater in amplitude. The second wave in the wave train has the same amplitude as the trough, but over time, it decays into three or four waves with decreasing wave periods. The initial inequality between the crest of the first wave and the succeeding trough enables landslide-generated tsunami to obtain greater run-up heights than those induced by earthquakes.

Wave generation by landslides depends primarily upon the volume of material moved, the depth of submergence, and the speed of the landslide. The volume of material can be determined knowing the height of the slide, its horizontal length, and the initial slope (Figure 6.3). The initial velocity of the landslide depends upon the trigger mechanism. If the mechanism is an earthquake, then the slide obtains its initial velocity instantaneously. If the slide is caused by slope instability, then it must slowly accelerate, obtaining a terminal velocity that depends upon its mass and density, and the angle of the slope. The delayed arrival time of a tsunami at shore is one of the clearest signals that is has been generated by a submarine landslide.

The initial and terminal velocities of a slide can be related to the duration of the slide and its travel distance as follows:

$$U_\infty = s_0 t_0^{-1} \tag{6.2}$$

$$U_i = s_0 t_0^{-2} \tag{6.3}$$

where U_i = the initial velocity of a submarine slide (m s^{-1})
   U_∞ = the terminal velocity of a submarine slide (m s^{-1})
   s_0 = travel distance of the slide (m)
   t_0 = duration of the slide (s)

Unless a slide cuts submarine cables, its duration will remain unknown. This applies to all slides in the geological record. In this case, the velocity of the slide must be determined hydrodynamically. If viscous drag effects at the base of the slide can be ignored, the initial and terminal velocities of a slide can be determined as follows:

$$U_i = (m_s - m_o) g (\sin \beta - \mu \cos \beta) (m_s + C_m m_o)^{-1} \tag{6.4}$$

$$U_\infty = [2g (m_s - m_o) (\sin \beta - \mu \cos \beta)]^{0.5} (C_d w l m_o \sin \beta \cos \beta)^{-0.5} \tag{6.5}$$

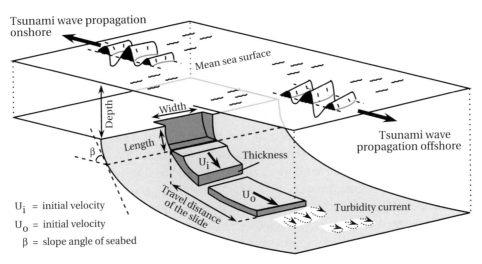

6.3 Schematic representation of a coherent submarine slide.

where m_s =   mass of the slide (kg)

m_o =   mass of displaced water (kg)

β =   slope angle (degrees)

C_d =   drag coefficient along the upper surface of the slide (~0.002)

C_m =   added mass coefficient (dimensionless)

μ =   Coulomb friction coefficient (~0.005)

w =   width of slide (m)

l =   length of slide parallel to the slope (m)

These equations assume that material fails as a coherent block. This may be the case for slumps, but any sustained movement causes disintegration of the slide into a debris avalanche and eventually a turbidity current. Turbidity currents are irrelevant in the generation of tsunami, because by the time sediment has become mixed with water and begun to stratify in the water column, the tsunami has been generated and is moving away from its source area. Turbidity currents are only important in laying down distinct deposits that are more than likely a signature of palaeo-tsunami.

The wavelengths and periods of landslide-generated tsunami range between 1 and 10 km and 1 and 5 minutes respectively. These values are much shorter than those produced by earthquakes. The wave period of a landslide-generated tsunami increases as the size of the slide increases and the slope decreases. It is independent of water depth, the depth of submergence, and the mass of the block. As a first approximation the height of the crest of the tsunami wave above still water can be approximated by the following formula:

$$H_t \quad = \quad ks_o \sin \beta \, Ha_o{}^{-n} \qquad\qquad\qquad 6.6$$

where H_t = the tsunami wave amplitude above mean sea level (m)

Ha_o = the Hammack number at termination (dimensionless)

$$= t_o (gd)^{0.5} (l \cos \beta)^{-1} \qquad\qquad 6.7$$

d = initial depth of submergence in the ocean (m)

k = a coefficient dependent upon depth of submergence and slope

n = an exponential ranging from $1 \leq n \leq 2$

The Hammack number is the ratio of the time scale of solid block motion to the time that it takes long waves to propagate away from the region of generation. Equation 6.6 can have many values depending upon the value of n. If the relationship is linear ($n = 1$), then the maximum wave amplitude is simply a function of the terminal slide velocity. However, if n equals two – its maximum value – then the maximum wave height depends upon the initial acceleration of the slide. There is a special case where the slope angle equals 45° and n lies midway between these two values (~1.5). Here, tsunami wave amplitude becomes simply a function of the thickness of the slide and the initial depth of submergence in the ocean. These relationships can be expressed by the following formulae:

If $n = 1$ $H_{max} = kU_\infty l \cos \beta (gd)^{-0.5}$ 6.8

If $n = 2$ $H_{max} = ka_o(l \cos \beta)^2 (gd)^{-1}$ 6.9

If $n \sim 1.5$ and $\alpha = 45°$ $H_{max} \propto c^{1.75} d^{-0.75}$ 6.10

where c = the thickness of the slide (m)

                       a_o = the initial acceleration of the slide (m s^{-2})

The Hammack number also defines the Froude number that quantifies wave behaviour and form. The Froude number for tsunami created by submarine landslides has the following form:

$$Fr = U^2 C^{-2} \qquad\qquad 6.11$$

where Fr =   the Froude number

U =   the velocity of the slide from Equations 6.4 or 6.5

C =   the velocity of the tsunami (Equation 2.2)

If the tsunami wave travels faster than the slide, both the Froude and Hammack numbers are very different from unity and nonlinear effects such as wave breaking and bore formation will not develop. Submarine landslides rarely exceed 50 m s^{-1}, while their associated tsunami travel at 100–200 m s^{-1}. Recent modelling indicates that these latter velocities may approximate 1,500 km hr^{-1}. Hence, tsunami generated by submarine landslides outpace the slide that forms them and produces several simple long waves in a wave train. In this case, that maximum tsunami wave height becomes independent of bottom slope and very exact as follows:

If $Fr << 1$ $H_{max} = 0.25\ \pi c^2 d^{-1}$ 6.12

The heights defined by the above formulae can be modelled using Navier–Stokes equations similar to those described in Chapter 2.

HISTORICAL TSUNAMI ATTRIBUTABLE TO LANDSLIDES

(Miller, 1960; Moore and Moore, 1988)

Both terrestrial and submarine landslides can produce tsunami. While historically rare, both are impressive. For example, the largest tsunami run-up yet identified occurred at Lituya Bay, Alaska, on 9 July 1958 following a rockfall triggered by an earthquake. Water swept 524 m above sea level opposite the rockfall, and a 30- to 50-m tsunami wave propagated down the bay towards the ocean. This tsunami will be described in detail subsequently. As pointed out in Chapter 1, steep-sided fjords in both Alaska and Norway are also subject to slides. There have been seven tsunamigenic events in Norway, which together have killed 210 people. The heights of these tsunami ranged between 5 and 15 m with run-ups of up to 70 m above sea level, close to the source. Five of these events occurred in Tafjord, in 1718, 1755, 1805, 1868, and 1934. In the 7 April 1934 event, 1×10^6 m^3 of rock dropped 730 m from an overhang into the fjord. The initial tsunami was 30–60 m high, decaying to 10 m several kilometres away. It travelled at a velocity of 22–43 km hr^{-1}. Historically, a slide sixteen times this volume has occurred in Tafjord. Loen Lake in Norway has also been subject to slide-induced tsunami, with three events occurring in 1936 alone. The largest of these produced run-ups of 70 m above lake level, killing seventy-three people. In the United States, slides have also generated tsunami in Disenchantment Bay, Alaska, on 4 July 1905 and several times in Lake Roosevelt, Washington. The Disenchantment Bay event produced maximum run-up of 35 m, 4 km from the source, while slides into Roosevelt Lake have generated run-ups of 20 m on at least two occasions. One of the more unusual tsunami produced by a rockslide occurred in the Vaiont Reservoir in Italy in October 1963. Here, 0.25×10^9 m^3 of soil and rock fell into the reservoir, sending a wave 100 m high over the top of the dam and down the valley, killing 3,000 people.

Tsunami generated by submarine slides have been common historical occurrences. For example, the Grand Banks earthquake of 18 November 1929 triggered a submarine landslide that is famous for the turbidity current that swept downslope into the abyssal plain of the North Atlantic. Less well known is the devastating tsunami that swept into the Newfoundland Coast. Similarly, the 11-m-high tsunami that swept the foreshores of Sagami Bay after the Great Tokyo earthquake of 1 September 1923 is now thought to have been produced by submarine landslides. Major tsunami generated by submarine slides triggered by earthquakes have also occurred at Port Royal, Jamaica, in June 1692; at Ishigaki Island, Japan, on 4 April, 1771; and at Seward, Valdez, and Whittier, Alaska, following the Great

Alaskan Earthquake of 1964. The Port Royal tsunami flung ships standing in the harbour inland over two-storey buildings and killed 2,000 inhabitants, while the Ishigaki Island tsunami carried coral 85 m above sea level and killed 13,486 people. The Sanriku tsunami of 1923, which took 2,144 lives, may also have been the product of a submarine slide triggered by an earthquake because a submarine canyon lying offshore was deepened by 590 m.

In the Pacific region alone, there have been sixteen tsunami events in the last 2,000 years in which submarine landslides were triggered by earthquakes, killing 68,832 people. Another sixty-five tsunami events involving either submerged or terrestrial landslides took a further 14,661 lives. Together, landslide-induced tsunami may have been involved in 18% of all deaths due to tsunami in the Pacific Ocean region. In the twentieth century, submarine slides have occurred off the Magdalena River, Colombia, and the Esmeraldas River, Ecuador, in the Orkdals Fjord, and off several Californian submarine canyons. In 1953, submarine landslides cut cables in Samoa and produced a 2-m-high tsunami. On 18 July 1979, a tsunami generated by a landslide, unaccompanied by any earthquake or adverse weather, destroyed two villages on the Southeast Coast of Lomblen Island, Indonesia, killing at least 539 people. Smaller tsunami were produced by submarine slides at Nice, France, on 16 October 1979; at Kitimat Inlet, British Columbia, on 27 April 1975; and in Skagway Harbour, Alaska, in November 1994. In the latter two cases, run-ups of 8.2 m and 11 m respectively were observed.

The Lituya Bay Landslide of 9 July 1958
(Miller, 1960; Pararas-Carayannis, 1999)

Lituya Bay is a T-shaped, glacially carved valley lying entirely on the Pacific Plate. Glaciers reach almost to sea level in Crillon and Gilbert inlets at the head of the Bay (Figure 6.4). The main bay measures 11.3 km long and 3 km wide, with a 220-m maximum depth. Cenotaph Island obstructs the centre of the bay, and La Chaussee Spit, which is the remnant of an arcuate terminal moraine left over from the last glaciation, blocks the entrance to the sea. The bay has been subject to giant waves geologically. For example, run-ups of 120 m and 150 m above sea level were produced by events in 1853 and on 27 October 1936 respectively. The 9 July 1958 event is the largest, however, reaching an elevation of 524 m above sea level. The trigger for the event was an earthquake with a surface magnitude, M_s, of between 7.9 and 8.3 that occurred around 10:16 P.M. along the Fairweather Fault at the junction of the Pacific and North American Plates. The earthquake's epicentre was 20.8 km southeast of the head of Lituya Bay. Vertical and horizontal ground displacements of 1.1 m and 6.3 m respectively occurred along the fault and reached the surface in Crillon and Gilbert inlets at the head of Lituya Bay. Vertical and horizontal accelerations reached 0.75 and 2.0 g respectively in the region. None of these disruptions was sufficient to gen-

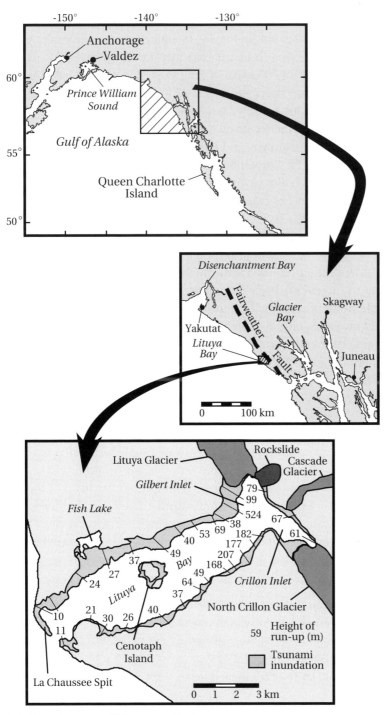

6.4   Location map of Lituya Bay, Alaska.

erate a giant wave. The earthquake triggered landslides in the northern part of the embayment. Terrestrial landslides are inefficient mechanisms for tsunami generation because only about 4% of their energy goes into forming waves. No known landslide has ever produced a wave the size of the Lituya Bay wave, which reached eight times the height above sea level of the largest slide-generated wave recorded in any Norwegian fjord. A Jökulhlaup or water bursting from an ice-dammed lake high up on Lituya Glacier could have also caused the wave. The water level in this lake dropped 30 m following the earthquake, and the hydraulic head was certainly sufficient to generate a giant wave. However, neither the volume of water nor the rate at which the lake drained was sufficient to allow such a high wave to develop. Besides, the maximum wave run-up did not occur near the discharge point for any Jökulhlaups or glacier burst from the glacier.

The tsunami appears to have been created by a sudden impulsive rockfall. Within 50 to 60 seconds of the earthquake, 30.5×10^6 m^3 of consolidated rock dropped 600–900 m from the precipitous northeast shoreline of Gilbert Inlet into the bay. The impact of this rockfall was analogous to that of a meteoroid crashing into an ocean – a phenomenon that will be described in Chapter 8. The impact not only displaced an equivalent volume of water, it also created a large radial crater in the bottom sediments of the lake that left an arcuate ridge up to 250 m from the shoreline. The sudden displacement of water and sediment sheared 400 m off the front of Lituya Glacier and flung it high enough into the air that a distant observer reported seeing the glacier lift above a ridge that had hidden it from view. On the opposite spur to the glacier, splash from the impact raced upslope to a height of 524 m above sea level, or more than three times the water depth of Gilbert Inlet. This splash has been mistaken for a giant wave. To distinguish the effects of this phenomenon from more coherent tsunami, the term *splash tsunami* will be used in this text. The rockfall, in combination with ground uplift at the head of the bay, generated a solitary wave 30 m high that swept down the bay to the ocean at a speed of 155–210 km hr^{-1}. Mathematical modelling using the same Navier–Stokes solutions as those described in Chapter 2 supports these figures. Run-up from the wave swamped an area of 10.4 km^2 on either side of the bay and penetrated as much as 1 km inland (Figure 6.5). Soil and glacial debris were swept away, exposing clean bedrock; however, little erosion of bedrock took place despite theoretical water velocities as high as 30 m s^{-1}.

Grand Banks Tsunami, 18 November 1929

(Heezen and Ewing, 1952; Long et al., 1989; Whelan, 1994; Dawson et al., 1996; Piper et al., 1999)

The Grand Banks tsunami was produced by a submarine landslide or slump triggered at 5:02 P.M. by an earthquake that had a surface magnitude, M_s, of 7.2 on

6.5 Aerial photograph of the foreshores of Lituya Bay swept by the tsunami of 9 July 1958 (from Miller, 1960). Height bars for tsunami run-up are scaled relative to each other.

the Richter scale. The earthquake had an epicentre in about 2,000 m depth of water (Figure 6.6). Numerous submarine slides occurred in a 120- to 260-km-wide swathe over a distance of 110 km along the slopes of the continental shelf. The slides occurred at two scales, as small rotational slumps 2–5 m thick and as larger ones 5–30 m thick. These coalesced over several hours into debris flows and then a turbidity current. The event is noteworthy because it was the first submarine debris flow detected. The turbidity current was hundreds of metres thick and swept downslope over the next eleven hours, cutting twelve telephone cables between Europe and North America. Based upon the time when each cable was broken, the turbidity current obtained a maximum velocity of 15–20 m s^{-1}. At its terminus, the resulting turbidite covered an area of 160,000 km^2 of seabed with 185 km^3 of sediment, to a maximum depth of 3 m. Sediment was deposited more than

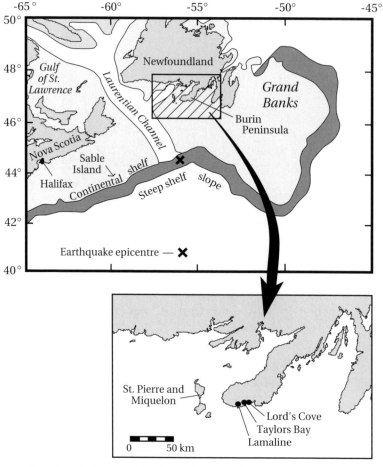

6.6 Location of the Grand Banks earthquake of 18 November 1929 and the Burin Peninsula of Newfoundland affected by the resulting tsunami.

400 km from the site of the slump in water depths of more than 2,500 m. This is one of the biggest turbidity currents yet identified either historically or in the geological record.

Little publicity has been given to the tsunami that followed the 1929 event. Two and a half hours after the earthquake a 3-m-high wave surged into the Burin Peninsula on the South Coast of Newfoundland, directly opposite the headwall of the slide (Figure 6.6). The wave was caused by backfilling of the depression left in the ocean by the removal of material that formed the turbidity current. It is difficult to attribute the wave to a single rotational slide because hundreds were involved. However, a cluster of large slides occurred in 600 m depth of water directly south of the Burin Peninsula. The wave travelled at a velocity of 140 km hr^{-1} and was concentrated by refraction into the coves along this coast, where forty isolated fishing communities were sheltered. The story in Chapter 1 refers to this wave as sheep riding a rising mountain of water in the moonlight (Figure 6.1). The tsunami wave train reached the coast on top of a high spring tide and surged up the steep shores as three successive waves, destroying boats and houses, and carrying sediment into small coastal lagoons. The characteristic tsunami signature of a fining-upwards sand layer, sandwiched between peats, has been identified in the lagoons at Taylors Bay and Lamaline. Run-up varied between 2 and 7 m with an extreme value of 27 m elevation at Taylors Bay. Two other waves in rapid succession followed the first wave. Twenty-eight people died in Newfoundland. So isolated were the communities that news of the disaster did not reach the outside world until two days later. The death toll was not restricted to Newfoundland. The wave also radiated out from the headwall of the slide and swept down the coast of Nova Scotia, where one person was killed. However by the time it reached Halifax, the wave was only 0.5 m high. The wave also spread throughout the North Atlantic Ocean and was measured on tide gauges as far away as South Carolina and Portugal, although no damage was reported.

The event is considered rare for this part of the North American Coast, with a recurrence interval of 1,000–35,000 years. However, a similar, but smaller, event occurred in the same region in 1884. This tsunami was also triggered by an earthquake and broke submarine cables. Since European settlement in North America, three other large earthquakes along the East Coast could have generated tsunami if they had happened at sea or triggered submarine landslides. These events occurred at Cape Anne, Massachusetts, in 1755; Charleston, South Carolina, in 1886; and Baffin Bay in 1934. While rare, the Grand Banks slide illustrates that the East Coast of North America is susceptible to deadly tsunami caused by slides. Large submarine slides are a common bathymetric feature of the eastern continental margin of North America and the Gulf of Mexico, and will occur again.

GEOLOGICAL EVENTS

Hawaiian Landslides

(Lipman et al., 1988; Moore and Moore, 1988; Moore et al., 1994a, b; Masson et al., 1996)

Some of the best evidence for giant submarine landslides has been discovered on the Hawaiian Islands using GLORIA (Geologic Long-Range Inclined Asdic), SeaBeam, and HMR-1 wide-swathe, side-scanning sonar systems. These systems sweep a swathe 25–30 km wide and measure features on the seabed down to 50 m in size. At least sixty-eight major landslides more than 20 km long have been mapped over a 2,200-km length of the Hawaiian Ridge, between the main island of Hawaii in the southeast and Midway Island in the northwest (Figure 6.7). The average interval between events is estimated to be 350,000 years. Because shield volcanoes typically have triple junction rift zones, amphitheatres are cut into the volcano at the headwall of giant landslides. If landslides have been ubiquitous on an island, then the island takes on a stellate shape termed a Mercedes star. On older volcanoes, these head scarps may be buried by subsequent volcanic activity. Some of the largest landslides occurred near the end of shield building when a volcano stood 2–4 km above sea level. The major trigger for the avalanches are earthquakes on the younger islands; however, on the older islands mass failures are even triggered by storm surges and internal waves. The length of the slides increases from 50 to 100 km for the older western volcanoes to 150–300 km for the younger eastern ones – making them some of the longest slides on Earth. The volume of material in each slide is as much as 5,000 km³. Smaller landslides having volumes of tens of cubic kilometres have also been detected in shallower waters but not mapped in detail because of the difficulty of using side-scan sonar at these depths. In total, about half of the volume of the Hawaiian Ridge consists of landslide material. The landslides can be grouped into slower-moving slumps and faster-moving debris flows. The latter have the potential for generating colossal tsunami wave heights.

Slumps have an internal consistency whereby large chunks slough off from a volcano along faultlines or rifting zones to a depth of 10 km. Parts of slumps may produce smaller debris avalanches. Slumps and their resulting tsunami are historically common in Hawaii. In 1868, slumps associated with an earthquake with a surface magnitude, M_s, of 7.5, produced a 20-m-high tsunami that killed eighty-one people. In 1919, a similar magnitude earthquake produced a submarine landslide off Kono that resulted in a 5-m-high tsunami run-up. In 1951, a small earthquake generated a local tsunami at Napoopoo. Two hours later, part of a cliff fell into Lealakakua Bay and generated another local tsunami. Most recently in 1975, an earthquake again with a surface magnitude, M_s, of 7.5 caused a 60 km section of the flank on Kilauea to subside 3.5 m and to move 8 m into the ocean. The

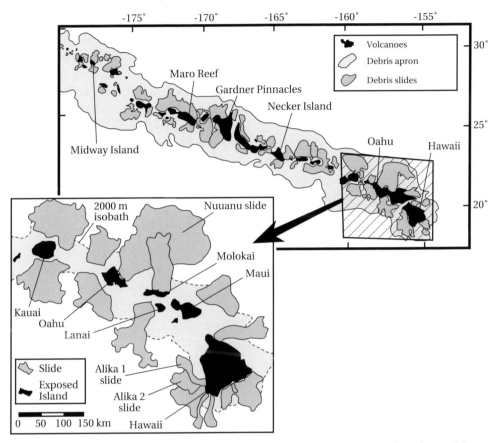

6.7 Location of submarine slumps and debris flows on the Hawaiian Ridge (based upon Moore et al., 1989, 1994b).

resulting tsunami had a maximum run-up height of 14.3 m and killed sixteen people.

Submarine debris avalanches are morphologically similar to those witnessed on volcanoes on land, where speeds of several hundred kilometres per hour have been inferred. If the Hawaiian debris avalanches obtained similar velocities, then they were major agents for tsunami generation in the Pacific. Unlike slumps, debris avalanches are thinner, having thicknesses of 0.05–2.0 m; however, they may travel as far as 230 km on slopes as little as 3°. The largest submarine avalanche is the Nuuanu debris avalanche on the northern side of Oahu Island. It is 230 km long, with a maximum thickness of 2 km at its source. It covers an area of 23,000 km² and has a volume of 5,000 km³ (Table 6.1). This represents one of the largest debris avalanches on Earth. It possibly occurred 200,000 years ago and would have sent a wave 20 m high crashing into the U.S. West Coast. Debris avalanches are characterised by hummocky terrain at their distal end, with some of the hummocks consisting of blocks of volcanic rock measuring 1–10 km in width. The largest individual block identified so far occurs in the Nuuanu deposit and measures 30 km × 17 km × 1.5 km.

The Alika debris avalanches on the western side of the main island of Hawaii are two of the youngest events in the island chain (Figure 6.7). The older avalanche, Alika 1, covers an area of 2,300 km²; the younger, Alika 2, is slightly smaller, covering an area of 1,700 km². Together, both failures incorporate 600 km³ of material, although the volume of material missing from the western side of Hawaii Island totals 1,500–2,000 km³ – much of it originating underwater down to 4,500 m depths. The second slide travelled more than 50 km from the coast in a northwesterly direction towards the small island of Lanai. Uncharacteristically steep slopes, reaching 2,000 m above sea level on the southwest side of Mauna Loa, appear to represent the headwall of the slides. Failure has occurred in the unbuttressed flanks of Mauna Loa and Hualalai volcanoes along rift zones infilled with dykes. Sedimentation rates on these slide deposits suggest that they are only a few hundred thousand years old. This age and the potential of the slides for tsunami generation links them to anomalously high beach deposits of a similar age on nearby islands.

Evidence that the Alika slides generated a major tsunami comes from gravel and boulder beds found on nearby islands to the northwest. These deposits reach elevations above sea level of 61 m on the southeastern tip of Oahu, 65 m on the South Coast of Molokai, 73 m on the West Coast of Maui, 79 m on the northwest corner of Hawaii, and 326 m on the South Coast of Lanai. On Lanai, catastrophic wave run-up may have deposited discontinuous dunelike boulder ridges, known as the Hulopoe Gravel, up gullies and on interfluves. The waves were also erosive removing a 2 m depth of weathered soil and basalt in a 2-km-wide strip parallel to the coast. The highest elevation of stripping is 365 m above sea level on Lanai and 240 m on the West Side of Kahoolawe. On the northeast sides of both islands, the height of stripping only reached 100 m above sea level. Because all islands are rising, the elevation of run-up would have been lower than these values at the time of the tsunami.

This stripping removed spherically weathered basaltic boulders up to 0.5 m in diameter from the soil and swept then downslope. These boulders were then reentrained by subsequent waves, together with coral gravel, and deposited in three inversely graded beds up to 4 m thick. The size of boulders and thickness of the deposit decreases upslope to a 150-m elevation. The larger particles support each other and are imbricated upslope – facts indicative of transport in suspension or as bedload. The voids between the large clasts are infilled with silt and pebbles that include marine foraminifera and sea-urchin spines. Fine calcareous material can be found filling crevices to an altitude of 326 m on Lanai. Mollusc shells are scattered throughout the deposit with species derived from 20 to 80 m depth of water characteristic of a slightly warmer environment than exists at present. The surface of the deposit consists of a branching network of ridges with a relief of 1 m and a spacing of 10 m. The ridges are asymmetric in profile with the steepest end facing downslope – a fact suggesting that backwash was the last process to mould the surface.

The hydrodynamics of the flow can be determined from this internal fabric and morphology. The inverse grading is suggestive of a thin, fast moving sheet of water typical of wave run-up or backwash. With a dune spacing of 10 m and a maximum boulder size of 1.5 m, the flow of water was 1.6 m deep (Equation 3.11) and obtained a minimum velocity of 6.3 m s^{-1} (Equation 3.1). These conditions are favourable for deposition of inversely graded deposits under run-up or backwash. The boulder-sized material and high elevation of deposition and stripping indicate that the flow velocity and run-up limits were an order of magnitude greater than that produced by storm waves or by tsunami generated by a distant tectonic event. The latter have a recorded maximum run-up height of no more than 17 m on these islands. The presence of multiple beds indicates a wave train that included three or four catastrophic waves, with the second wave being most energetic and reaching a maximum elevation of 365 m above sea level. If run-up height is set at ten times the tsunami wave height approaching shore, then the tsunami wave was 19–32 m high when it reached the coast. This is a conservative estimate, given the steep slopes of the islands and the fact that the swash was sediment-laden. Only waves generated by submarine landslides or meteorite impacts with the ocean are this big. The first wave in the tsunami wave train picked up material from the ocean, washed over and dissected any offshore reefs, and carried all this material in suspension, at high velocity, up the slopes of the islands. The first wave was so powerful on Molokai that it cracked the underlying bedrock to a depth of 10 m. These crevices were then subsequently infilled with fluidised sediment. The first wave also picked up boulders from the weathered surface as it swept upslope. These were then mixed with the marine debris in the backwash and laid down by subsequent waves into graded beds. The last backwash moulded the surface of the deposits into dune bedforms.

The asymmetry in the elevation of maximum stripping around Lanai and Kahoolawe suggests that the source of the tsunami had to be from the southeast. Uranium–thorium dates from the coral on Lanai show that the event occurred during the Last Interglacial around 105,000 years ago. However, the deposit on Molokai represents an older event that occurred at the peak of the Penultimate Interglacial 200,000–240,000 years ago. The most likely source of the tsunami was one or more of the Alika submarine debris avalanches off the West Coast of the main island of Hawaii. Slides to the southwest of Lanai cannot be ruled out, but they are older than the age of the deposits. If this latter source caused the tsunami, then the wave must have been generated by water backfilling the depression left in the ocean. This mechanism is similar to that proposed for the tsunami that swept the Burin Peninsula on 18 November 1929. The tsunami could also have been generated by a meteorite impact in the Pacific Ocean off the Southeast Coast of Hawaii. Meteorites as agents of tsunami generation will be discussed in Chapter 8. Unfortunately, the Hulopoe Gravel on Lanai has been questioned as a tsunami deposit. Some locations covered in Hulopoe Gravel also contain soil profiles between layers, and it is difficult to envisage deposition up gullies that should have then been scoured by backwash. Finally, the imprint of human activity cannot be ignored in the deposition of either some of the boulder piles or molluscs on Lanai.

The Canary Islands
   (Masson, 1996; Carracedo et al., 1998)

The Canary Islands have formed over the last 20 million years as shield volcanoes that rise from depths of 3,000–4,000 m in the ocean to heights of 1,000 m above sea level. The islands are similar to the Hawaiian Islands in that their origin has been linked to hot-spot activity. Because of structural control and scarring by fourteen landslides, the islands have taken on Mercedes star outlines (Figure 6.8). The scars consist of large amphitheatres or depressions that are backed by high cliffs. The most prominent of these lie on the island of El Hierro, where landslides have left a Matterhorn-like peak. As well, there are two prominent scars on Tenerife to the east. Giant landslides occur more frequently on the younger islands that are less than 2 million years old. Seven turbidites, ranging in volume from 5 to 125 km^3 and linked geochemically to the islands, have been mapped on the abyssal plain west of the Canary Islands. All are less than 650,000 years old. The slides associated with these turbidites reach volumes of 1,000 m^3 and were most likely induced by oversteepening of the volcanoes through successive eruptions or by the emplacement of vertical dykes.

Most of the debris flows have formed on the west side of the islands. Two of the oldest occur on the west sides of El Hierro and La Palma Islands, and are more than 550,000 year old. The largest debris flow on Tenerife Island occurred as recently as 170,000 years ago. El Hierro – the youngest island – has undergone the most recent activity. Two slides exist on the northwest corner of El Hierro. The older slide is 133,000 years old and has been proposed as a possible source for the tsunami that deposited chevron ridges on the Bahamas during the Last Interglacial. The younger slide contains by far the largest mass of debris identified as originating from the Canary Islands. Its headwall scarp is at least 8 km long and up to 900 m high. A debris avalanche, deposited closest to shore, contains individual blocks 1.2 km in

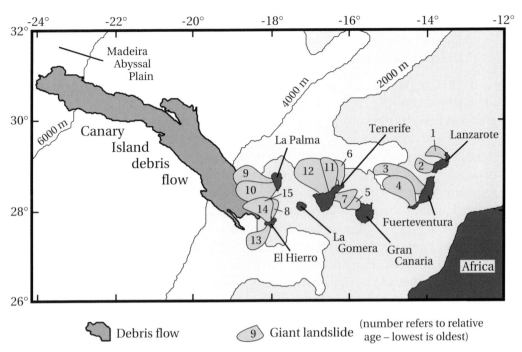

6.8 The giant landslides of the Canary Islands (based upon Masson, 1996, Carracedo et al., 1998). The debris flows are numbered sequentially according to decreasing age.

diameter and 200 m high. Further seaward, a debris flow 50–75 m thick was laid down. It is known as the Canary Island debris flow and covers an area of 1,500 km³ (Figure 6.8). This debris flow merges into a turbidite that is spread out on the seabed 600 km to the west of the islands. All the deposits are linked to a single cata-strophic failure, involving 700–800 km³ of material, that occurred 13,000–17,000 years ago when global sea levels were over 100 m lower than present.

Tsunami deposits have now been identified in the Canary Islands and linked by their juxtaposition to landslides on Tenerife. None has been dated, however, so they cannot be associated with any specific debris avalanche. One deposit has been mapped on the Jandia Peninsula on the south side of Fuerteventura Island. The deposit is unstratified and consists of well-rounded boulders and pebbles, marine molluscs, and angular pieces of basalt. It lies at an elevation of 35–75 m above present sea level up to 1 km inland. Fuerteventura Island is tectonically sta-ble, so the elevation of the deposit cannot be attributed to tectonic uplift. A similar deposit, consisting of two upwards-coarsening beds of beach boulders and shell, has also been discovered up to 90 m above present sea level at Agaete, on the Northwest Coast of Gran Canaria. A deposit containing shell also exists more than 200 m above sea level on La Palma.

The Storegga Slide of 7,950 BP

(Hall, 1812; Dawson, 1994; Long et al., 1989; Harbitz, 1992; Henry and Murty, 1992; Bondevik et al., 1997a, b)

In 1815, Sir James Hall postulated that a range of bedrock sculpturing features that included hairpin erosion marks several metres in length and the crag-and-tail hills that make up the city of Edinburgh were shaped by a tsunami similar to the one that had destroyed Lisbon in 1755. It was one of the first attempts in geology to explain the evolution of a landscape by invoking a single physical process. Hall of course was wrong. He knew nothing about continental glaciation that was subsequently used to explain the landforms. He was also unfortunate in picking the wrong landforms as examples. Had he examined the raised estuary plains (carseland) or the rocky headlands that dominate the East Coast of Scotland, he would have found not only his signatures for catastrophic tsunami, but also many of the additional ones presented in Chapter 3 (Figure 3.1). The tsunami did not originate in the North Atlantic as Hall envisaged, but from the region of known submarine slides at Storegga off the East Coast of Norway (Figure 6.9). While not large enough to swamp the hills of Edinburgh, they were certainly a significant factor in moulding the coastal landscape of Eastern Scotland.

There were three Storegga (great edge) slides involving a total of 5580 km³ of sediment that collapsed along 290 km of the continental slope of Norway. The slides travelled over 500 km across the sea floor at velocities up to 50 m s⁻¹ (Figure 6.9). The average thickness of the resulting deposits is 88 m, with values as high as 450 m. The slides contain blocks measuring 10 km × 30 km that are up to 200 m thick. Sediment moved into water more than 2,700 m deep and formed turbidity currents over 20 m thick. This thickness is seven times more than that of the turbidite deposited by the Grand Banks slide of 18 November 1929. The triggering mechanism for the Storegga slides is uncertain, but earthquakes and decomposition of gas hydrates within sediments have been suggested as causes.

The first slide occurred over 30,000 years ago, while the last two occurred close together between 6,000 and 8,000 years ago. The first slide was the biggest, involving 3,880 km³ of sediment; however, its tsunami happened at lower sea levels and had little visible effect on today's coastline. The last two slides moved or remobilised 1,700 km³ of sediment. Most of the evidence for catastrophic tsunami comes from the second Storegga slide, which has been radiocarbon-dated as being 7,950 ± 190 years old. The modelled deep-water wave height of the tsunami was 8–12 m at its source and 2–3 m where it swept into the North Sea and the open Atlantic Ocean. Equation 6.10 yields a tsunami height of 2.3 m in the open ocean based upon the geometry of the slide. The tsunami had a wave period of two to three hours – values that are much longer than those normally associated with earthquake-induced tsunami. The first wave reached the Northeast Coasts of Iceland and Scotland in just over two hours and took another eight hours to propa-

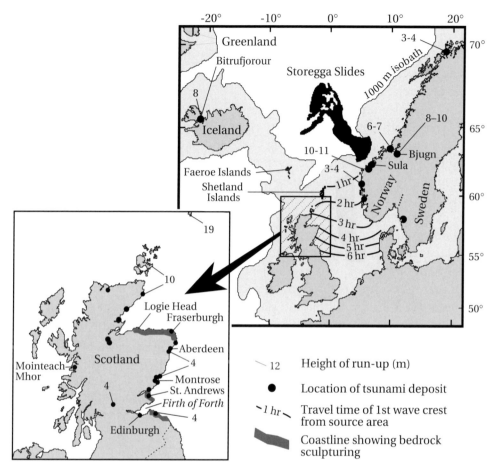

6.9 Location of Storegga slides near Norway and coastlines in the North Atlantic affected by the resulting tsunami from the second event. Run-ups based on Dawson (1994) and Bondevik et al. (1997a). Travel times in the North Sea are from Henry and Murty (1992).

gate through the North Sea (Figure 6.9). Most of the tsunami's energy was focussed towards Greenland and Iceland. Based on shallow-water long-wave equations, the tsunami was 3 m high approaching the coast of Greenland and Iceland, and 1 m high approaching Scotland. Maximum run-up heights were 10–15 m for the first slide and 5–8 m for the second. Along the East Coast of Scotland, modelled run-ups ranged from 3 to 5.5 m. These values may be conservative. If the slide moved at a maximum theorised velocity of 50 m s⁻¹ or initially underwent rapid acceleration, then the calculated run-up heights of the second slide were 13.7 m along the East Coast of Iceland and Norway, 11.5 m along the East Coast of Greenland, and 18 and 5.3 m along the North and East Coast of Scotland respectively. These latter figures agree with field evidence in Northeast Scotland showing that the wave reached 4 m above contemporary sea level at most locations in Northeastern Scotland, 10 m at the northern tip of Scotland, and 19 m on the Shetland Islands. Along the Norwegian Coast, tsunami deposits indicate that run-ups reached 10–11 m above sea level along the coastline adjacent to the headwall of the slide, and up to 4 m elsewhere.

The most prominent signature of the Storegga tsunami is the presence of thin sand layers sandwiched between silty clays and buried 3–4 metres below the surface of the raised estuarine plains or carseland of Eastern Scotland (Figure 6.9). These deposits have been described at over seventeen sites and are best developed in the Firth of Forth region, where they can be found more than 80 km inland. This long penetration up what was then a shallow estuary is beyond the capacity of even the largest storms in the North Sea and requires tsunami wave amplitudes at the maximum range of those modelled. Generally, the tsunami deposited these sand layers 4 m above the high-tide limit. The presence of these buried sand layers has been known since 1865; however, it wasn't until the late 1980s that researchers realised that the sands were evidence of tsunami originating from the Storegga submarine slides. The basic characteristics of the sands have already been described in Chapter 3. The sands, some of which are gravelly, include marine and brackish diatoms and peat fragments from the underlying sediments. The most common diatom species is *Paralia sulcata* (Ehrenberg) Cleve, which constitutes over 60% of specimens. Many of the diatoms are broken and eroded – features indicative of transport and deposition under high-energy conditions. Generally, the anomalous layer comprises grey, micaceous, silty fine sand less than 10 cm thick, although thicknesses of 75 cm have been found. In places, multiple layers exist, consisting of a series of layers of moderately sorted sand. Grain size fines upwards both within individual units and throughout the series. Detailed size analysis indicates that as many as five waves may have reached the coast, with the first and second waves having the greatest energy (Figure 3.3).

This sequence of sands is also well preserved in raised lakes along the Norwegian Coast. As the tsunami raced up to 11 m above sea level, it first eroded the seaward portion of coastal lakes around Bjugn and Sula, located adjacent to the headwall of the Storegga slides (Figure 6.9). The wave then laid down a graded or massive sand layer containing marine fossils. As the wave lost energy upslope, it deposited a thinner layer of fining sand. Only one wave penetrated the upper portions of the lakes, but closer to the sea, up to four waves deposited successively thinner layers of sediment in deposits 20–200 cm thick (Figure 6.10). As each wave reached its maximum limit inland, there was a short period of undisturbed flow during which larger debris such as rip-up clasts, water-logged wood fragments; and fine sand and organic debris, torn up and mulched by the passage of the tsunami wave, accumulated over the sands. Fish bones from the marine species *Pollachius vierns* (coalfish) and buds from *Alnus spp.* (alder) were

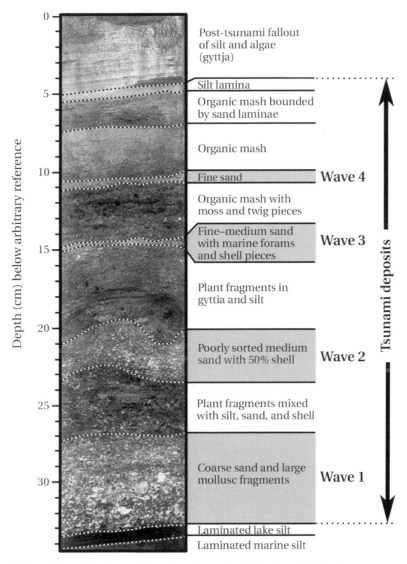

6.10 Alternating layers of sand and organic debris deposited by four successive tsunami waves of diminishing height in Kvennavatnet lake basin near Bjugn, Norway, following the second Storegga slide (from Figure 6 in Bondevik et al., 1997b). The photograph is ©Blackwell Science Ltd. and is reproduced by permission.

found in the organic mash. Both the length of the fish bones and the stage of development of the buds indicate that the event occurred in the late autumn. Each backwash eroded into the freshly deposited organic layer, but because water was ponded in lakes, velocities were not as high as in the run-up. Finally, when the waves abated, fine silt and sand settled from suspension in the turbid lake waters and capped the tsunami deposit with thin, muddy layers of silt termed *gyttja*.

The buried sand is not the only evidence of the Storegga tsunami. The 1- to 3-m open-ocean amplitude of the wave ensured that its effects were widespread throughout the North Atlantic. Besides the buried sands, a raised dump deposit of sand and cobbles has been found at Bitrufjorour, Iceland (Figure 6.9). Large aligned boulders have also been found along the Skagerrak Coast of Sweden pointing towards the slides. Finally, a buried splay of sand, sandwiched between two peat layers, rises to the surface of an infilled embayment on the West Coast of Scotland at Mointeach Mohr. The extent and sheltered location of this site rules out storm surge as the mechanism of deposition. The contact between the sand and the older, lower peat unit is erosional, with pieces of the lower peat incorporated into the sand. It appears that the sands in the tsunami deposit originated from the beach and dunes at the mouth of the embayment and were deposited rapidly landward in a similar fashion to the model proposed for tsunami-swept barriers in Chapter 4 (Figure 4.2).

This evidence certainly indicates that the tsunami was large enough to have been very erosive along exposed rocky coasts. Indeed, bedrock-sculptured features and tsunami-generated landscapes of the type described in Chapters 3 and 4 respectively are present along the East Coast of Scotland in the Edinburgh area and prominently along the Grampian coastline north of Aberdeen (Figure 6.9). Ironically, many of these features are similar to those originally described by Hall in his ill-fated hypothesis. For example, a raised fluted rock drumlin is cut into Carboniferous sandstone on the low-tide platform at St. Andrews (Figure 6.11). The crest of the feature lies 2 m above high tide and over 150 m from the backing cliff. Although weathered, the sides of the flute still preserve the distinct form of cavettos, while the upper surface is imprinted with muschelbrüche (scalloped-shaped depressions) aligned parallel to the alignment of features. Transverse troughs, formed by roller vortices, have been carved out from the surrounding platform and are not only aligned with the strike of the sandstone beds, but also perpendicular to the rock drumlin alignment. The platform surface rather than being planed by storm waves has a relief of 0.4–0.6 m, dominated by rock plucking that forms smaller flutes aligned parallel to the main feature.

Similar, but more effective, erosion, with large-scale development of fluted promontories, is most prominent along the Grampian coastline between Fraserburgh and Logie Head. The sand layer deposited by the Second Storegga tsunami reaches its maximum thickness – 75 cm – along the coast at Fraserburgh. Platforms and promontories have been cut into sandstones and metamorphic slates, phyllites and schists. One of the better examples of large-scale sculpturing occurs at MacDuff. Here, bedrock has been shaped into large rock drumlins or flutes rising 7–8 m above the high-tide line (Figure 6.12). The drumlins are dominated by profuse rock plucking, are detached from each other and the backing cliff, and rise *en echelon* landward, where they end abruptly in a well developed cliff whose base is slightly raised, about 2–3 m above the berm line of the modern beach. While

6.11 Raised rock drumlin or flute on the platform at St. Andrews, Scotland. The rock drumlin preserves sculptured s-forms in the form of muschelbrüche and cavettos. Weathering has subsequently modified these features. Roller vortices have cut transverse troughs into the platform surfaces.

6.12 Detached rock drumlins or flutes lying *en echelon* along the coastline at MacDuff, Scotland. Plucking dominates as the main mechanism of erosion, but smooth sculptured features such as potholes and muschelbrüche can be distinguished at a smaller scale. The porthole has formed through vortex formation on the front and back of the stack. The porthole and islet in the middle point towards the headwall of the Storegga slide.

6.13 Fluted stacks about 4 m high on the beach at Cullen, Scotland. Although weathered, each stack still preserves numerous cavettos on its flanks. The flutes align towards the headwall of the Storegga slide.

plucking dominates, the sides of some flutes have been carved smoothly and the *en echelon* arrangement is suggestive of helical flow under catastrophic flow. Despite the irregular nature of the eroded bedrock surfaces, smooth potholes appear on the seaward sides of some flutes. These potholes lie above the present active wave zone. In one case, vortex formation on the front and back of a stack has carved out a porthole aligned towards the headwall of the Storegga slides (Figure 6.12). Similar forms have been linked to tsunami in Eastern Australia, where vortices peel off the ends of headlands (Figure 3.22). Further north, at Gardenstown and Cullen, isolated fluted stacks over 4 m high are present in the middle of embayments. The features are so remotely linked to the adjacent rock coastline that in some cases they appear as stranded erosional remnants in the middle of beaches (Figure 6.13). Their upper parts often lie above the limit of present storm waves. Many flutes still show evidence of cavettos along their sides. The alignment of the flutes is structurally controlled, but the features are best developed where the strike of the bedrock points towards the headwall of the Storegga slides.

Tsunami also have moulded headlands. For example, at Logie Head, the end of the headland, which rises over 15 m above present sea level, is separated from the main cliffline by an erosional depression (Figure 6.14). The toothbrush-shaped form is similar to the eroded headlands of Southeast Australia interpreted as a prominent signature of tsunami-eroded bedrock terrain (Figure 4.5). Concentrated high-velocity flow or even wave breaking over the back of the headland generates such erosional forms. Large storms can be ruled out as a mechanism for moulding these bedrock-sculptured features. Although North Sea storms can generate 15-m-high waves superimposed on a 2-m-high storm surge, such waves are short in wavelength and break offshore of headlands. Both fluted terrain and the chiselled headland at Logie Head require sustained, high-velocity, unidirectional flow that only a catastrophic tsunami resulting from a megaslide or meteorite impact into the sea could produce.

It might also be conceivable to attribute such features to glacial activity or to subglacial flow. After all, the features Hall described were later attributed to just such a process. Indeed, the orientation of flutes along the Grampian coastline cor-

6.14 The toothbrush-shaped headland at Logie Head. The larger depression has formed either by catastrophic wave breaking near the shoreline or by concentrated high-velocity flow.

respond to flow lines for the Late Devensian ice sheet that covered this region. However, the fluted features at St. Andrews (and also east of Edinburgh at Dunbar), which also point in the same general direction as those along the Grampian Coast, are not aligned with the direction of ice sheet movement. Finally, if the flutes are the products of glaciation or catastrophic subglacial water flow, then the features should not be limited just to the immediate coastline. Their similarity to features in Southeastern Australia, which at no time has been affected by glacial ice during the Pleistocene, implies a common mechanism for both localities. It is only fitting that after 150 years, evidence can be found in Eastern Scotland for Hall's 1812 hypothesis for bedrock sculpturing by catastrophic tsunami – albeit on a smaller scale than he envisaged.

THE RISK IN THE WORLD'S OCEANS

Other Volcanic Islands
(Stoddart, 1950; Holcomb and Searle, 1991; Keating, 1998)

Large debris avalanches are now known to be associated with at least two other midocean volcano complexes, La Réunion in the West Indian Ocean and Tristan da Cunha in the South Atlantic Ocean (Figure 6.2). On Réunion, the Grand Brule slide fans out from the East Coast. It consists of four nested submarine landslides that have reached depths of 0.5 km, 1.1 km, 2.2 km, and 4.4 km below sea level. The Tristan da Cunha complex consists of three volcanoes rising 3,500 m from the sea floor. All three of these islands are bounded by near-vertical cliffs 150–500 m high. In places, these cliffs form amphitheatres that appear to be the headwalls of former landslides. The largest of these occurs on the northwest side of Tristan da Cunha itself and is associated with a debris avalanche more than 100 m thick, covering an area of about 1,200 km^2 and having an estimated volume of 150 km^3. The slide has been tentatively dated as younger than 100,000 years.

If amphitheatre forms in cliffs or a stellate-shape island are the signatures of former landslides, then this process could have removed between 10 and 50% of the exposed portion of most volcanic islands. For example, in the Pacific Ocean such features appear on American and Western Samoa, Tahiti, the Society chain, and the Marquesas Group (Figure 6.2). On these islands, the highest seacliffs do not face the prevailing winds or swell but are protected from marine erosion by reefs. On the island of Tutuila in the Samoan Islands and on the islands of the Manua Group up to half of the volcanic complex is missing. On the Samoan islands, SeaMARC II side-scan sonar reveals profuse slump blocks, chaotic slumps, landslide sheet flows, turbidites, and avalanche debris flows. Landslides have also removed large portions of the upper parts of Guam in the Mariana Islands and of Rarotonga, Mangaia, and Aiutaki in the Cook Islands. As well, the western half of Volcán Ecuador in the Galápagos Islands is missing. In the Atlantic Ocean, the Cape Verde group and the Island of St. Helena also have marked sea cliffs, while in the Caribbean Sea large headwall scars are evident on the westward sides of Dominca, St. Lucia, and St. Vincent Islands in the Lesser Antilles. The Azores Islands are also faceted with amphitheatre scars. In the Indian Ocean, large landslides can be inferred from Gough, Marion, Prince Edward, Amsterdam, St. Paul, Bouvetoya, Possession, and Peter I Islands.

Nor do volcanoes have to emerge above sea level to have undergone failure. The seas are pockmarked by numerous atolls, guyots (eroded volcanic islands), and seamounts evincing amphitheatre and stellate forms similar to the above. Mass wasting is one of the major mechanisms reducing high volcanic islands to guyots. Mapping of submerged guyots on the Hawaiian Ridge shows the same density of landslides as present around the main islands. There are approximately a thousand seamounts higher than 1,000 m in the Pacific Ocean. Over 300 of these seamounts lie along the Mariana Island arc in the West Pacific (Figure 6.2). Many seamounts show extensive turbidites on their flanks and have the potential to generate landslides 20–50 km^3 in size. The perceived view of an uneroded circular or elliptical atoll is also illusionary. For example, on Johnston Atoll in the Line Islands south of Hawaii, one or more major landslides have removed much of the southern margin. Blocks of carbonate up to a kilometre in size have been detected on the adjacent seabed, which in places has been infilled to a depth of 1,500 m. Ninety-five percent of atolls are in fact polygonal in shape, with deep embayments cut into at least one seaward flank. If the aprons around these features are signs of past landslides, then such deposits cover 10% of the ocean. In each case of a failure, there has been the potential for a tsunami.

Other Topography
(Moore, 1978; Kenyon, 1987; Lipman et al., 1988; Carlson et al., 1991)

Other topography besides volcanoes can also produce submarine landslides. These include river deltas, passive continental margins, submarine canyons, deep-sea fans, the walls of deep trenches near subduction zones, and the slopes of midocean ridges. Some of the sites where slides have been identified as originating from these types of topography are mapped in Figure 6.2. Major river deltas are prone to landslides because of the volume of sediment continually being built up on their submerged distal ends. The rivers with the largest sediment loads – the Amazon, Mississippi, Nile, and Indus – have built up relatively steep fans more than 10 km thick. The Amazon Fan consists of several major slide deposits, the largest of which covers an area of 32,000 km^2. Two debris flows have been mapped in 1,000–3,000 m depth of water off the Mississippi delta. The larger of the two is 100 km wide and 300 km long.

Passive continental margins lie along tectonically inactive edges of crustal plates. Many of the slides emanating from these margins are derived from sedimentary units that are only 10–100 m thick. Sediment has accumulated over time along these margins through subaerial erosion. Failure occurs on slopes parallel to bedding planes. While the size of these slides is small compared to the Hawaiian ones, the widespread nature of the evidence is worrisome. For example, the eastern seaboard of the United States has at least four large submarine canyons cutting through the shelf edge, leading to distributary fans on the abyssal plain. Levees on the fans indicate that large debris or gravity flows have occurred often. Slides are numerous off the West Coast of Africa, where hummocky slides, block fields, debris flows, and turbidity deposits have all been mapped. Few submarine slides have yet been detected off the West Coast of North America, mainly because bathymetry has not been mapped in detail. However, a 6.8- km^2 slide with a volume of 1 km^3 occurred on 27 April 1975 off the fjord delta at Kitimat, British Columbia. The resulting tsunami had a run-up height of 8.2 m. A 75-km-long slide also has been detected off the Monterey Fan in California. One area that has been mapped well is the continental slope on the north side of the Aleutian Islands facing the Bering Sea. The area is relatively quiet seismically but is underlain by gas hydrates or a zone of sedimentary weakness. Here, mass failures up to 55 km long and containing blocks 1–2 km across have been identified emanating from some of the largest canyons in the world on low slopes of 0.5–1.8°. The volume of the landslides ranges between 20 and 195 km^3. In Northwestern Europe, the Storegga slides are not unique. The 1,000-m bathymetric contour is the site of at least fifteen other slides on the continental slope between West Ireland and Northern Norway. The largest of these is equivalent in size to the smallest Storegga slide. At least seven of these slides exist along the coast of Ireland within a few hundred kilometres of the coast. Remnant slides also exist on underwater platforms between the British Isles and Iceland with a debris flow, again as large as the smaller Storegga slide, situated at the base of the Rockall Trough.

The sides of deep ocean trenches are only susceptible to submarine slides if ocean sediment has accumulated here as part of a tectonic process. Thick sediment layers can pile up on oversteepened slopes as the result of tectonic off-scraping. Slides from trenches have been reported in the Sunda, Peru–Chile, Puerto Rican, and eastern and western Aleutian trenches. One of the largest slides occurs in the Sunda Trench off the Bassein River in Burma. The slide covers an area of 3,940 km^2 and has a volume of 960 km^3. The slide was triggered either by an earthquake or by overloading of sediments brought down the Irrawaddi River at lower sea levels. Along other trenches, localised slides of 10 km length appear to be a common feature. Where crustal plates are sliding past each other, slides can develop in subduction zones. The Ranger slide off the coast of California is one of the biggest of this type identified to date. It covers an area of 125 km^2 and incorporates 12 km^3 of material. Even midocean ridges can generate slides. Along the Mid-Atlantic Ridge, one such slide, comprising 19 km^3 of material, was caused by the failure of a 4-km × 5-km block on the flank of a mountain bordering the rift.

More worrisome are the coasts where no mapping has been carried out. Most of the coastline surrounding the Indian Ocean has not been mapped in enough detail to identify individual slides. Even in a developed country such as Australia, parts of the East Coast have only been mapped since 1990 using side-scan sonar. This coastline is passive and assumed to have low seismicity. However, a slide measuring 10 km × 20 km was detected 50 km off the coast south of Sydney. The age of the slide is unknown, but there is now substantial evidence for the presence of recent large tsunami along the adjacent coastline. As described in Chapter 4, the signatures of tsunami are common elsewhere around Australia, but unfortunately any link to submarine slides remains speculative without detailed mapping.

CHAPTER SEVEN

Volcanic Eruptions

7.1 An artist's impression of the tsunami from the third explosion of Krakatau hitting the coast of Anjer Lor at about 10:30 A.M. on 27 August 1883. ©Lynette Cook

INTRODUCTION

(Latter, 1981; Intergovernmental Oceanographic Commission, 1999)

Historically, volcanoes cause 4.6% of tsunami and 9.1% of the deaths attributable to this hazard, totalling 41,002 people. Two events caused this disproportionately large death toll: the Krakatau eruption of 26–27 August 1883 (36,000 deaths) and the Unzen, Japanese eruption of 21 May 1792 (4,300 deaths). Tsunami account for 20–25% of the deaths attributable to volcanic eruptions. The eruption of Santorini around 1470 B.C. is not included in these statistics because of a lack of written record. Santorini and the Krakatau eruption of 1883 will be discussed in more detail subsequently in this chapter. The main locations of the sixty-five tsunami linked to eruptions historically are plotted in Figure 7.2. The vast majority of these are restricted to the Japanese-Kuril Islands and the Philippine-Indonesian Archipelagoes. Both of these regions form island arcs where one plate is being subducted beneath another. Explosive volcanism with caldera formation is a common occurrence in these regions. Other isolated cases of eruptions that have generated tsunami are associated with hot spots beneath the Pacific Plate. Unfortunately, volcano-induced tsunami have neither been well recorded nor described except for a few events such Krakatau in 1883.

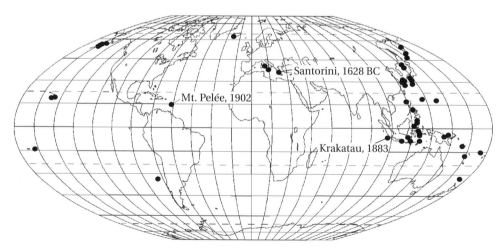

• Location of volcanoes generating tsunami historically

7.2 Location of volcanoes that have generated tsunami in recorded history (based on Latter, 1981, and Intergovernmental Oceanographic Commission, 1999)

CAUSES OF VOLCANO-INDUCED TSUNAMI
(Latter, 1981; Blong, 1984; Lockridge, 1988a, 1990)

There are ten mechanisms whereby volcanic eruptions can generate tsunami. These together with their major events are summarised in Table 7.1. Submarine landslides sloughed off from nonerupting volcanoes are not included in this table because they were dealt with in the preceding chapter. Many of the events listed in Table 7.1 were catastrophic. The majority of volcanic eruptions are accompanied by seismic tremors. If these are substantial enough and the volcano lies near or in the ocean, the tremors can trigger tsunami. For example, the eruption of Vesuvius in the southeast corner of the Bay of Naples on the West Coast of Italy, in August A.D. 79, was preceded by a tsunami induced by seismic activity. Pliny the Elder, the commander of the Roman fleet at Misenum, sailed to the coast at the base of the mountain to rescue inhabitants five days before the final eruption. He could not get near the shore because of a sudden retreat of the shoreline. Two of the largest events due to seismic activity occurred on the 10 January 1878 and 8 January 1933 with the eruptions of Yasour Volcano in the New Hebrides and Severgin Volcano in the Kuril Islands, respectively. The respective tsunami reached 17 and 9 m above sea level.

Pyroclastic flows, or *nuées ardentes,* are generated by the collapse under gravity of hot vertically ejected ash clouds. When these reach the ocean surface, they spread out rapidly as density flows that can either displace water or transfer energy to the ocean and generate a tsunami. The size of the resulting tsunami depends upon the density of the flow. If the density is less than seawater, then the ash cloud rides the surface of the ocean, generating a small wave. However, if the flow is denser than seawater, the cloud will sink to the bottom of the ocean and will displace water pistonlike in front of it. In some cases, these flows can travel tens of kilometres along the seabed. Pyroclastic flows have the potential to generate devastating tsunami remote from the source of the eruption. For example, Tambora in 1815 generated a tsunami 2–4 m high in this manner despite lying 15 km inland. The 7 May 1902 eruption of Mt. Pelée, Martinique produced a *nuée ardente* that swept into the harbour of St. Pierre and generated a tsunami that travelled as far as Fort de France, 19 km away. Two of the largest of these types of events occurred in Indonesia during the eruption of Ruang on 5 March 1871 and Krakatau on 26–27 August 1883, producing tsunami 25 and 10 m high, respectively.

Submarine eruptions within 500 m of the ocean surface can disturb the water column enough to generate a surface tsunami wave. Below this depth, the weight and volume of the water suppress surface wave formation. Tsunami from this cause rarely propagate more than 150 km from the site of the eruption. One of the largest such tsunami occurred during the eruption of Sakurajima, Japan, on 9 September 1780, when a 6-m-high wave was generated. More significant are submarine explosions that occur when ocean water comes in contact with the magma chamber.

TABLE 7.1 Causes of Historical Tsunami Induced by Volcanoes

Mechanism	Percentage of Events	Examples	Date	Height (m)
Volcanic earthquakes	22.0	New Hebrides	10 January 1878	17
Pyroclastic flows	20.0	Ruang, Indonesia	5 March 1871	25
		Krakatau, Indonesia	26–27 August 1883	>10
Submarine explosions	19.0	Krakatau, Indonesia	26–27 August 1883	42
		Sakurajima, Japan	9 September 1780	6
Caldera formation	9.0	Ritter Island	13 March 1888	12–15
		Krakatau, Indonesia	26–27 August 1883	2–10
Landslides	7.0	Unzen Volcano, Japan	21 May 1792	6–9
Basal surges	7.0	Taal Volcano, Philippines	Numerous	?
Avalanches of hot rock	6.0	Stromboli, Italy	Numerous	?
Lahars	4.5	Mt. Pelée, Martinique	5 May 1902	4.5
Atmospheric pressure wave	4.5	Krakatau, Indonesia	26–27 August 1883	<0.5
Lava	1.0	Matavanu Volcano, Samoa	1906–1907	3.0–3.6

Source: Based on Latter, 1981.

This water is converted instantly to steam and in the process produces a violent explosion. Krakatau during its third explosive eruption in 1883 produced a tsunami 40 m high in this manner.

Formation of a caldera during the final stages of an explosive eruption near the sea can permit water to flow rapidly into the depression. This sets up a wave train that can propagate away from the caldera within five minutes. Stratovolcanoes are particularly prone to collapse. In the Ring of Fire subduction zone around the Pacific Ocean there are hundreds of these types of volcanoes. Krakatau's numerous eruptions produced calderas that may have been responsible for some of the tsunami observed in the Sunda Strait. A comparable event occurred on Ritter Island, Papua New Guinea, on 13 March 1888. Here, the formation of a caldera 2.5 km in diameter produced a 12- to 15-m-high tsunami. While the initial wave heights radiating out from the caldera can be large, the actual volume of water displaced may be small. In addition, because the height of the wave decays inversely to the square root of the distance travelled, the effect of the tsunami diminishes rapidly away from the point source.

The slopes of a volcano are inherently unstable during eruptions because of earthquakes, inflation, or collapse. Collapsing material can form a debris avalanche that can travel at speeds of 100 m s^{-1}. Horseshoe-shaped scars are left behind as evidence of the failure. There have been 200 such avalanches over the past 2 million years. Some of these exceeded 20 km^3 in volume and travelled 50–100 km from their source. For small avalanches of 0.1–1.0 km^3 in volume, the travel distance ranges between 6 and 11 times the elevation of the initial avalanche. For higher volumes, the travel distance can increase to eight to twenty times the vertical drop. For example, the collapse of a 2,000-m-high volcano could generate a debris avalanche that can travel 16–22 km from its source. Avalanching is a significant hazard associated with volcanoes in Alaska, Kamchatka, Japan, the Philippines, Indonesia, Papua New Guinea, the West Indies, and the Mediterranean Sea. In most cases, the resulting landslide is localised enough to generate small tsunami that are highly directional in their propagation away from the volcano. However, some of the greatest death tolls have been caused by such events. For example, during the eruption on 21 May 1792 of Unzen Volcano, in Japan, 0.34 km^3 of material sloughed off its flank. The landslide travelled 6.5 km before sweeping into the Ariake Sea, where it generated a tsunami with run-ups 35–55 m above sea level along 77 km of coastline on the Shimabara Peninsula and along the opposite side of the sea, 15 km away. Six thousand houses were destroyed, 1,650 ships were sunk, and a total of 14,524 people lost their lives. Similar-magnitude waves were generated by landslides on Paluweh Island, Indonesia, on 4 August 1928 (160 dead), and recently on Ili Werung Volcano, Indonesia on 18 July 1979 (500 dead). Mt. St. Augustine at the entrance to Cook Inlet, Alaska, also has the potential to generate 5- to 7-m-high tsunami through sloughing. Eleven major debris avalanches, occurring at 150- to 200-year intervals, have originated from this volcano. One of the largest occurred on 6 October 1883. The avalanche swept 4–8 km into Cook Inlet on the north side of the volcano. Within half an hour, a 9-m-high tsunami flooded settlements at English Bay, 85 km up the inlet.

Basal surges or lateral blasts are formed when a volcano erupts sideways. Taal Volcano in Lake Bombon, Philippines, has been subject to at least five basal surge events since 1749. All of the resulting tsunami took lives. Avalanches of hot rock can generate tsunami when ejecta, piled up on the side of an erupting volcano, collapses into the sea. These events occur frequently on Stromboli volcano in Italy. The resulting tsunami are usually localised and have not been associated with any deaths. Lahars are ash deposits that fail after becoming saturated with water. Failures occur during an eruption because of the displacement of ground water, mixing of ash with water in a crater lake, or melting of snow and ice at the crest of the volcano. When the lahar reaches the ocean, it can generate significant tsunami; however, these are usually localised because the lahar tends to flow down valleys. For example, on 5 May 1902, a 35-m-high lahar from Mt. Pelée swept down Rivière Blanche north of the nearby town of St. Pierre. When it reached the sea, it generated a 4.5-m high tsunami that only affected the lower part of the town, killing one hundred people. Large explosive volcanoes generate a pressure pulse through the atmosphere. Krakatau, in 1883, generated tsunami in the Pacific Ocean, and in Lake Taupo in the middle of the North Island of New Zealand, via this mechanism. However, nowhere did the tsunami exceed more than 0.5 m in height. The 1955–1956 eruption of Bezymianny on the Kamchatka Peninsula in Russia also generated a global pressure wave, but the resulting tsunami in the Pacific Ocean did not reach more than 0.3 m in height. Finally, if lava reaches the ocean en masse it can generate tsunami. These events are rare and have only been noted at one or two locations, with no widespread destruction being produced. For example, the lava from the eruption of Matavanu Volcano on Savaii, Samoa in 1907 generated a tsunami 3.0–3.6 m high when it poured into the sea. No deaths were reported.

KRAKATAU, 26–27 AUGUST 1883
(Verbeek, 1884; Latter, 1981; Self and Rampino, 1981; Blong, 1984; Myles, 1985; Nomanbhoy and Satake, 1995)

Krakatau was one of the largest explosive eruptions known to humanity. It is the only eruption for which detailed information exists on volcano-induced tsunami. To date, the mechanisms generating tsunami during its eruption are still debated. The volcano lies in the Sunda Strait between Sumatra and Java, Indonesia (Figure 7.3). The Javanese *Book of Kings* describes an earlier eruption, referring to Krakatau as Mount Kapi. Then, the volcano exploded and created a sea wave that

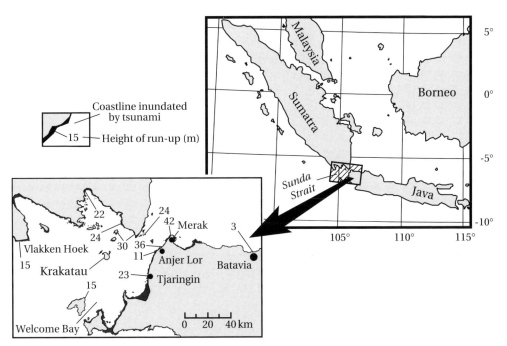

7.3   Coastline in the Sunda Strait affected by tsunami following the eruption of Krakatau on 26–27 August 1883 (based on Verbeek, 1884, Blong, 1984, and Myles, 1985).

inundated the land and killed many people throughout the northern part of Sunda Strait. Krakatau had last been active in 1681, and during the 1870s the volcano underwent increased earthquake activity. In May 1883, one vent became active, throwing ash 10 km into the air. By the beginning of August, a dozen Vesuvian-type eruptions had occurred across the island. On 26 August, loud explosions recurred at intervals of 10 minutes, and a dense tephra cloud rose 25 km above the island. The explosions could be heard throughout the islands of Java and Sumatra. In the morning and later that evening, small tsunami waves 1–2 m in height swept the strait, striking the towns of Telok Betong on Sumatra's Lampong Bay, Tjaringin on the Java Coast north of Pepper Bay, and Merak (Figure 7.3). On the morning of 27 August, three horrific explosions occurred. The first explosion at 5:28 A.M. destroyed the 130 m peak of Perboewatan, forming a caldera that immediately infilled with seawater and generated a tsunami. At 6:36 A.M., the 500-m-high peak of Danan exploded and collapsed, sending more seawater into the molten magma chamber of the eruption and producing another tsunami. The third blast, at 9:58 A.M., tore the remaining island of Rakata apart. Including ejecta, 9–10 km³ of solid rock was blown out of the volcano. About 18–21 km³ of pyroclastic deposits spread out over 300 km² to an average depth of 40 metres. Fine ash spread over an area of

2.8×10^6 km², and thick pumice rafts impeded navigation in the region up to five months afterwards. A caldera 6 km in diameter and 270 m deep formed where the central island had once stood. This third blast was the largest sound ever heard by humanity and was recorded 4,800 km away on the island of Rodriguez in the Indian Ocean, and 3,200 km away at Elsey Creek, Northern Territory Australia. Windows 150 km away were shattered. The atmospheric shock wave travelled around the world seven times. Barometers in Europe and the United States measured significant oscillations in pressure over nine days following the blast. The total energy released by the third eruption was equivalent to 200 megatons of TNT. (Kinetic energy for volcanic eruptions and meteorites exploding in the atmosphere or impacting with the ocean is expressed in megatons of TNT. One megaton of TNT is equivalent to 4.185×10^{15} joules.)

The two predawn blasts each generated tsunami that drowned thousands in the Sunda Strait. The third blast-induced wave was cataclysmic and devastated the adjacent coastline of Java and Sumatra within 30 to 60 minutes (Figure 7.3). The coastline north of the eruption was struck by waves with a maximum run-up height of 42 m. The tsunami penetrated 5 km inland over low-lying areas. The largest wave struck the town of Merak. Here, the 15-m-high tsunami was increased to 40 m because of the funnel-shaped nature of the bay. The town of Anjer Lor was swamped by an 11 m high wave (Figure 7.1), the town of Tjaringin by one 23 m in height, and the towns of Kelimbang and Telok Betong were each struck by a wave 22–24 m high. In the latter town, the Dutch warship *Berouw* was carried 2 km inland and left stranded 10 m above sea level. Coral blocks weighing up to 600 tonnes were moved onshore. Within the Strait, eleven waves rolled in over the next fifteen hours, while at Batavia (now Jakarta) fourteen consistently spaced waves arrived over a period of thirty-six hours. Between 5,000 and 6,000 boats in the strait were sunk. In total, 36,417 people died in major towns and 300 villages were destroyed because of the tsunami.

Within four hours of the final eruption, a 400-m-high tsunami arrived at Northwest Cape West Australia, 2,100 km away. The wave swept through gaps in the Ningaloo Reef and penetrated 1 km inland over sand dunes. Nine hours after the blast, 300 riverboats were swamped and sunk at Kolkata (formerly Calcutta) on the Ganges River 3,800 km away. The wave was measured around the Indian Ocean at Aden on the tip of the Arabian Peninsula, Sri Lanka, Mahe in the Seychelle Islands, and on the Island of Mauritius. The furthest this tsunami wave was observed was 8,300 km away at Port Elizabeth, South Africa. Tsunami waves were measured over the next thirty-seven hours on tide gauges in the English Channel, in the Pacific Ocean, and in Lake Taupo in the centre of the North Island of New Zealand, where a 0.5-m oscillation in lake level was observed. Around the Pacific Ocean, tide gauges in Australia and Japan and at San Francisco and Kodiak Island measured changes of 0.1 m up to twenty hours after the eruption. Honolulu recorded higher

oscillations of 0.24 m with a periodicity of 30 minutes. Smaller, subsequent eruptions of Krakatau generated lesser tsunami throughout the strait until 10 October.
The last tsunami was observed in Welcome Bay, where it surged 75 m inland
beyond the high-tide mark.

The tsunami in the Pacific have been attributed to the atmospheric pressure
wave because many islands that would have effectively dissipated long-wave
energy obscured the passage from the Sunda Strait into this ocean. The atmospheric pressure wave also accounts for seiching that occurred in Lake Taupo,
which is not connected to the ocean. Finally, it explains the long waves observed
along the coast of France and England when the main tsunami had effectively
dissipated its energy in the Indian Ocean. The generation of tsunami in Sunda
Strait and the Indian Ocean has been attributed to four causes: lateral blast,
collapse of the caldera that formed on the north side of Krakatau Island, pyroclastic flows, and a submarine explosion. Lateral blasting may have occurred
to a small degree on Krakatau during the third explosion; however, its effect
on tsunami generation is not known. During the third explosion Krakatau collapsed in on itself, forming a caldera about 270 m deep and with a volume of
11.5 km^3. However, modelling indicates that this mechanism underestimates
tsunami wave heights by a factor of three within Sunda Strait. Krakatau generated
massive pyroclastic flows. These flows probably generated the tsunami that
preceded the final explosion. At the time of the third eruption, ash was ejected
into the atmosphere towards the northeast. Theoretically, a pyroclastic flow
in this direction could have generated tsunami up to 10 m in size throughout
the strait; however, the mechanism does not account for measured tsunami
run-ups of more than 15 m in height in the northern part of Sunda Strait. The
pyroclastic flow now appears to have sunk to the bottom of the ocean and travelled 10–15 km along the seabed before depositing two large islands of ash. The
40-m-high run-up measured near Merak to the northeast supports this hypothesis. The tsunami's wave height corresponds with the depth of water around
Krakatau in this direction. As well, the third explosion of Krakatau at 9:58 A.M.
more than likely produced a submarine explosion as ocean water came in contact
with the magma chamber. Van Guest's description of the eruption (presented in
Chapter 1) indicates that the magma chamber was visible in the strait before the
third explosion at 10 o'clock in the morning. This explosion can be modelled in a
similar fashion to the caldera collapse, but so as to incorporate outward movement of water immediately after the blast. A submarine explosion could have
generated tsunami 15 m high throughout the Strait. If the explosion had a lateral
component northwards, as indicated by the final configuration of Krakatau
Island, then this blast, in conjunction with the pyroclastic flow, would account for
the increase in tsunami wave heights towards the northern entrance of Sunda
Strait (Figure 7.3).

SANTORINI, AROUND 1470 B.C.

(Yokoyama, 1978; Pichler and Friedrich, 1980; Kastens and Cita, 1981;
Bryant, 1991; LaMoreaux, 1995; Cita et al., 1996; Johnstone, 1997;
Pararas-Carayannis, 1998c)

The prehistoric eruption of Santorini around 1470 B.C., off the island of Thera in
the southern Aegean Sea north of Crete (Figure 7.4), is probably the biggest volcanic explosion witnessed by humans. It is also one of the most controversial
because legend, myth, and archaeological fact frequently get intertwined and dis-
torted in the interpretation of the sequence of events. The eruption has been
linked to the lost city of Atlantis described by Plato in his *Critias,* to the destruction
of Minoan civilisation on the island of Crete 120 km to the south (Figure 7.4), and
to the exodus of the Israelites from Egypt in the Bible. Certainly, Greek flood myths
refer to this or similar events that generated tsunami in the Aegean. Plato's story of
Atlantis is based on an Egyptian story that has similarities with Carthaginian and
Phoenician legends.

The great Minoan empire was a Bronze Age maritime civilisation centred
on the island of Crete that flourished from 3000 B.C. to 1400 B.C. The Minoan
seafarers dominated trade in the Eastern Mediterranean, and on this basis were
able to accumulate great wealth and prevent the development of any other
maritime power that could threaten them. The Minoans were noted for their
cities and great palaces at Ayía, Knossos, Mallia, Phaestus, Triáda, and Tylissos –
all decorated by detailed and lively frescoes. By far the largest and best-known
palace was that of the legendary king Minos at Knossos, rivalling any other
Middle Eastern structure in size. The eruption of Santorini, also known as
Stronghyli – the round island – did not destroy Minoan civilisation, but it certainly weakened it. The tsunami from the eruption is believed to have sunk most
ships near the coast and in harbours, and to have greatly disrupted sea trade that
was pivotal to the stability of the civilisation. Ash falls also disrupted agriculture.
Within fifty years of the eruption of Santorini, the Mycenaean Greeks, who had
escaped its effects, were able to conquer the Minoans and take over their cities
and palaces.

The timing of the Santorini eruption has also been linked to the plagues of
Egypt (Exodus 6:28–14:31) and the exodus of the Israelites from that country. In 1
Kings 6:1 the exodus is dated as occurring 476 years before the rule of Solomon.
Scholars believe that Solomon began his rule in 960 B.C., putting the Exodus
around 1436 B.C. Other evidence indicates that the exodus occurred in 1477 B.C.
Both dates encompass the reign in Egypt of either Hatshepsut or her son Tuthmo-
sis III of the eighteenth dynasty. The transition of rule between the two rulers is
known as being a time of catastrophes. In the biblical account, the river of blood
may refer to pink pumice from the Santorini eruption preceding the explosion.
This pumice, after it was deposited, would easily have mixed with rainwater and

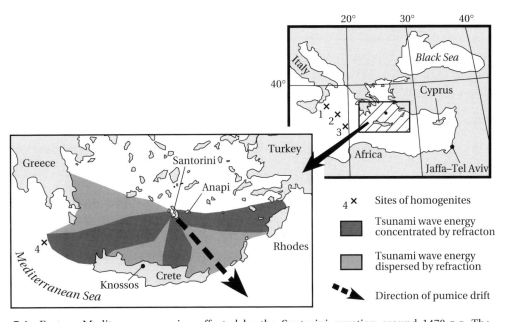

7.4 Eastern Mediterranean region affected by the Santorini eruption around 1470 B.C. The numbers refer to sites where homogenites exist: 1, Calabrian Ridge; 2, Ionian Abyssal Plain; 3, Bannock Basin; and 4, Mediterranean Ridge. Refraction patterns based on Kastens and Cita (1981).

flowed into any stream or river, colouring it red. The three days of darkness possibly refer to tephra clouds blowing south across Egypt at the beginning of the eruption. The darkness was described as a "darkness, which could be felt". Egyptian documents around 1470 B.C. refer to a time of prolonged darkness and noise, to a period of nine days that "were in violence and tempest: none . . . could see the face of his fellow", and to the destruction of towns and wasting of Upper Egypt. There is also direct reference to the collapse of trade with Crete (Keftiu). Volcanic shards have been found in soils on the Nile delta with the same chemical composition as tephra on the Santorini Islands. The parting of the Red Sea most likely occurred in the marshes at the northern end of the sea. The Bible attributes the parting to wind (Exodus 14:21); however, the wind may refer to the atmospheric pressure wave produced by the explosion of the volcano. Such waves, akin to that generated by Krakatau in 1883, can generate seiching or tsunami in enclosed basins or distant oceans.

Santorini volcano is part of a volcanic island chain extending parallel to the coast of Asia Minor (Figure 7.4). The Aegean is the only zone in the Eastern Mediterranean where subduction of plate boundaries is active. Of all these volcanic islands, only two, Santorini and Nisyros, have erupted in recorded times. Santorini forms a complex of overlapping shield volcanoes consisting of basaltic and andesitic lava flows. The volcano has erupted explosively at least twelve times during the last 200,000 years. Its height has been reduced over this time from a single mountain 1,500 m high to three islands less than 500 m in height surrounding a submerged caldera. The eruption around 1470 B.C. was the most recent of these and was one of the largest eruptions on Earth in the last 10,000 years. The timing is debatable. Acidity in Greenland ice cores suggest that the major eruption occurred in 1390 ± 50 B.C., although radiocarbon dating on land suggests an age around 1450 B.C. or 1470 B.C. Dendrochronology based on Irish bog oaks and Californian bristlecone pine puts the age of the event as old as 1628 B.C. At this latter time, Chinese records report a dim sun and failure of cereal crops because of frost. Large volcanic events cool temperatures globally by as much as 1° C over the space of several years. The range of dates may not be contradictory because there is evidence that Thera may have erupted several times over a time span of 200 years.

The eruption around 1470 B.C. had four distinct phases. The first was a Plinian phase with massive pumice falls. This was followed by a series of basal surges producing profuse quantities of pumice up to 30 m thick on Santorini. The third phase was associated with the collapse of the caldera and production of pyroclastic flows. About 4.5 km^3 of dense magma was ejected from the volcano, producing 10 km^3 of ash. The volume of ejecta is similar in magnitude to that produced by the Krakatau eruption in 1883. The ash drifted to the east-southeast, but did not exceed 5 mm thickness in deposits on any of the adjacent islands, including Crete. The largest thickness of ash measured in marine cores appears to originate from pumice that floated into the Eastern Mediterranean. It is possible at this stage that ocean water made contact with the magma chamber and produced large explosions, which generated tsunami in the same way that the eruption of Krakatau did. The final phase of the eruption was associated with the collapse of the caldera in its southwest corner. The volcano sunk over an area of 83 km^2 and to a depth of between 600 and 800 m. According to the Krakatau model, this final event produced the largest tsunami, directing most of its energy westwards (Figure 7.4). It is estimated that the original height of the tsunami was 46–68 m in height, and maybe as high as 90 m. The average period between the dozen or more peaks in the wave train was 15 minutes.

Evidence of the tsunami is found in deposits close to Santorini. On the island of Anapi to the east, sea-borne pumice was deposited to an altitude of 40–50 m above present sea level. Considering that sea levels at the time of the eruption may have been 10 m lower, this represents run-up heights greater than those produced by Krakatau in the Sunda Strait. On the Island of Crete, the wave arrived within 30 minutes, with a height of approximately 11 m. Refraction focussed wave energy on the northeast corner of Crete, where run-up heights reached 40 m above sea level. In the region of Knossos, the tsunami swept across a 3-km-wide coastal plain, reaching the mountains behind. The back-wash concentrated in valleys and watercourses, and was highly erosive. Evidence for the tsunami is also found in the Eastern Mediterranean on the western side of Cyprus, and further away at Jaffa–Tel Aviv in Israel. At the latter location, pumice has been found on a terrace lying 7 m above sea level at the time of the eruption. However, the tsunami wave here had already undergone substantial defocussing because of wave refraction as it passed between the islands of Crete and Rhodes. The greatest tsunami wave heights occurred west of Santorini. Based upon linear wave theory, the wave in the central Mediterranean Sea was 17 m high, while closer to Italy over the submarine Calabrian Ridge, it was 7 m high. Bottom current velocities under the wave crest in these regions ranged between 20 and 50 cm s^{-1} – great enough to entrain clay to gravel sized particles. The maximum pressure pulse produced on the seabed by the passage of the wave ranged between 350 and 850 kdyne cm^{-2}. Spontaneous liquefaction and flow of water-saturated muds is known to occur under pressure pulses of 280 kdyne cm^{-2}and greater.

Some of the evidence for a large tsunami comes from the discovery of unusual deposits on the seabed of the central Mediterranean Sea, where wave heights were highest. These deposits – labelled homogenites – formed in the deep ocean as the result of settling from suspension of densely concentrated, fine-grained sediment. This process produced homogeneous units up to 25 m thick with a sharp basal contact. Homogenites can be linked hydrodynamically to the passage of a tsunami wave. As sediment fails via liquefaction due to the pressure pulse, oscillatory flow under the wave suspends finer particles, creating turbulent clouds of sediment. It is estimated that the slurries exceeded concentrations of 16,000 mg l^{-1}. In comparison, the highest measured sediment concentrations on the ocean seabed and in muddy tidal estuaries rarely exceed 12 mg l^{-1} and 300 mg l^{-1} respectively. Gravity sorting occurred under this extreme concentration. Sand-sized particles settled first to the bottom and were deposited at the erosional contact with the seabed as a fining upward unit whose thickness ranged from a few centimetres to several metres. Finer clay-sized sediment was deposited over the next few days as a massive undifferentiated clay deposit that was up to 20 m or more thick. Homogenites differ from turbidites described in Chapter 3 by their greater thickness, lack of laminations, and undifferentiated particle size. Homogenites differ from debris flows by the absence of large clasts or rock pieces derived from continental sediments.

Four types of homogenites can be differentiated. In the Western Mediterranean, on the Ionian Abyssal Plain, a 10- to 20-m-thick deposit, with an estimated volume of 11 km^3, was laid down on the seabed over an area of 1,100 km^2. It appears that the tsunami wave slammed into the continental shelf of North Africa and either directly or indirectly triggered a mega-turbidity current. This current carried terrigenous and shelf sediment into the deep Mediterranean Sea, eroding flanks of undersea ridges and depositing homogenites with an erosional base on upslopes. In one location this turbidity current rode up a ridge 223 m above the abyssal plain and deposited sediment. In the eastern part of the Mediterranean, bottom velocities and the related powerful pressure pulse liquefied sand into depressions, forming uniform deposits several metres thick with a sandy base overlying an erosional contact. These deposits form in cobblestone-shaped basins with a vertical relief of 200 m. Finally, in the Bannock Basin, the passage of the wave destabilised evaporites. The resulting deposits are 12 m thick and consist of 3 m of sand overlain by 9 m of graded mud deposited from suspension in high-density brines trapped at the bottom of 100-m-deep depressions in the seabed. All of the homogenites found in the Mediterranean are derived from a single event and date around the time of the Santorini eruption. Homogenites are not found in the Eastern Mediterranean Sea, where tsunami wave heights were insufficient to cause resuspension or liquefaction of bottom sediment.

Since the Santorini eruption around 1470 B.C., there have been many others. Since 197 B.C. at least eleven eruptions have formed the two islands that presently exist in the centre of the caldera. Eruptions in 1650, 1866, and 1956 have given rise to tsunami with damaging consequences. An earthquake preceded the 1650 eruption and generated a 50-m-high tsunami that swept 4 km inland in places. The 1866 event generated two tsunami that had run-up heights of 8 m along nearby coasts. Earthquakes associated with the latest eruption on 9 July 1956 produced a tsunami that had a run-up height of 24 m and killed fifty-three people. The Santorini volcano remains one of the most dangerous in terms of tsunami in the world today.

The last three chapters have summarised how geophysical processes originating from the Earth generate tsunami. When reviewed, the magnitudes of tsunami associated with these processes are indeed impressive. Tsunami have dispersed across the Pacific after numerous historical earthquakes. Five events since 1600 have produced run-up heights of 51–115 m in elevation. Volcanic eruptions, while rarer in terms of tsunami, have generated similar magnitude run-ups, but these have been localised. The Santorini eruption of around 1470 B.C. may hold the record for the biggest volcano-induced tsunami, with an initial wave height of 90 m. Tsunami generated by submarine landslides may be bigger yet. The Lituya Bay landslide of 9 July 1958 generated a splash tsunami that achieved a run-up height of 524 m above sea level. However, in terms of area affected, the Storegga slide of 7,950 ± 190 years ago may have been the biggest – considering that some suspicion hangs over some of the evidence attributed to the Lanai slide in Hawaii. Many of the tsunami induced by these processes produced some of the signatures of tsunami outlined in Chapters 3 and 4. However, only the Storegga event can be linked to the full range of signatures that includes bedrock-sculpturing features. A dichotomy thus exists in that observable tsunami have not commonly generated bedrock sculpturing features that exist so widely along rocky coasts, especially those in Australia. One mechanism, meteorite impact with the ocean, is capable of generating tsunami equivalent to or bigger than the largest tsunami produced by other mechanisms. The nature of meteorite-induced tsunami will be discussed in the next chapter.

CHAPTER EIGHT

Comets and Meteorites

8.1 An artist's impression of the Chicxulub impact tsunami as it crossed the coastal plain of the United States. The impact event was responsible for the extinction of the dinosaurs, 65 million years ago. The figure appeared originally in Alvarez (1997). ©Princeton University Press 1997 and reprinted by permission.

INTRODUCTION

Comets originate from the Oort Cloud, which is an area of icy debris beyond the orbit of Pluto. Comets are captured and brought into the inner solar system mainly by the gravitational attraction of Jupiter. Here, these comets tend to disintegrate over successive orbits, producing a large number of smaller active comets and inactive asteroids. Many asteroids orbit close to the Earth and in a significant number of cases intersect the Earth's orbit. The population of Near Earth Objects (NEOs) is replenished discontinuously over time. When an asteroid impacts with the Earth it is called a meteorite. If it explodes in the atmosphere it is called a bolide. Our present era is under the influence of a large comet that entered the inner solar system within the last 20,000 years. This chapter looks at the formation of Near Earth Objects, the probability of their impact with the Earth, the effect of resulting tsunami, and evidence for their occurrence in recent times.

NEAR EARTH OBJECTS (NEOS)

What Are They?
(Steel, 1995; Verschuur, 1996; Lewis, 1999)

There are two main classes of celestial objects – asteroids and comets – that can cross the Earth's orbit and eventual impact with the Earth. Asteroids mainly orbit the sun between Mars and Jupiter. Because of Jupiter's gravitational influence, debris in this region was prevented from agglomerating into a planet, although there is suffi-cient evidence to suggest that many objects in the asteroid belt have collided with each other. Over 20,000 objects ranging in diameter from several metres to over 1,000 km have been discovered in the asteroid belt. The density of these objects is similar to that of igneous rock. Jupiter still exerts an attraction on asteroids and has the potential to destabilise them into Earth-crossing orbits. Four groups of asteroids have near Earth orbits. The Apollos, the first group, cross the Earth's orbit but spend most of their time just beyond it. Their orbital period around the sun is greater than one year. The Amors, the second group, orbit further out, crossing the orbit of Mars as well as that of the Earth. These objects have unstable orbits, and they are affected the most by the gravity of Jupiter. The Atens, the third class of asteroids, spend most of their time inside Earth's orbit. Their orbital period about the sun is less than one year. These latter objects were only discovered in the 1970s. A fourth group of aster-oids is suspected. These asteroids lie close to the Earth and are thought to originate from debris remaining from meteorite impacts with the moon.

Comets enter the inner solar system from the Kuiper belt lying outside the orbit of Neptune or from the Oort belt lying beyond the outer boundary of the solar sys-tem. The gravitational attraction of Jupiter and Saturn can force a comet as large as 200 km in diameter to traverse the inner part of the solar system once every t200,000 years. There are three types of comets: short, intermediate, and long period. Short period comets orbit the sun with periods of less than twenty years. They are called Jupiter family comets because they come under the gravitational influence of Jupiter and have unstable orbits. The best example is Comet P/Encke. Intermediate-period or Halley-type comets have orbital periods between 20 and 200 years. Long-period comets have orbital periods greater than 200 years and tend to appear in the inner solar system only once.

The best-known comet to impact with the Earth occurred at Chicxulub on the Yucatan Peninsula of Mexico at the Cretaceous–Tertiary boundary. This event threw up large volumes of sediment into the stratosphere, attenuated solar radiation significantly for months, if not years, and led to the extinction of the dinosaurs. Comets consist mainly of ice and stony or iron material ranging in size from sand grains to boulders hundreds of metres in diameter. As comets approach the sun, they slowly disintegrate as ice is vaporised and discharged with other gases in a tail that can extend millions of kilometres into space. There is growing evidence that comets are responsible for much of the debris orbiting the inner solar system. Some main-belt asteroids may be the remnants of comets, and most of the Earth-crossing asteroids have short-lived orbits about the sun, suggesting that they too originated from the breakup of comets. The Earth is known to cross through at least twelve comet debris trails that form prominent meteorite showers.

The most important comet stream is the Taurid complex. It is theorised that a large comet, about 100 km in diameter, initially entered the inner solar system 20,000 years ago and began to break up about 14,000 years ago. Further disintegration events occurred around 7500 B.C. and 2600 B.C. The latter break-up happened during the Bronze Age and was associated with meteorite strikes within the next few centuries that impacted dramatically upon civilisations. Debris from the disintegration forms the Taurid complex, which consists of a dozen asteroid objects 0.5–2.0 km in diameter. There may be another hundred objects in this size range yet to be discovered in the complex. As well, probably only 1–2% of Taurid objects smaller than 0.5 km in size have been spotted to date. The break-up of the original comet also generated four prominent meteor showers appearing annually with a major peak in October–December and a minor peak in April–June. One prominent reactivated comet, P/Encke, is also part of the Taurid complex. Comet P/Encke first appeared in 1786 following the last large display of meteorites associated with the Taurids. Comet P/Encke is about 5 km in diameter, has a mass of 10^{13} tonnes, and orbits the sun every 3.5 years in an Earth crossing orbit. The Taurid complex also contains other large objects that presently have orbits that intersect the Earth in the last few days of June each year. The Tunguska airburst of 30 June 1908 was a Taurid object, as was a 1-km-sized object that struck the far side of the Moon on 19 June A.D. 1178. This latter event produced the 20-km-diameter Giordano Bruno crater as well as spreading dust across the face of the New Moon, splitting the upper horn with flame, and causing the moon to "throb like a wounded snake". The Monk Gervasse, in Kent, who saw the sequence repeatedly several times, witnessed the event. This latter description also appears in Maori legends in New Zealand. The event itself may have sent debris crashing into the Earth several days latter.

The perceived view that relatively small NEOs hundreds of metres in size are innocuous is misleading. For example, a 100- to 500-m-diameter object moving at a velocity of 30 km s^{-1} will release energy equivalent to 100–30,000 megatons of TNT. The large asteroid 1989FC, which was 500 m in diameter and missed the Earth by 650,000 km, would have destroyed civilisation if it had hit our planet. It would have created a crater about 10 km in diameter, released energy equivalent to the world's total nuclear arsenal, triggered an earthquake with a surface magnitude, M_s, of 9–10, spread a hot plume burning tens of thousands of square kilometres of forest, ejected billions of tonnes of rocks and dust into the atmosphere with consequent lowering of temperatures and partial blocking of photosynthesis for several years, induced acid rain due to atmospheric reaction of nitrogen and oxygen, and depleted ozone in the stratosphere.

How Frequent Have Been Comet and Meteorite Impacts?
(Rasmussen, 1991; Hasegawa, 1992; Asher et al., 1994; Steel, 1995;
Verschuur, 1996; Lewis, 1999)

It is estimated that, since hominids evolved, between 200 and 500 extraterrestrial bodies several hundred metres in size have impacted with the Earth. About 70% of NEOs are asteroids, of which 50% are derived originally from comets. The best approximation of the probability of impact with the Earth of various sizes of objects is presented in Figure 8.2. There are large uncertainties, spanning an order of magnitude, on the return period of these objects because the total population of NEOs is presently unknown. Observations of the more frequent, smaller iron meteorites and bolides are also uncertain because historical records beyond a few centuries are very limited globally. The range in kinetic energy presented in Figure 8.2 reflects the range in speed of objects striking the Earth – typically between 10 and 45 km s^{-1}. Astronomical observations indicate that one to three Near Earth Objects with a diameter of 1 km could impact the Earth every 100,000–200,000 years. This size object is undoubtedly more common then present observations indicate and may occur every 50,000 years. An object 50 m in diameter crashes into the Earth every century, while a Tunguska-sized object of 60 m diameter occurs every 200–300 years.

These probabilities may be conservative because they are based upon recorded events rather than the probable population of NEOs. Simulations have been made of the number of type of impacts that could randomly be expected over a period of 10,000 years. Over this time span, there could be 110 impact events, 285 Tunguska style air bursts over land, 680 over the ocean and 12 ocean impacts that could produce oceanwide tsunami. Only four events would be big enough to leave a crater on land, and all of these would have a high probability of being eroded or buried.

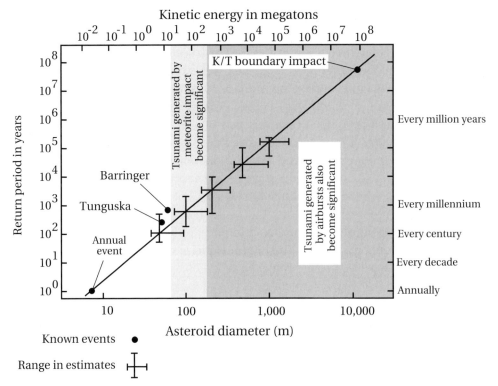

8.2 Probability of comets or asteroids of given diameter striking the Earth. The sizes generating significant tsunami are shaded (based upon Verschuur, 1996, and formulae presented in the text). Estimates derived from Michael Paine's web page at **http://www1.tpgi.com.au.users/tps-seti/Impact.jpg.**

Of the Tunguska-sized events, one or two impacts would have been equivalent to 500–1,000 megatons of TNT. Their impacts must have affected global climate and the course of human history. All of the Tunguska-sized objects that strike the ocean could have generated significant tsunami. Gerrit Verschuur makes the interesting comment that, historically, few population centres existed around the shores of the Atlantic or Pacific Oceans. The cradles of civilisation emerged in relatively sheltered river valleys around smaller seas – the Mediterranean, Red Sea, and Persian Gulf, or in mountain regions such as the Andes and the south Indian highlands. Our ancestors may have been wiser than us and avoided coastal areas because of the potentially devastating effects of meteorite-generated tsunami that were occurring at regular intervals in circumoceanic regions.

These probabilities also assume a random but constant flux of objects intersecting the Earth. However, the appearance of comets and meteorites in the inner solar system, and their impact with the Earth, is clustered in time in a phenomenon astronomers call coherent catastrophism. This is logical when it is realised that cometary disintegration leads to the production of objects 10–100 m in size, with some kilometre-sized objects orbiting about the sun within the confines of a narrow stream or trail. These objects tend to be clustered within this stream. Resonant interaction by Jupiter and the inner planets upon a stream such as the Taurid complex periodically allows objects within the stream to intersect the Earth's orbit, leading to multiple bombardments of Tunguska-sized objects (or larger) over periods of one to four centuries. In terms of global catastrophes, it is not the random impact of celestial objects greater than 1 km in diameter occurring on average every hundred thousand years that is important, but rather the occurrence of clusters of Tunguska-sized objects during epochs of high activity. The latter can affect civilisations deleteriously through direct impact, the generation of tsunami, or modifications to the atmosphere leading to sudden periods of global cooling.

Some measure of coherence in meteorites and comets can be obtained from Chinese, Japanese, and European records of meteor, comet, and fireball sightings gathered over the last 2,000 years. The accumulated record, up to the beginning of the nineteenth century when scientific observations began in earnest, is plotted in Figure 8.3. The Asian records are the most complete, with European sightings accounting for less than 10% of the record over the last thousand years. The comet record from Asia is also plotted in Figure 8.3. A quasicyclic pattern is evident in these records that can be linked to the dominance of the Taurid complex in the inner solar system. Peak occurrences of cosmic input to the atmosphere occurred between 401 and 500, 801 and 900, 1041 and 1100, 1401 and 1480, 1641 and 1680, and 1761 and 1800. These intervals have been shaded. The first period corresponds with the last extended epoch of nodal intersection with the Taurid complex, while the prominent fluxes in the eleventh and fifteenth centuries correspond to nodal intersections with parent objects in the complex. The fifteenth century represents the last phase of coherent catastrophism associated with the Taurid complex. In addition, there is a preference for sightings to occur in July–August and October–November. Some astronomers believe that many of these peaks were responsible for climate changes and direct impacts that have affected the course of human history.

HOW DO EXTRATERRESTRIAL OBJECTS GENERATE TSUNAMI?

Mechanisms for Generating Tsunami
(Chyba et al., 1993; Nemtchinov et al., 1996; Verschuur, 1996)

There are four types of extraterrestrial objects based upon density: comets (~1.0 g cm^{-3}), carbonaceous bodies (2.2 g cm^{-3}), stony asteroids (3.5 g cm^{-3}) and iron asteroids (7.9 g cm^{-3}). Comets are generally considered to be dirty snowballs; however, larger comets, in breaking up in successive orbits within the inner solar system, may produce many of the meteorites of various densities that strike the Earth.

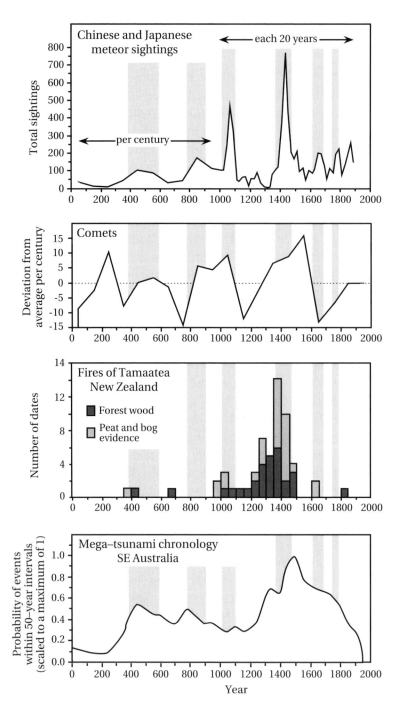

8.3 Incidence of comets and meteorites, and related phenomena, between A.D. 0 and 1800. The meteorite records for China and Japan are based upon Hasegawa (1992), while meteorite records for Europe come from Rasmussen (1991). Peak occurrences are shaded. The Asian comet record is based upon Hasegawa (1992). The calibrated radiocarbon dates under the Fires of Tamaatea are from Mooley et al. (1963) for forest wood and from McGlone and Wilmshurst (1999) for peats and bogs. The radiocarbon dates of prehistoric tsunami events in Australia are based upon the author's published work and other acknowledged research. See text for more details.

These four classes of objects also have different yield strengths that can vary over several orders of magnitude. The yield strength determines how easily the objects will fragment when they hit the atmosphere. Finally, these objects travel through space at different speeds. Comets travel at 25–50 km s^{-1}, while the Near Earth Objects move at slower speeds of 15 km s^{-1}. If objects fragment and explode in the atmosphere as bolides before striking the Earth's surface, they can still generate tsunami. In this case the size of the wave depends very much upon the height of the explosion in the atmosphere.

Any object greater than 1 km in diameter tends to intersect the Earth's atmosphere without fragmenting or exploding. In effect, these large objects travel fast enough that they do not have time to see the atmosphere before striking the surface. Comets less than 580 m in diameter, and stony and iron asteroids less than 320 and 100 m in diameter respectively, begin to distort or fragment travelling through the atmosphere. Any object entering the atmosphere at a shallow angle is more likely to reach the ocean without breaking up; however, this doesn't necessarily lead to bigger tsunami. Distortion without fragmentation leads to a pancake-shaped body of greater diameter impacting into the ocean. Even if a meteorite fragments, the fragments can hit the ocean as a hollow shell, creating a cavity that can be ten times greater than the radius of the original asteroid or comet. Theoretically, an iron meteorite with a diameter of less than 30 m could generate a tsunami by this mechanism. The initial waves formed in this case are technically not tsunami, as they are formed by the air blast. The real tsunami comes about 5 seconds later when the cavity in the water collapses. These fragmentation and distortion aspects have not yet been considered in the generation of tsunami, but may be very important.

Under fragmentation, a tsunami wave can be created by a minimum expenditure of 2–4 megatons of energy. The Tunguska airburst occurred at 7:40 A.M. on 30 June, 1908, flattening radially 2,150 km^2 of forest near the Podkamennaya Tunguska River, in central Siberia (60°55' N, 101°57' E). The explosion was equivalent to 10–20 megatons of TNT released into the atmosphere at an elevation of 8–9 km. This was one thousand times more than the Hiroshima atomic bomb. The airblast that reached the ground had only one percent of the bolide's kinetic energy – equivalent to about 0.1–0.2 megatons of TNT. The object must have been a small comet with a diameter of 60 m travelling at a velocity of 15 km s^{-1}. If the object had been an iron meteorite, then it must have been travelling at a higher speed of 30–40 km s^{-1} in order to explode at this height. Such an event is too rare, because this high velocity excludes about 90% of all known Earth-crossing asteroids. If the object had been a stony meteorite, then some debris would have reached the ground. None did.

The impact of a meteorite into an ocean can be modelled using the same Navier–Stokes equations described in Chapter 2. In fact, meteoritic tsunami are similar to those generated by rockfalls such as the Lituya Bay tsunami of

9 July 1958 discussed in Chapter 6. For example, a small meteoroid of only 500 m diameter falling in the ocean at 20 km s^{-1}, at a low angle of entry, could carve a path at least twelve miles long across the ocean. The horizontal and vertical accelerations of the resulting seismic waves would be much greater than any known earthquake. An important similarity between low-angled meteorite impacts and the Lituya rockfall-induced tsunami is the role played by splash. The Lituya run-up height of 524 m was technically not caused by a tsunami, but by splash. Low-trajectory meteorites can eject into the atmosphere enormous volumes of water that travel at high velocities, travel hundreds of kilometres and fall back to the surface of the Earth over equivalent distances (Figure 8.4). Such effects have only been mimicked in a small way on the Earth by rockfalls such as that produced by the Lituya event. Splash tsunami may be a far more important geomorphic process of meteorite impacts than any tsunami generated by cratering in the ocean. The significance of such events will be described later in this chapter.

Size of Tsunami

(Shoemaker, 1983; Huggett, 1989; Nemtchinov et al., 1996; Verschuur, 1996; Goodman, 1997; Hills and Mader, 1997; Toon et al., 1997; Crawford and Mader, 1998; Paine, 1999; Ward and Asphaug, 2000)

As a rough approximation, the height of a tsunami generated by a meteorite impact with the ocean can be determined by relating its kinetic energy to that of known tsunami generated by earthquakes and volcanoes. For example, the Chilean earthquake of 1960 generated a tsunami with kinetic energy equivalent to 2–5 megatons of TNT, while the Alaskan earthquake of 1964 generated a tsunami equivalent to 0.14 megatons of TNT. The eruption of Krakatau in 1883 was equivalent to 200 megatons of TNT. While not all of this latter energy went into the tsunami, run-up of over 40 m occurred within a 100-km radius. The airblast of the Tunguska bolide at the ocean's surface produced only 0.1–0.2 megatons of energy. This would only have generated a localised tsunami of around 0.2 m in height had it occurred over the ocean. However, had the comet reached the ocean's surface, it would have created a tsunami with an initial wave height of over 900 m. Thus most meteorites must be large or dense enough to crash into the ocean without fragmenting in order to produce a significant tsunami.

Of the meteorites or comets that reach the Earth's surface, about 70% will strike the ocean. At velocities of over 20 km s^{-1}, these objects burst upon contact with the ocean, sending out splash that can reach the top of the atmosphere and creating a tsunami wave train. The height of the tsunami wave depends upon the displacement of water from the pseudocrater blasted into the ocean. The volume of water absorbs most of a meteorite's energy so that no imprint or crater is left behind on the seabed.

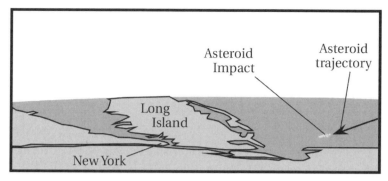

Just before impact

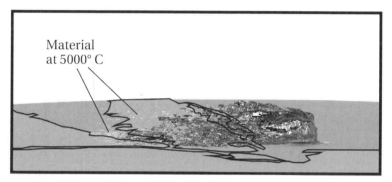

2.9 seconds after impact

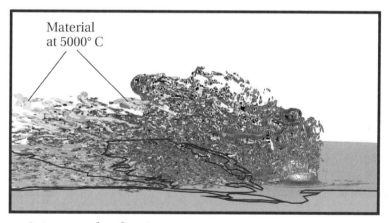

8.4 seconds after impact

8.4 Computer simulation of the splash from an asteroid striking the ocean off the coast of Long Island. The asteroid is 1.4 km in diameter and is travelling northwards at a speed of 20 km s^{-1}. Note the vapour heated over 5,000° C. Dark material is water vapour; white material is water. This phenomenon is called a splash tsunami. The simulations were performed at Sandia National Laboratories using an Intel Teraflops super computer. They took eighteen hours to complete. Images are used courtesy Dr. David Crawford. More images and movies are located at **http://sherpa.sandia.gov/planet-impact/asteroid/.**

The displaced water piles up around the crater and forms a ring whose width equals the radius of the crater. The tsunami's wave height can be approximated initially by the following formula:

$$H_t = r d R_t^{-1} \qquad\qquad 8.1$$

where H_t = height of the tsunami wave above mean sea level (m)
r = the radius of the pseudocrater in the ocean (m)
R_t = the distance from the centre of impact (m)
d = water depth (m)

As this ring moves out from the centre of impact, it causes water to oscillate up and down, forming about four ringlets that propagate outwards as a tsunami wave train. This process is similar to the ripples that form when a pebble is thrown into a pond. As the wave moves away from the impact site, its height then becomes dependent upon the distance from the centre of impact. Shoemaker proposed a simple formula, based upon analogies to nuclear explosions, to estimate the size of the crater generated by any meteorite as follows:

$$D = SpW^{0.3} \qquad\qquad 8.2$$

where D = crater diameter (m)
Sp = $90\,\rho_e^{-0.3}$ $\qquad\qquad 8.3$
= density correction
ρ_e = density of material ejected from an impact crater (g cm^{-3})
W = $0.12 \times 10^{-12}\, m v_m^2$ $\qquad\qquad 8.4$
= kinetic energy of impact (kilotons of TNT)
v_m = impact velocity of meteorite (m s^{-1})
m = $1.33\,\pi\rho_m\,r_m^3$ $\qquad\qquad 8.5$
= mass of meteorite (kg)
r_m = radius of meteorite (m)
ρ_m = density of meteorite (g cm^{-3})

Using Equation 8.4, the kinetic energy of an iron meteorite 40 m in diameter with a density of 7.8 g cm^{-3}, travelling at a velocity of 20 km s^{-1} – such as that which created the 1.2-km-wide Barringer crater in Arizona – is equivalent to 12.5 megatons of TNT. Equations 8.2 and 8.3 indicate that this object would have produced a pseudocrater 1.5 km in diameter in the ocean.

When Equation 8.2 is substituted into Equation 8.1, the size of the tsunami is overestimated for small meteorites. Equation 8.1 also does not account adequately for tsunami generated by stony meteorites or meteorites in shallow seas. Stony meteorites smaller than 100 m in diameter tend to fragment in the ocean, while larger ones tend to dissipate some of their energy in the atmosphere. For example, a 200-m-diameter stony meteorite travelling at 25 km s^{-1} would impact with a force equivalent to 940 megatons of TNT. If this meteorite exploded as an airburst, then only 20% of its energy would reach the surface of the ocean. This latter component is more than thirty times greater than the energy of the Chilean tsunami of 1960 and approximately equal to the largest Krakatau eruption of 1883. The diameter of the smallest object that can reach the Earth's surface virtually intact is 40 m for an iron meteorite, 130 m for a stony meteorite, and 380 m for a short-period comet. The recurrence interval for a stony meteorite, 130 m in diameter, is about every thousand years. The tsunami created by large meteorites impacting in the ocean are also depth limited. As a rule of thumb, depth becomes a limiting factor when it is less than twelve times the diameter of the meteorite. For example, meteorites greater than 167 m in diameter will be depth-limited if the meteorite falls into water less than 2,000 m deep. In this case, the resulting tsunami is 60% smaller than if the meteorite had fallen into deeper water. Stony meteorites also produce waves that are 60% smaller than those produced by denser asteroids. The following equations more realistically model these three conditions:

Iron meteorite $\quad\quad H_t = 1.87\ W^{0.54}\ R_t^{-1}$ $\hfill$ 8.6

Stony meteorite $\quad\quad H_t = 3900[(0.005\ r_m)^3\ (0.05\ v_m)^2\ (0.33\ \rho_m)]^{0.54}\ R_t^{-1}$ $\hfill$ 8.7

Shallow water $\quad\quad H_t = 0.0229\ W^{0.25}\ dR_t^{-1}$ $\hfill$ 8.8

These relationships are tabulated for meteorites of between 0.1 and 1.0 km in diameter in Table 8.1. The heights of tsunami produced by an iron meteorite are also plotted in Figure 8.5. This size range covers those objects that could realistically strike the Earth's ocean in the near future. Tsunami wave height quickly attenuates away from the site of impact. It also increases sizeably as the diameter of the meteorite increases. For example, a 100-m-diameter iron meteorite would produce tsunami of 27.1 m, 2.7 m, and 0.7 m in height, within 50 km, 500 km, and 2,000 km respectively of the centre of impact. This size meteorite is the minimum limit presently proposed for the detection of NEAs and one equivalent in energy to the Krakatau eruption of 1883. The wave heights are equivalent to tsunami generated by the Krakatau eruption over similar distances. The 0.7-m height is much higher than the 0.2 to 0.4-m open ocean height postulated for the Chilean tsunami of 1960, 2,000 km away from its source. As mentioned in Chapter 5, this latter tsunami's run-up measured 3 m high on the Hawaiian Islands and 1–2 m in Japan, and killed 61 and 190 people respectively in each area.

If run-up height of a tsunami is approximately ten times its open ocean height, then according to Equation 2.25, the tsunami wave produced by a 100-m-diameter iron meteorite would penetrate inland 890 m on any flat coastal plain within 2,000 km of the impact. An iron meteorite, 500 m in size can generate a tsunami wave that is approximately 35 m high leaving the impact site. Two thousand kilometres away, this tsunami would still be approximately 10 m high. If this size meteorite landed in the middle of the Pacific Ocean, its tsunami would be just under 5 m in height approaching any of the surrounding coastlines. Theoretically, this wave could sweep inland over 12 km across any flat coastal plain in the Pacific Ocean

TABLE 8.1 Tsunami Heights Generated at Distances of 50, 500, and 2,000 km from the Impact Site of Iron or Stony Meteorites with the Ocean

Iron Meteorite Characteristics				Tsunami Height (m)					
				Iron Meteorite			Stony Meteorite		
Meteorite Diameter (m)	Kinetic Energy (megatons TNT)	Kinetic Energy (joules × 10^{18})	Crater Diameter (km)	50 km Distance	500 km Distance	2,000 km Distance	50 km Distance	500 km Distance	2,000 km Distance
100	198	1	3.4	27.1	2.7	0.7			
150	667	3	4.9	52.2	5.2	1.3	15.5	1.6	0.4
200	1,581	7	6.3	83.3	8.3	2.1	24.8	2.5	0.6
250	3,089	13	7.7	119.5	12.0	3.0	35.6	3.6	0.9
300	5,337	22	9.1	160.6	16.1	4.0	47.8	4.8	1.2
350	8,475	35	10.5	206.1	20.6	5.2	61.3	6.1	1.5
400	12,651	53	11.8	255.9	25.6	6.4	76.1	7.6	1.9
450	18,013	75	13.1	309.7	31.0	7.7	92.1	9.2	2.3
500	24,710	103	14.4	367.3	36.7	9.2	109.3	10.9	2.7
550	32,889	138	15.7	428.7	42.9	10.7	127.5	12.8	3.2
600	42,699	179	17.0	493.6	49.4	12.3	146.9	14.7	3.7
650	54,288	227	18.3	561.9	56.2	14.0	167.2	16.7	4.2
700	67,804	284	19.5	633.6	63.4	15.8	188.5	18.9	4.7
750	83,396	349	20.8	708.5	70.8	17.7	210.8	21.1	5.3
800	101,212	424	22.0	786.6	78.7	19.7	234.0	23.4	5.9
850	121,400	508	23.2	867.7	86.8	21.7	258.2	25.8	6.5
900	144,108	603	24.5	951.9	95.2	23.8	283.2	28.3	7.1
950	169,485	709	25.7	1,039.1	103.9	26.0	309.2	30.9	7.7
1000	197,679	827	26.9	1,129.1	112.9	28.2	336.0	33.6	8.4

Note: Velocity of impactor = 20 km s^{-1}; density of iron meteorite = 7.9 g cm^{-3}; density of stony meteorite = 3 g cm^{-3}; ocean depth at impact = 5,000 m.

Source: Based on Shoemaker, 1983, and Hills and Mader, 1997.

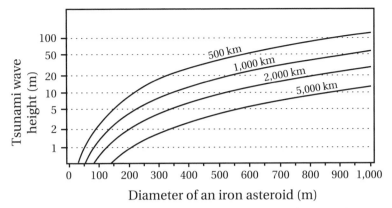

8.5   The size of tsunami generated by iron meteorites of various diameters striking the ocean. The heights are at distances of 500 km, 1,000 km, 2,000 km, and 5,000 km from the centre of impact. The meteorites have a density of 7.8 g cm^{-3} and impact with a velocity of 20 km s^{-1}.

region. Stony meteorites are less effective at generating tsunami than iron meteorites, but the impact of their run-up is just as impressive. A 500-m stony meteorite would generate a tsunami that is 2.7 m high, 2,000 km from the impact site. This is bigger than any historical tsunami in the Pacific. Such a wave could sweep inland, 5.3 km over any flat coast surrounding the Pacific Ocean.

Meteorites larger than 1 km in diameter will produce catastrophic tsunami in any ocean. Simulations have been performed on the effects of a meteorite 5 km in diameter falling into the mid-Atlantic or West Pacific Ocean. Such objects have a return period of once every 10 million years. At 1,000 km from the impact site, the resulting tsunami would be over 45 m high. At 2,000 km distance, the wave would begin to shoal onto continental shelves with a height of 22 m. Modelling, using incompressible shallow-water long-wave equations indicates that a significant amount of this wave's energy would be reflected from the front of flat continental shelves. Hence, coasts protected by wide shelves, such as those found along the East Coast of the United States and Northern Europe, are less affected by cosmogenic tsunami than are steep coasts such as those found along the coasts of Australia or Japan. However, even on the most protected coastline, the wave formed by a meteorite more than 5 km in diameter would be 30 m or more high.

Recent mathematical and computer modelling indicates that there is a wide range in the calculated heights of tsunami emanating from meteorite impacts. Figure 8.6 presents a simulation of the wave height produced by a stony meteorite 200 m in diameter travelling at a speed of 20 km s^{-1}. The leading edge of the resulting tsunami travels 25 m from the centre of the impact within 5 minutes and is over 300 m high. At distances of 50 km, 500 km, and 2,000 km the wave is still over 100 m, 11 m, and 6 m high respectively. These heights are more than four times greater than the calculated heights based upon nuclear explosions (Table 8.1).

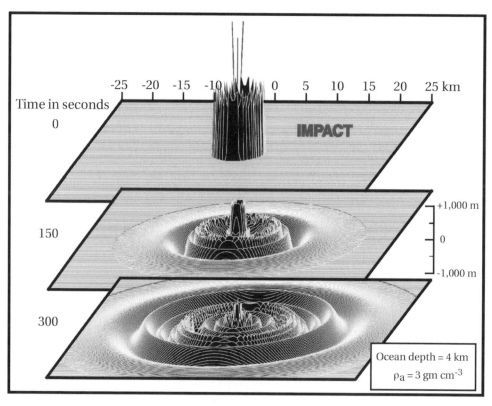

8.6 Modelled results of the initial development of a tsunami created by the impact of a 200-m part diameter stony meteorite with a density of 3 g cm^{-3} travelling at 20 km s^{-1}. Within 300 seconds, a tsunami propagates outwards more than 50 km. The leading wave in the bottom panel is about 325 metres high (based on Ward and Asphaug, 2000). Dr. Steven Ward, Institute of Tectonics, University of California at Santa Cruz, kindly provided a digital copy of this simulation.

At the other extreme, simulations of meteorite impacts into an ocean have been performed on a supercomputer at the Sandia National Laboratories in New Mexico, using algorithms that modelled accurately the impact of the Shoemaker–Levy 9 comet into Jupiter's atmosphere in July 1994. These are the same calculations used in the simulation of splash effects shown in Figure 8.4. Despite an uncertainty factor of two, the computer modelling yields tsunami heights that are smaller than those derived using Equation 8.6 by a factor of almost ten. These differences are summarised in Table 8.2 for a small range of meteorite sizes. The differences have been calculated 500 km from the centre of impact for an iron meteorite striking an ocean 5,000 m deep at a velocity of 20 km s^{-1}. The modelled tsunami wave heights span one and a half orders of magnitude – a fact indicating that there is not a broad consensus amongst researchers on the heights of meteorite-generated tsunami.

TABLE 8.2 Crater and Tsunamic Characteristcs Modelled Using Analogues to Nuclear Explosions and the Shoemaker-Levy Comet Impact into Jupiter

	Asteroid Diameter (m)		
	250	500	1,000
ANALOGUE TO NUCLEAR EXPLOSIONS			
Crater size (km)	7.7	14.4	26.8
Tsunami height (m)	11.9	36.5	112.1
SANDIA COMPUTER SIMULATIONS			
Crater size (km)	5	10	20
Wave velocity (m s^{-1})	166	166	900
Wave period (s)	150	180	300
Tsunami height (m)	1.3	5.0	12.0

Note: Latter modelling is the same as that used in Figure 8.4. Wave heights are taken 500 km from the impact site.
Source: Based on Crawford and Moder, 1998.

GEOLOGICAL EVENTS

Hypothesised Frequency

As of 1998, 160 impact craters have been identified geologically on the surface of the Earth, excluding the polar icecaps. Only seven of these are submarine. While some of the preserved craters may have occurred on the margins of oceans, no deep-ocean-basin impact structure has been recognised. Comet and meteor impacts were common in the early history of the Earth, but since the end of the Carboniferous era 225 million years ago, the Earth has only been struck randomly by debris flung into the inner solar or by asteroids orbiting between the Earth and Jupiter. Ninety craters have been identified since the Carboniferous. By far the largest is Chicxulub crater, which is 180 km in diameter and is buried beneath the Yucatan Peninsula in Mexico. Chicxulub is one of the few submarine impacts now preserved under a continent. This event has been linked to the Cretaceous–Tertiary boundary extinction of the dinosaurs and will be described in more detail subsequently. If the distribution of meteorite impacts was spread evenly over the Earth's surface, and if it can be assumed that the oceans have continually occupied at least 70% of the world's surface area, then a minimum of 210 impacts should have occurred in the oceans since the Carboniferous. This represents conservatively one large impact event approximately every million years.

Using Equation 8.2 and the size of known impact craters, it is possible to calculate the energy released by each impact, and the diameter of the impacting object.

In this analysis, it is assumed that all of the objects that struck the continents were iron asteroids with a density of 7.8 g cm^{-3}, and that they struck crustal material with a density of 2.65 g cm^{-3}. In addition, the diameter of craters larger than 3 km has been adjusted upwards by a factor of 1.3 to account for collapse due to gravity. Using Equations 8.6 and 8.8, it is possible to calculate the hypothetical distribution of tsunami wave heights in the ocean produced by this theorised population of meteorites since the Cretaceous. This distribution is plotted in Figure 8.7 at a distance of 1,000 km from each impact site. All objects are assumed to have fallen into an ocean that is 5,000 m deep. Shallow water corrections have been applied where required because they make the theorised tsunami wave heights more realistic. For example, the Chicxulub impact object, which had a postulated diameter as great as 15 km, would have generated a tsunami 4.6 km high without this correction, as opposed to one 104 m high with it. The latter value is closer to estimates derived from analysis of sedimentary deposits laid down by the wave.

Excluding Chicxulub, the average height of meteorite-generated tsunami waves in the ocean over the past 225 million years is 12.2 m. Of these, 35.6% or seventy-five, of the hypothesised impacts with the ocean generated tsunami that were less than 2 m in height. If the cut-off value for mega-tsunami approximates 5 m, then 57.8% of impacts generated this type of event. The probability of a meteorite producing a tsunami more than 5 m in height over the next million years is 0.26%. Since civilisation began, the probability of such a large event has only been 0.0026%. Mega-tsunami greater than 20 m in elevation are rare and represent in total 11.1% of hypothesised events. While the effect of mega-tsunami in any ocean would be catastrophic, geologically, they appear not to have been a dominant force shaping world coastal landscape.

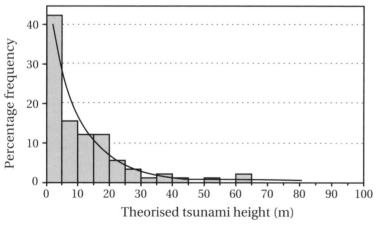

8.7 Theoretical distribution of tsunami wave heights generated by iron meteorite impacts with the world's oceans over the last 225 million years. Density of the meteorites is 7.8 g cm^{-3}

Chicxulub, the Cretaceous–Tertiary (K/T) Extinction Event

(Bourgeois et al., 1988; Smit et al., 1996; Verschuur, 1996; Alvarez, 1997)

In the last ten years, one of the great mysteries of geology – the extinction of the dinosaurs – has been unravelled. In 1980, Walter Alvarez and his colleagues proposed that a comet had caused a global winter that not only killed the dinosaurs, but also wiped out 67% of all species at the boundary between the Cretaceous and the Tertiary. That hypothesis was speculative until the crater for the impact was eventually found underlying Chicxulub, on the Yucatan Peninsula in Mexico. Finally, between 1988 and 1992, tsunami deposits were found in the region that linked the Chicxulub crater to what is now called the K/T extinction event.

The comet responsible for the extinction event was about 10–15 km in diameter, and produced a crater 180 km in diameter in anhydrite-rich limestones and dolomites. Two fireballs issued from the impact site. The first consisted of a cloud of extremely hot vapourised rock, followed closely by a superheated cloud of CO_2 gas released from the carbonates. The heat was so intense that all vegetation burned out to a radius of several thousand kilometres, loading the atmosphere with soot. Seismic waves with surface magnitudes, M_s, of 9–11 on the Richter scale caused faulting in shallow waters around the Gulf. Huge submarine landslides were triggered on steeper slopes. Following the impact, sunlight was blocked by an estimated 100×10^9 tonnes of dust thrown into the atmosphere – a process that eliminated photosynthesis for a two- to six-month period. Large amounts of CO_2 and SO_2 were released into the atmosphere together with the formation of equivalent amounts of nitrogen oxides. Subsequent scavenging of these molecules from the atmosphere produced an acid rain rich in sulphuric and nitric acids. All animals dependent upon photosynthesising organisms became extinct. In total, 38.5% of all genera and 67% of all species disappeared. Only the end of the Permian 250 million years ago saw the extermination of more species.

The evidence for extinction is one of the main signatures of the K/T event left in the geological record. However, the distribution of tsunami deposits up to 9 m thick on continental shelves and the seafloor provide the conclusive proof of an impact event of enormous magnitude. Equations 8.1 and 8.2 indicate that, had the Chicxulub comet fallen into the deep ocean, the tsunami wave would have been over 400 m high 1,000 km from the centre of impact. However, the impact occurred on the continental shelf, close to the southern shore of the proto–Gulf of Mexico, which formed an enclosed, shallow sea, 1,500 km in diameter (Figure 8.8). The ejecta curtain, the initial blast wave of compressed air from the impact, and finally a atmospheric blast driven by the release of CO_2 from limestone bedrock, produced a tsunami 100 m high that rolled into the southern United States (Figure 8.1). Within an hour of the impact, the burning forests that had been flattened and then ignited by the initial blast wave were picked up, mixed with uncompacted sediment and ripped-up bedrock, and then driven inland hundreds of kilometres over the flat coastal plains adjacent to the proto-Gulf. Backwash reentered the Gulf as a wave tens of metres high, accompanied by channelised backflow that eroded channels across the wide shelf. This slurry spread along the seabed as a turbidity current, depositing a chaotic mixture of ejecta, bedrock clasts, sand, and terrestrial vegetation across the abyssal plain. Reflection of this wave back and forth across the Gulf ensured that landmasses were repetitively swept by tsunami diminutively over the next few days. Slowly, sediment suspended in the water column and dust put into the atmosphere settled to the seabed over the next few weeks as quiescence returned to a lifeless ocean. This latter sediment contained the iridium-rich signature of the cosmogenic source. Millions of years later as the modern Gulf formed, remnants of the waves' passage were uplifted and then exposed around the Chicxulub impact centre – along the Mexican Gulf Coast, in exposures on the Brazos River in Texas, at several sites in Alabama, and on the Island of Haiti (Figure 8.8).

The nature of the turbidite signature varies depending upon its proximity to the initial impact and its relative location to the shoreline of the proto-Gulf (Figure 8.8). The deposits have been attributed to turbidity currents generated by submarine slides or to bottom currents generated by the passage of tsunami waves. The enormous height of the wave would have ensured that even the deepest part of the Gulf was in shallow water and that cobble-sized material was moved. However, the passage of the wave over the seabed was only capable of deforming bottom sediment or rearranging the sediment that was already there. In quiescent deep water, this sediment was most likely mud. However, no homogenites similar to those deposited following the eruption of Santorini have been found. All deposits contain sand that must have been brought there from the shelf by backwash generated by the tsunami or by turbidity currents. Because tsunami have wavelengths tens if not hundreds of kilometres long, current velocities at the seabed generated by the oscillatory nature of the wave must have been unidirectional, pointing away from the impact site for several minutes under the crest. This would have been followed by a longer period of reverse flow that decreased in intensity until the next wave in the tsunami wave train approached. Thus, sandy sediment once deposited in the deep ocean could have been reworked by other waves in the tsunami wave train and by the seiches that followed. In addition, ejecta – in the form of spherules – is absent or sparse at the base of some deposits. Ejecta present at the base of deposits in the ocean did not necessarily fall there in situ from the atmosphere. Because ejecta were thrown high into the atmosphere, it took time for it to settle back to the Earth's surface. During this time the Earth rotated eastwards. Thus most ejecta tended to fall on the shelf to the west of the impact site, forming a layer 1 m thick. Strong tsunami backwash swept these spherules seaward and deposited them first below the subsequent turbidites. Shelf sands then covered the turbidites, leading to a reversal in stratigraphy.

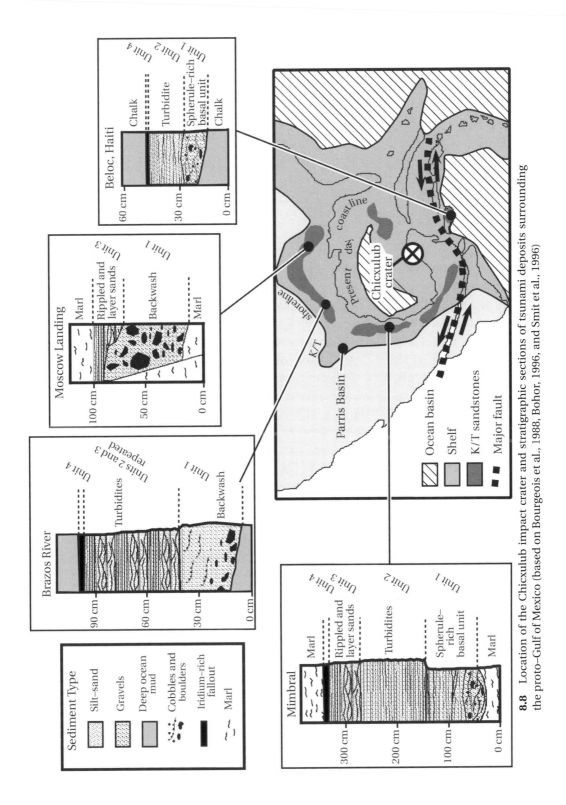

8.8 Location of the Chicxulub impact crater and stratigraphic sections of tsunami deposits surrounding the proto–Gulf of Mexico (based on Bourgeois et al., 1988, Bohor, 1996, and Smit et al., 1996)

Both the hydrodynamics of the tsunami and the relative abundance of ejecta are reflected in the deposits. At Beloc, on the Island of Haiti, which was in water about 2 km deep at the time of the impact, deposits are thinnest because they are furthest from sources of sand on the shelf. A turbidity current first deposited a layer of ejecta 15–70 cm thick. This grades upwards into a deposit of fine sand and silt 20 cm thick with low-angled cross-bedding (Figure 8.8). Thin iridium-rich layers of sandy silt 1–2 mm thick occur at the top of the sequence. This upper segment in many locations is disturbed, implying that a second tsunami may have affected the area afterwards. This could have been generated by volcanism triggered by the impact or by subsequent landslides on an unstable seabed. In slightly shallower water, but still offshore of any shelf, west of the impact site at Mimbral where still water conditions existed on the seabed, fine limestone-rich clays called marls were overlain by 1 m of ripped-up limestone, mixed together with the ejecta debris from the impact zone. These sediments are best preserved in the channels cut by tsunami backwash and are overlain by at least 2 m of sand derived from the distant shoreline and brought to the seabed by turbidity currents. This unit can be traced over a distance of 2,000 km, from Alabama through Texas to the southern border of Mexico. This unit is overlain by a metre of crosscurrent beds of sequential rippled sand and fine clay. In many respects, the deposit has all of the characteristics of a Bouma sequence as described in Chapter 3. Finally, a thin layer, several centimetres thick, caps the sequence. This uppermost layer contains the iridium-rich dust fallout that settled out of the atmosphere over several weeks following the impact. Closer to shore, about 100 km landward of the shelf edge at a site like the Brazos River in Texas, the tsunami wave swept over the muddy, flat shelf, scouring out swales with a relief of 0.7 m. Backwash dominates these sites. The bottom part of the sand deposits is about 1.3 m thick and consists of rounded calcareous cobbles, shell, fish teeth, terrestrial wood debris, and angular pieces of mudstone. At Moscow Landing, Alabama, the shelf was only 30 m deep at the time of impact. Seismic waves preceded the tsunami creating normal faults in a north–south direction. These are paralleled by grooves, flute casts, scour features, and lineations created by the passage of the wave over undular topography moulded into soft bottom muds. Here, backwash in the form of undertow wedged the bottom sediments against fault blocks. Boulders are also present in the deposit. The bottom unit fines upwards into parallel-laminated and symmetric, ripple-laminated sandstone, siltstone, and mudstone. In places, the sequence is repeated up to three times, indicating the passage of more than one wave. The upward part of the sequence, which is iridium-rich, shows evidence of seiching.

The size of clasts at the Brazos River site gives some indication of the bottom shear velocities that must have been generated over the shelf. These values are as high as 1 km s⁻¹, far exceeding those that could be generated by storms on a shelf in 50–100 m depth of water. The velocities allude to a tsunami wave between 50 and 100 m high with a period of 30–60 minutes. Boulder deposits indicative of a mega-tsunami are rare; however, they have been found close to the shoreline of the proto-Gulf at Parras Basin in Northeastern Mexico and in the mid-south of the United States. At the latter location, anomalous sandstone boulders up to 15 m in diameter have been found 80 m above floodplains on hills that would have been coastal headlands at the time.

Other Events
(Gersonde et al., 1997; Mader, 1998)

Chicxulub tends to dominate the public's perception of what a large meteorite or comet impact can do. While this is the biggest event in the last 225 million years, it is not the only one to have generated an impressive tsunami. For example, an event known as the Eltanin Meteorite Event occurred during the Pliocene 2.15 million years ago. This meteorite, estimated to have been 4 km in diameter, plunged into the Pacific Ocean 700 km off the southwest corner of South America and exploded, sending ejecta into the atmosphere. If the object had a density of 3.6 g cm⁻³ and struck at a speed of 20 km s⁻¹, it potentially generated a mega-tsunami – according to Equation 8.6 – at least 30 m in height along nearby coasts in South America and Antarctica. Spreading waves from the impact site would have extended into the North Pacific and Southern Oceans with a deep-water wave height of 10 m reaching the most distant coasts. Conservatively, the resulting tsunami would have been 4 m, 10 m, and 20 m high off the coasts of Japan, California, and New Zealand respectively. Run-ups would have been amplification by a factor of two to three times as the wave travelled across continental shelves.

Evidence for this tsunami consists of unusual skeletal deposits of marine and land mammals found mixed together on the Peruvian Coast, corresponding to the time of the Eltanin impact. In coastal Antarctica, late Pliocene deposits of marine sediments containing continental shelf diatoms have been found several tens of metres above sea level, also dating from this period. The sediment layers are less than 0.5 m thick and are analogous to the splays of shelf sediment described in Chapter 3, onlapped onto coasts by modern tsunami. Finally, the splash from the impact may have lobbed marine diatoms and other microfossils thousands of kilometres into the ice-free Transantarctic Mountains of Antarctica. If this is so, then this evidence may resolve one of the discrepancies between theory and field evidence for the size of cosmogenically generated mega-tsunami – namely, that splash may be a potent force generating some of the geomorphic evidence attributed to catastrophic flows.

MORE RECENT EVIDENCE FROM LEGENDS AND MYTHS

Deluge Comet Impact Event 8,200 ± 200 Years Ago
(Kristan-Tollmann and Tollmann, 1992)

If cosmogenically generated tsunami are so rare, certainly within the time span of human civilisation, then a paradox exists because evidence for such events certainly appears often in the geological record and in human legends. Traditionally, the difficulty in discriminating between fact and fiction, between echoes of the real past and dreams, has discouraged historians and scientists from making inferences about catastrophic events from myths or deciphered records. Yet, common threads appear in many ancient tales. Stories told by the Washo Indians of California and by the Aborigines of South Australia portray falling stars, fire from the sky, and cataclysmic floods unlike any modern event. Similar portrayals appear in the *Gilgamesh* myth from the Middle East, in Peruvian legends, and in the Revelations of Saint John and the Noachian flood story in the Bible. Victor Clube of Oxford University and William Napier of the Royal Observatory of Edinburgh have pieced together consistent patterns in ancient writings, which they interpret as representing meteoritic showers 3,000–6,000 years ago. One of the more disturbing accounts has been compiled from these legends by Edith and Alexander Tollmann of the University of Vienna, who believe that a comet circling the sun fragmented into seven large bodies that crashed into the world's oceans 8,200 ± 200 years ago. This age is based on radiocarbon dates from Vietnam, Australia, and Europe. The impacts generated an atmospheric fireball that globally affected society. This was followed by a nuclear winter characterised by global cooling. More significantly, enormous tsunami swept across coastal plains and, if the legends are to be believed, overwashed the centre of continents. The latter phenomenon, if true, most likely was associated with the splash from the impacts rather than with conventional tsunami run-up. Massive floods then occurred across continents. The event may well have an element of truth. Figure 8.9 plots the location of the seven impact sites derived from geological evidence and legends. Two of these sites, in the Tasman and North Seas, have been identified as having mega-tsunami events around this time. The North Sea impact centre corresponds with the location of the Storegga slides described in Chapter 6. Here, the main tsunami took place 7,950 ± 190 years ago. One of the better dates comes from wood lying above tektites in a sand dune along the South Coast of Victoria, Australia. The tektites are associated with the Tasman Sea impact and date at 8,200 ± 250 years before present. These dates place the Deluge Comet impact event – a term used by the Tollmanns – around 6200 B.C. This event does not stand alone during the Holocene. It has been repeated in recent times – a fact supported by Maori and Aborigine legends from New Zealand and Australia.

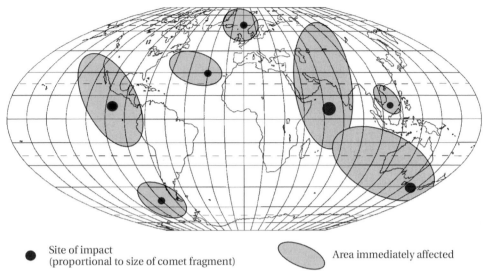

8.9 Reconstruction of the impact sites of fragments of the Deluge comet 8,200 ± 200 years ago based upon geological evidence and legends (from Kristan-Tollmann and Tollmann, 1992).

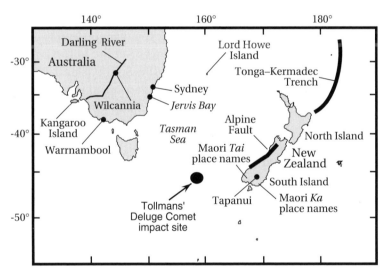

8.10 Location map of Southeast Australia and New Zealand where physical and legendary evidence exists for tsunami.

Mystic Fires of Tamaatea

(Mooley et al., 1963; Steel and Snow, 1992; McGlone and Wilmshurst, 1999)

One of the more intriguing legends associated with the Taurids is the New Zealand Maori legend known as the Mystic Fires of Tamaatea. The legend originates in the North Island, but ethnographic evidence is best chronicled in the

Southland and Otago regions of the South Island, centred on the town of Tapanui (Figure 8.10). Here there appears to be evidence for an airburst that flatten trees similar to the Tunguska event. The remains of fallen trees are aligned radially away from the point of explosion out to a distance of 40–80 km. Maori legends in the area tell about the falling of the skies, raging winds, and mysterious and massive firestorms from space. The Sun was screened out, causing death and decay. Maori names in the region refer to a Tunguska-like explosion. Tapanui, itself, translates as "the big explosion", while Waipahi means "the place of the exploding fire". Place names such as Waitepeka, Kaka Point, and Oweka contain the southern Maori word ka, which means fire. Some place names put the timing of the fires in the Southern Hemisphere winter around June at the timing of the Taurids. A deluge then followed the widespread fires. One legend states that the Aparima Plains west of Invercargill were flooded. Dimpling on the plain suggests that trees were toppled landward by water from the sea, and Maori place names such as Tainui, Tairoa, and Paretai, inland from the ocean, suggest a tsunami was involved because the affix tai translates as wave. The Maori also attribute the demise of the Moas, as well as their culture, to an extraterrestrial event. The extinction of the Moa is remembered as Manu Whakatau, "the bird felled by strange fire". One Maori song refers to the destruction of the Moa when the horns of the Moon fell down from above. On the North Island, the disappearance of the Moa is linked to the coming of the man/god Tamaatea who set fire to the land by dropping embers from the sky. Remains of Moa on the South Island can be found clustered in swamps as if these flightless birds fled en masse to avoid some catastrophe. Southern Maori legends tell of stones falling from the sky that caused massive firestorms that not only annihilated the Moa, but also Maori culture.

The age when these fires occurred can be determined by radiocarbon-dating wood debris from the fires. The dating evidence comes from two sources: buried wood and carbon derived from unconformal layers in swamps and bogs that have been interpreted as fire-induced. These dates traditionally have been interpreted as reflecting the time of deforestation due to Maori occupation in New Zealand. However, many of the dates come from uninhabitable high country that was burnt on a vast scale. The distribution of dates is plotted in Figure 8.3 and spans at least two centuries, with the ages peaking at the beginning of the fifteenth century. This wide range in dates is logical knowing that mature trees, already hundreds of years old, burnt. The crucial point is that few ages occur after the fifteenth century. The Fires of Tamaatea legend may well have a cosmogenic origin. The peak in dates is synchronous with the highest number of meteor sightings by Chinese and Japanese astronomers for the past 2,000 years (Figure 8.3). More important, the timing of the fires is also coherent with the occurrence of mega-tsunami along the nearby coastline of Southeast Australia.

EVIDENCE FROM AUSTRALIA
(Oliver, 1988; Bryant et al., 1996)

The evidence from Australia for cosmogenic mega-tsunami is based upon the magnitude of geomorphic features and their contemporaneous occurrence over a wide region that includes the Tasman Sea and the East Coast of New Zealand. This chronology coincides with the timing of legends and the influx of comets and meteorites over the last millennium. As pointed out in Chapter 1, Australia historically has not been affected significantly by large tsunami. The closest sources for earthquake-generated tsunami lie along the Tonga–New Hebrides Trenches and the Indonesian Archipelago. An earthquake with a surface magnitude greater than 8.3 on the Richter scale can be generated in the Southwest Pacific every 125 years. The highest tsunami recorded at Sydney since 1870 occurred on 10 May 1877 and had a height of 1.07 m. The Chilean earthquake of 1868 produced a tsunami height of 1.0 m, while the Chilean earthquake of 22 May 1960 generated a tsunami height of 0.85 m. On the West Coast, the biggest tsunami run-up measured 6 m at Cape Leveque, Western Australia, on 19 August 1977 following an Indonesian earthquake.

Palaeo-tsunami generated by conventional mechanisms and larger than these historical events are possible. The proximity of the northwest coastline to the volcanically and seismically active Indonesian Archipelago makes large tsunami with run-ups of 10 m a distinct possibility. Additionally, the East Coast lies exposed to tsunami generated by earthquakes on seamounts in the Tasman Sea and along the Alpine Fault running down the West Coast of the South Island of New Zealand. This latter fault last ruptured in the fifteenth century before European colonisation of the region. Nor can volcanic activity be ruled out along the East Coast. Active volcanoes lie in the Tonga–Kermadec Trench region north of New Zealand. In A.D. 1453, a volcanic eruption in Tonga created a crater 18 km long, 6.5 km wide, and 0.8 km deep. The volcano erupted with a force equivalent to 20,000 megatons of TNT and produced a tsunami wave 30 m high. Finally, local slides off the Australian continental shelf cannot be ignored. A very large submarine landslide, mentioned in previous chapters, lies 50 km offshore from the coast south of Sydney. This slide is a prime candidate for the tsunami-deposited barrier described in Chapter 4 along the adjacent coast (Figure 4.1).

Mega-Tsunami Evidence for a Cosmogenic Source
(Bryant and Young, 1996; Jones and Mader, 1996; Bryant et al., 1997; Bryant and Nott, 2001)

Some of the Australian evidence for tsunami is on a scale much bigger than could possibly be generated by the geophysical processes described in Chapters 5–7. Very few tsunami attributed to these latter types of events have generated anything approaching the bedrock-sculptured s-forms outlined in Chapter 3 (Figure 3.1). For example, while the Lituya Bay landslide of 1958 generated a tsunami that surged 524 m above sea level and obtained velocities of 210 km hr^{-1}, it only cleared soil and glacial debris overlying bedrock on nearby slopes. Only the Storegga slide – and this may be one of the Tollmanns' meteorite impacts – produced s-forms similar to those profusely blanketing the rocky headlands of the New South Wales South Coast. Meteorite impacts with the ocean can unequivocally generate the large tsunami necessary for the formation of s-forms. Modelling results, using the SWAN code described in Chapter 2, indicate that a 6-km-diameter asteroid impacting into the central Pacific would produce a tsunami 15 m high along the New South Wales Coast. As shown in Table 8.1, much smaller impacts near Australia could also produce waves with this height. Four other signatures also stand out as unique features of cosmogenic tsunami: whirlpools bored in bedrock, imbricated boulders fronting cliffs, mega-ripples, and overwashing of headlands up to 130 m high.

Whirlpools bored into bedrock, of the type shown in Figure 3.23, are rare. Isolated bedrock plugs of the type shown on the frontispiece of this book are rarer still. The kolks and tornadic flow necessary to form them have been described in detail in Chapter 3 (Figure 3.25). Suffice it to say here that kolks involve enormous hydraulic lift forces produced by turbulent bursting and steep pressure gradients across vortices. Tornadic flow involves the breakdown of a wide, parent vortex with secondary vortices developing around its circumference. The current speed around the vortex is so high that bedrock can be bored in a matter of minutes. These types of flow can only be produced by cosmogenic tsunami.

Imbricated and aligned boulders were also depicted in Chapters 3 and 4 as a signature that uniquely separates the presence of tsunami from storms. While the boulders perched on top of 33-m-high cliffs at Jervis Bay are impressive evidence of the high velocity flow that only tsunami can produce (Figure 3.11), their magnitude pales in comparison to other boulder deposits found in the region – namely at Gum Getters Inlet and Mermaids Inlet. At Gum Getters Inlet, angular boulders 6–7 m in diameter have been stacked up to 30 m above sea level into a small indent in the cliffs (Figure 8.11). It would be tempting to attribute this debris to cliff collapse but for the fact that the imbricated blocks rise to the top of the cliffs. The deposit is all the more unusual in that the indent is virtually protected from dominant southeast storm swell. Imbricated blocks of similar size choke the entrances of two narrow and deep gulches at Mermaids Inlet (Figure 8.12). Some of the largest blocks, which are over 5 m in length, have not simply dropped from the cliff faces but have been rotated 180° and shifted laterally in suspension flow. Not even the 26.2-m high run-up at Riang–Kroko, following the Flores Island tsunami of 12 December 1992, produced the magnitude and degree of organisation of these deposits (Figure 3.5). The depth of overland tsunami flow in the Jervis Bay region has been theorised at 9.5 m. The boulder features at Gum Getters Inlet

8.11 Imbricated boulders stacked against a 30-m-high cliff face on the south side of Gum Getters Inlet, New South Wales, Australia. Some of these boulders are the size of a boxcar. Note the person circled for scale.

are suggestive of even greater flow depths of 15–20 m, which only a cosmogenic tsunami could generate.

Just as dramatic are the dunes at Crocodile Head, Jervis Bay, and Sampson Point, Western Australia. Both of these features were described in Chapters 3 and 4. The former lie atop 80-m-high cliffs, have a relief of 6.0–7.5 m, and are spaced 160 m apart. They are akin to the undulatory-to-lingoidal giant ripples that are features of catastrophic flow such as that observed in the scablands of Washington State. The flow over the dunes at Crocodile Head is theorised to have been 7.5–12.0 m deep and to have obtained velocities of 6.9–8.1 m s^{-1}. The Sampson Point megaripples are gravelly (Figure 4.13) and have a wavelength approaching 1,000 m and an amplitude of about 5 m. Flow depth here is theorised to have been as great as 20 m with velocities of over 13 m s^{-1}. More important, the megaripples occur up to 5 km from the coast. These megaripples have never been described for conventional tsunami and could only have been produced by a cosmogenic event.

8.12 Boulder pile blocking the mouth of Mermaids Inlet, New South Wales, Australia. The largest blocks are more than 5 m in length. The gulch on the right contains a shelly beach radiocarbon dated at A.D. 1790 ± 70 years.

Finally, there is evidence of tsunami run-up higher and further inland than produced by conventional processes. The largest run-up produced historically by a volcano was 90 m on 29 August 1741 on the West Coasts of Oshima and Hokkaido Islands, Japan. Santorini may also have had a tsunami wave height of 90 m, but confirmed evidence for its run-up does not exceed 50 m above sea level. The largest tsunami run-up generated by an earthquake was 100 m on Ambon Island, Indonesia, on 17 February 1674. In recent times, an earthquake or submarine landslide off the Sanriku Coast of Japan produced run-up of 38.2 m on 15 June 1896. The highest palaeo-tsunami run-up identified in Australia so far is 130 m at Steamers Beach, Jervis Bay, on the crest of a chevron dune. This site has been referred to often in this text. However, this limit is underestimated because the wave still had enough force not only to flow over the headland and into Jervis Bay, but also to transport large boulders along a ramp inside the bay. The estimated flow velocity derived from these boulders using Equation 3.8 is 7.9 m s^{-1}. The potential for higher run-up may have been exceeded at Sampson Point. Here a palaeo-tsunami originating from the Indian Ocean overran hills, 60 m high, lying 5 km inland.

Aboriginal Legends of Comets
(Peck, 1938; Jones and Donaldson, 1989; Johnson, 1998)

This book began with a story based upon Aboriginal legends about a meteorite impact. Many of these legends are concentrated in the southeast corner of Australia, where some of the best signatures of large tsunami are preserved. As with Gervasse's description of the meteor impact with the Moon on 19 June A.D. 1178, the Aboriginal legend in Chapter 1 mentions that the moon rocked. There are also similarities with the Maori legend of the Fires of Tamaatea. In both, stars, fire, and stones fell from the sky, and there was a thunderous explosion. Further inland in New South Wales, the Paakantji tribe, near Wilcannia on the Darling River, also tell of the sky falling. A great thunderous ball of fire descended from the sky, scattering molten rock of many colours. As in the Maori legend of New Zealand, floods then followed this event. The floods may have been the consequence of millions of tonnes of seawater, vapourised by a meteorite impact with the ocean, condensing and falling as rain. In South Australia, another legend tells of stars falling to Earth to make the circular lagoons fringing the coast. Finally, it is curious that when Europeans made contact with Aboriginal coastal tribes in Western Australia, they noted that the Aborigines avoided the coast and made little attempt to use it for food, even though there was evidence of past usage in the form of large shell kitchen middens. As described in Chapter 4, the biggest mega-tsunami to affect Australia occurred on the West Australia Coast within the last thousand years, before European occupation.

Perhaps the most intriguing legend along the Southeast Coast of Australia is the story of the eastern sky falling. Aborigines south of Sydney believed that the sky was held up on supports and that these gave way on the eastern side. One version refers to the ocean as belonging to the sky. The ocean had fallen down, wiping out Aboriginal culture. Some tribes were even requested by others to send tribute to the east to be given to the spirit people in charge of holding up the sky so that it could be repaired. Archaeological evidence for tsunami and their impact on Aboriginal culture also exists along this coast. One of the deposition signatures of tsunami mentioned in Chapter 3 was the presence of disturbed Aboriginal kitchen middens, which form a special case of *dump* deposit more than 10 m above sea level on some rocky headlands. At Atcheson Rock, 60 km south of Sydney (Figure 3.20), tsunami overwashed a 20- to 25-m-high headland, boring whirlpools into the sides. The wave was travelling so fast that it separated from the headland and made contact with the sea 100–200 m on the lee side in a bay. Flow separation caused profuse amounts of coarse sediment to drop from the flow under gravity and be deposited on the lee side of the headland. On the far side of the bay, a dump deposit contains numerous silcrete hand axes and shaped blades that came from an Aboriginal camp at the head of the embayment. Aborigines in this camp initially would have heard, but not seen, the tsunami approaching. Their first indication of disaster would have been when they looked up and saw the ocean dropping down on them from the sky as the tsunami wave surged over the headland.

Dating of the deposits at Atcheson Rock indicates that the meteorite-induced tsunami occurred within the last 600 years, rather than in some distant Dreamtime. Archaeological research has shown that Aboriginal culture changed dramatically along this coast about 500 years ago. Instead of continuing their profuse gathering of marine shells for food, Aborigines switched to fishing. If a tsunami wave had the force to sweep over 130-m-high headlands in the region, then it would have been powerful enough to clear all marine shells from rock platforms. The event necessitated a change in lifestyle by Aborigines simply to survive starvation. There is also evidence from increased usage of rock shelters that Aborigines moved inland around this time. While interpreted as an indication of increasing population, it could also indicate abandonment of a dangerous coast similar to that observed in West Australia.

More physical and legendary evidence of tsunami comes from South Australia. Here, mainland Aborigines tell about Ngurunderi, who was a great, moody ancestral figure who lived in the sky. Long ago his two wives left him, and he came down from the sky to find them. He eventually found his wives wading in the water between Kangaroo Island and the mainland of South Australia. He was so angry that he decided to punish his wives. He ordered the sea to rise up as an enormous tidal wave and drown them. Noisily, the water rushed in so fast that it quickly drowned his wives, who were turned into stone. Their remains can be seen off the coast of Cape Jervis as rocks called the Two Sisters. The history of Aboriginal occupation of Kangaroo Island remains enigmatic. The island shows extensive evidence of Aboriginal occupancy, but when the first European, Matthew Flinders, landed on the island in 1802, it was totally unoccupied. Mainland Aborigines call Kangaroo Island, Kanga – the Island of the Dead. The coastline also evinces signatures of cosmogenic tsunami. Most significant are enormous, bored whirlpools on the northern coast of the island, where the Aboriginal legend is set. The features are larger than those found at Atcheson Rock. In addition, there are vortex-carved caves and massive piles of imbricated boulders, some over 4 m in diameter, near promontories. The Island of the Dead may be just that – evidence of another, tragic, cosmogenic tsunami witnessed by Australian Aborigines before European occupation, and then documented by the few survivors in legend form.

Chronological Evidence

(Asher et al., 1994; Steel, 1995; Young et al., 1997; Estensen, 1998; Bryant and Nott, 2000)

At present, no evidence has been found of a meteorite or comet impact linked to the signatures of mega-tsunami along the South Coast of New South Wales. Nor may any be found because it does not take a large meteorite impact in the ocean to produce the size of tsunami responsible for the observed evidence. Meteorite impacts also tend not to leave a crater on the seabed. For example, no crater for the Eltanin Meteorite has yet been found despite its 4-km diameter. However, the timing of tsunami events can be approximated using radiocarbon dating of marine shells deposited in dump deposits and sand layers, and attached to boulders transported by tsunami. Radiocarbon dating is only accurate for events that are older than 460 years. At least fifty-four dates have been obtained from the New South Wales South Coast. In addition, three samples related to tsunami were obtained from Lord Howe Island situated in the Tasman Sea halfway between Australia and New Zealand (Figure 8.10). At least ten additional dates were too young to be plotted. Each radiocarbon age is reported as an age with an error term. This information can be used to construct a probability distribution of dates for that sample. An overall time series was then constructed by summing these probabilities for all samples. For presentation purposes, this time series has been standardised to a maximum value of one. The resulting time series spanning the last 10,000 years is plotted in Figure 8.13, while that for the last 2,000 years is plotted in more detail in the bottom panel of Figure 8.3.

Six separate tsunami events can be recognised over the past 8,000 years, with peaks at 7500 B.C., 5000 B.C., 3300 B.C., 500–2000 B.C., A.D. 500, and A.D. 1500. There may be more events than this, but until further dating, it is impossible to know whether or not the broad sequence of dates between 500 and 2000 B.C. represents a single event or many. This later time span includes an impact event in the Middle East dated around 1600 B.C. Reference to fire and stones falling from the sky appear in the Bible and other manuscripts written around this time.

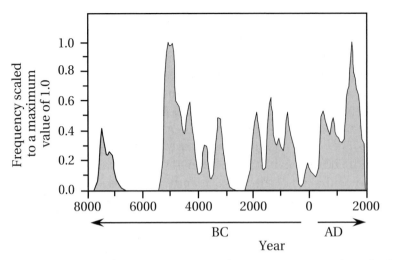

8.13 Chronology of tsunami events over the past 10,000 years along the South Coast of New South Wales. See text for details about the construction of this time series.

However, the record does not show any evidence for a Bronze Age event around 2350 B.C. that is believed to have destroyed civilisations simultaneously in Europe, the Middle East, India, and China. Nor do any of the dates cluster around the time of the Tollmanns' Deluge Comet impact event 8,200 ± 200 years ago. This may be due to the poor preservation potential of shell material this old or to the removal of such material by subsequent tsunami. However, thermoluminescence dating of sand layers deposited by tsunami on the New South Wales South Coast, indicates that a major discontinuity in sedimentation occurred 8,700 ± 800 years ago. This hiatus is within the time span of the Tollmanns' Deluge Comet impact event. The New South Wales event peaking in A.D. 1500 appears to be the largest, as it is associated with overtopping of the headland, 130 m high, at Steamers Beach, Jervis Bay. Because no large tsunami has been reported along the New South Wales Coast since European settlement in 1788, the shell samples that are too young for radiocarbon dating allude to a small, but significant, tsunami event in the early eighteenth century.

The peak of the A.D. 1500 tsunami event corresponds with the largest number of meteorite observations for the past two millennia (Figure 8.3). In addition, the peak at A.D. 500 corresponds with a clustering of meteorite sightings that is believed by astronomers to be one of the most significant over this time span in the Northern Hemisphere. Both of these clusterings are associated with the Taurid complex. Furthermore, the event around A.D. 1500 coincides with the calibrated ages for the Fires of Tamaatea across the Tasman Sea on the South Island of New Zealand. As well, the tsunami event at Atcheson Rock that accounts for the Aboriginal legend of the ocean falling from the sky occurred at this time, as does the age of the meandering backwash channels on the Shoalhaven Delta 40 km to the south (Figure 4.3). Other main sightings of meteorites from the Northern Hemisphere correspond with minor peaks in the Southeast Australian tsunami chronology. It would appear that meteorites, rather than comet impacts, correspond to the Australian chronology for palaeo-tsunami. The two minor clusters of meteorite activity between 1640 and 1800 may have produced cosmogenic tsunami that account not only for evidence of a pre-European event in New South Wales, but also for tsunami identified in Chapter 4 along the Coasts of North West Australia and Northeast Queensland. The lack of any mega-tsunami event since A.D. 1788 – the time of first European settlement – may only be fortuitous. Based upon the data for the last two millennia, there is a 50% probability that such an event could occur again in the next half century.

The events between the fifteenth and eighteenth centuries preceded European colonisation in Australia; however, they coincide with European exploration around the continent and Dutch colonisation in Indonesia. In the eighteenth century, without the means of determining longitude, merchant ships of the Dutch East India Company made their way to the colonial city of Batavia in Indonesia by sailing straight across the Indian Ocean until they sighted the Australian coastline, and then turning north. They would have sailed by the Northwest Coast of West Australia around the time a cosmogenic tsunami struck that coast. Many ships in pursuit of exploration and commerce were lost and presumed shipwrecked, but without hard evidence, it is best to put these losses down to storms.

Two shipwrecks in Australia stand out as unusual. The first relates to the *Mahogany* ship, now buried in sand dunes well above sea level at Warrnambool, Victoria. In 1521, three Portuguese caravels under the leadership of Cristovão de Mendonça sailed on a secret mission from Malacca, East Indies, to explore the Australian coastline. The reason for the secrecy was the intense competition between Spain and Portugal for world domination. Only one of Mendonça's ships made it back. Any record of his expedition disappeared into the secret Portuguese archives in Lisbon, where no one has seen them since. It is unlikely that they survived the earthquake and subsequent fire of 1755. In 1836, a mahogany ship was discovered, washed inland well above the limit of storm waves, near Warrnambool, on an isolated part of the Victorian Coast of southern Australia (Figure 8.10). The first Europeans known to have landed on this coast made the discovery. Unfortunately, the stranded ship was buried in shifting sands by 1880, never to be seen again. Intriguingly, evidence suggests that Mendonça did reach and map the South Coast of Australia. This evidence comes from the Dieppe maps, first published in the mid-1500s. They show remarkably detailed coastline down the East Coast of Australia and across the South Coast of Australia. The maps terminate at Warrnambool, Victoria. How the *Mahogany* ship managed to get into the sand dunes has remained an arcanum ever since.

The second shipwreck involves the *Zuytdorp,* a Dutch East India Company merchant vessel that was part of a convoy supplying the Dutch East Indies at regular intervals. In June–July 1712, the *Zuytdorp* crashed into the cliffs off Northwest Cape, Western Australia (Figure 3.2). Debris, including the ship's bell, was scattered amongst masses of boulders up cliffs rising 70 m above sea level – well above the limits of storm waves. The ship struck the reefs at the base of the cliffs suddenly because all six of its anchors were found intact without having been set, as would have been the case if the ship had been caught in a storm. Interestingly, the top of the cliffs is covered in a dump deposit of shell, sand, and angular gravels that has been misinterpreted by many anthropologists as an Aboriginal kitchen midden. The dates for both the *Mahogany* and *Zuytdorp* shipwrecks fit within temporal windows for two cosmogenic tsunami around the Australian Coast based upon radiocarbon dating.

There is controversy about the size of tsunami that can be generated by meteorite impacts. Also, if one examines the geological record, the theorised distribution of tsunami wave heights, calculated using one of the formulae leaning towards higher estimates, shows that cosmogenic tsunami have not been big enough to be a dominant force shaping the world's coastal landscape. On the other hand, there is plenty of evidence to indicate that some coastlines – mainly around Australia – have been affected by sufficient depth and velocity of water to transport boulders to the tops of cliffs 33 m high, deposit sandy bedforms on cliffs 80 m high, overwash headlands up to 130 m above sea level, and breech hills 60 m high lying 5 km inland. Similar evidence in the form of bedrock sculpturing can also be found along the coastlines of New Zealand and Northeastern Scotland. Two factors involving meteorite impacts with the ocean may account for the discrepancy between theory and fact. For example, meteorites of varying density and less than 1 km in diameter can fragment and undergo distortion before striking the ocean. If this is the case, craters ten times larger than the radius of the original asteroid or comet may dimple the ocean, creating a tsunami larger than could be produced by an unaltered meteorite. Second, large amounts of water and heated vapour can be flung into the ocean and tossed significant distances away from the centre of an impact (Figure 8.4). This high-velocity airborne splash may not only explain the inland flooding mentioned in many comet legends but also account for erosion of bedrock and emplacement of dump deposits on headlands and clifftops. Research on this aspect is in its infancy. Finally, the threat of splash or impact-related tsunami from meteorites may be alarmist. If coherent catastrophism is associated with the Taurid complex, then apart from the odd random Earth-crossing meteor or comet, the next large influx of meteorites will not occur until around the year A.D. 3000. The overall risk of all types of tsunami and society's mitigation of the threat will be discussed in Chapter 9.

PART FOUR

Modern Risk of Tsunami

CHAPTER NINE

Risk

9.1 Drawing of a tsunami breaking on the Japanese coast. The drawing probably represents the 1 September 1923 tsunami, which affected Sagami Bay following the Great Tokyo Earthquake. While this earthquake is noted for its subsequent fires and appalling death toll, it generated a tsunami 11 m in height around the bay. The drawing is by Walter Molino and appeared in *La Domenica del Corriere*, 5 January 1947. Source: Mary Evans Picture Library Image No. 10040181/04.

INTRODUCTION
(Bak, 1997)

There are three problems in assessing the risk of modern tsunami to any coast-line. All three suffer from popular misconceptions. The first problem involves the construction of probability of exceedence curves for the occurrence of tsunami based upon historical records. Such an approach is logically and scientifically flawed. It runs into a problem at the extreme end because many coastlines are devoid of credibly documented events. Second, it is often assumed that a coast-line is immune from the threat of large tsunami if it has not recorded any. Any study attempting to show otherwise is assumed to be fundamentally wrong. This concept treats the occurrence of tsunami as being some stochastic or random process mainly generated by earthquakes. Because the sea is flat, so is the seabed. Any idea of submarine landslides is discarded. So too is any consideration of vol-canoes as a cause of tsunami unless the smoke can be seen on the horizon. The idea that meteorites could cause tsunami is considered erroneous because no such phenomenon has been observed in the European-based historical record. Third, legends about tsunami, for whatever reason, are dismissed as myths, and, if any legend in a tsunamigenic region is bigger than the historic record, it is dis-missed as hyperbole by a primitive culture. Such attitudes are naive and ignore the fact that nature is critically self-organised. The laws for tsunami cannot be understood just by documenting tsunami that have occurred in the historical record. They must be set within the context of such events over hundreds of mil-lions of years. Large catastrophic tsunami tend to occur as a consequence of the same processes that produce small ordinary tsunami. Conversely, if catastrophic phenomena such as meteorites have generated large tsunami, then they can be implicated in some of the smaller, more frequent events. Finally, tsunami events of all sizes tend to be clustered in time. The latter concept incorporates the notion of coherent catastrophism involving meteorite impacts.

Much of the world's coastline has never experienced a large tsunami in its his-torical record, be it European or otherwise. This is clearly illustrated in Figure 9.2, which shows that portion of the world's coastline subject to historical observation in A.D. 1500 and A.D. 1750. The boundaries in Figure 9.2 are very liberally defined and, outside of Eastern Asia, reflect the presence of Europeans rather than any other society. In Chapter 8, the period around A.D. 1500 was identified as a possible time when a major meteorite struck the Southwest Pacific. No one who could put pen to paper was there to observe any such event. A second event may have occurred in the Australian region in the early part of the eighteenth century. Again, the event would have gone unrecorded except by the odd Dutch merchant ship or ship of adventure.

There are various methods for assessing the vulnerability of coastal popula-tions to the threat of tsunami. One of the simplest is to map population densi-ties. Such population density maps are readily available over the Internet and

AD 1500

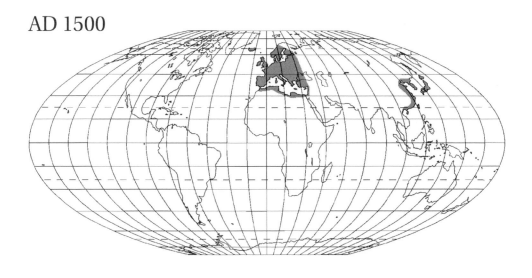

AD 1750

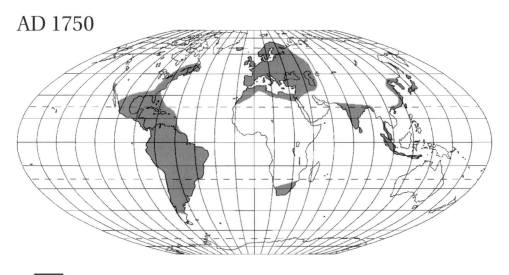

███████ Area with possible historical record of tsunami

9.2   The world's coastlines having historical records of tsunami in A.D. 1500 and A.D. 1750.

indicate that the most vulnerable regions of the world are the coastlines of
China, India, Indonesia, Japan, the Philippines, the eastern United States, the
Ivory Coast of Africa, and Europe. This approach ignores the economic impact of
tsunami. Without belittling the death tolls due to tsunami that have occurred
along isolated coastlines such as the Aitape Coast of Papua New Guinea or the
Burin Peninsula of Newfoundland, tsunami will have their greatest impact along
densely populated coasts of developed countries or where large cities are

located. For example, the next earthquake to strike Tokyo would have a world-wide economic impact. Any associated tsunami would destroy the shipping infrastructure so vital to that city's economy. It is possible to evaluate similar vulnerable coastlines in two ways. First, densely populated, economically developed coastlines can be detected by the amount of light they emit at night. The United States Air Force operates the Defense Meteorological Satellite Program (DMSP) that has a sensitive Operational Linescan System (OLS) that can detect visible and near-infrared light sources of 9–10 watts cm^{-2}. Maps of stable light sources, with a nominal spatial resolution of 2.8 km, are readily available. These maps exclude transient fires. One such global map current to 1997 is presented in Figure 9.3A. It clearly shows that the developed coastlines of the world lie in Western Europe, Japan, and the eastern United States. Tsunami would have the greatest economic impact along these coastlines.

Second, the largest coastal cities in the world require the greatest response irrespective of their role in the global economy. These cities are plotted in Figure 9.3B. Their populations are current to the year 2000. There are five coastal conurbations with populations of over 15 million people: Tokyo, New York, Osaka, Mumbai (Bombay), and Los Angeles. Were a major tsunami to strike any of these coasts, the impact would be severe. There are nine cities with populations of 10 to 15 million people. The majority of these are situated in poorly developed countries. Thirty-eight cities have populations of 2 to 5 million inhabitants. Over 60% of these are situated in Third World countries. It is only a matter of time before one of our world's major cities is crippled by a major tsunami.

WHAT LOCATIONS ALONG A COAST ARE AT RISK FROM TSUNAMI?

(International Tsunami Information Center 1998; National Oceanic and Atmospheric Administration, 1998)

A perusal of the chapters in this book will show that some locations along a coast are more susceptible to tsunami run-up, flooding, and inundation than others. Nine types of topography or coastal settings are particularly prone to tsunami. First and most obvious are exposed ocean beaches. Figure 7.1, which is an artist's impression of the tsunami generated by the eruption of Krakatau in 1883 hitting the coast of Anjer Lor, shows this clearly. If you live by the seaside, you are at risk from tsunami. This fact is clearly recognised for earthquakes by the United States National Oceanic and Atmospheric Administration (NOAA). In its publication "Tsunami! The Great Waves", it states "If you are at the beach or near the ocean and you feel the earth shake, move immediately to higher ground. DO NOT wait for a tsunami warning to be announced". Sometimes a tsunami causes the water near the shore to recede, exposing the ocean floor. Anyone who frequents the ocean should be aware that a rapid withdrawal of water from the shore is overwhelm-

A) Night lights of urban centres

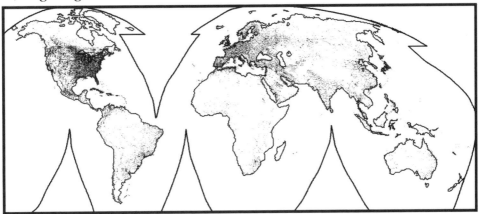

B) Major coastal cities

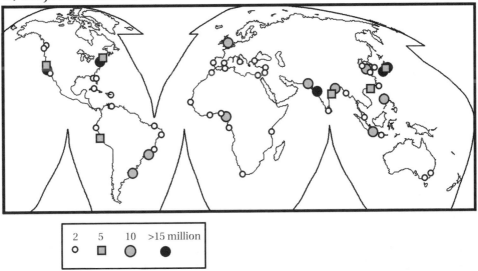

2	5	10	>15 million
○	▢	◯	●

9.3 Indicators of coastline where tsunami will impact the most. **A.** Night lights from major eco-nomically developed urban centres. Data based on satellite measurements using the United States Defense Meteorological Satellite Program (DMSP) Operational Linescan System (OLS). Source: **http://julius.ngdc.noaa.gov:8080/production/html/BIOMASS/night.html.** The darker the shad-ing, the greater is the concentration of people. **B.** Large coastal cities with over 2 million inhabi-tants. Data are current to the year 2000. Source: **http://www.citypopulation.de/World_j.html?E.**

ingly a clear signature of the impending arrival of a tsunami wave crest. The time until arrival may be less than a minute or, in the case of the coast near Concep-ción, Chile, following the Great Chilean Earthquake of 22 May 1960, up to 50 min-utes later.

Second, tsunami travel best across cleared land because frictional dissipation is lowest. This is shown mathematically by Equation 2.24 where the distance of inland penetration is controlled by the value of Manning's n, which is lower for smooth topography such as pastured floodplains, paved urban landscapes dominated by parking lots, and wide roads. The residents of Hilo, Hawaii, were dramatically made aware of this fact following the Alaskan earthquake of 1 April 1946 (Figure 9.4) and the Chilean earthquake of 22 May 1960 (Figure 5.11). On many flat coastlines that have been cleared for agriculture or development, authorities are now planting stands of trees to minimise the landward penetration of tsunami. The effect is clearly shown in Figure 9.5 at Riang–Kroko, on the island of Flores, following the 12 December 1992 Indonesian tsunami. The tsunami bore had sufficient energy to move large coral boulders; but these were deposited once the wave penetrated the forest and rapidly lost its energy through dissipation. If you like to live on the coast and are worried about tsunami, become green. Don't chop down the trees for the view, and be gracious to the neighbours that build in front of you, especially if they have an architecturally designed house with lots of corners and rough textured walls.

Third, tsunami flood across river deltas, especially those that are cleared and where the offshore bathymetry is steep. On these coasts – and they are numerous, for example the East Coast of Japan and the Southeast Coast of Australia – tsunami waves approach shore rapidly and with most of their energy intact. Delta surfaces lying only a few metres above sea level can allow tsunami to penetrate long dis-

9.4 People fleeing the third and highest tsunami wave that flooded the seaside commercial area of Hilo, Hawaii, following the Alaskan earthquake of 1 April 1946. Photograph courtesy of the United States Geological Survey. Source: Catalogue of Disasters #B46D01-352.

9.5 Coral boulders deposited in the forest at Riang–Kroko, Flores, Indonesia following the tsunami of 12 December 1992. Note the person circled for scale and the abrupt termination of debris upslope. Photo Credit: Harry Yeh, University of Washington. Source: NOAA National Geophysical Data Center.

tances inland, because once the wave gets onto the surface it propagates as if it was still travelling across shallow bathymetry. There are records of tsunami in small seas travelling 10 km inland across a delta for this reason.

Fourth, because of their long wavelengths, tsunami become trapped in harbours and undergo resonant amplification along steep harbour foreshores. As pointed out in Chapter 1, *tsunami* is a Japanese word meaning "harbour wave", and when they get into harbours, especially ones where the width of the entrance is small compared to the length of the harbour's foreshores, they become trapped and can't escape back out to sea easily. Inside a harbour or bay, long waves such as tsunami tend to travel back and form for hours dissipating their energy, not across the deeper portions but against the infrastructure built on the shoreline. Rapid changes in sea level and dangerous currents can be generated. Ria coastlines, such as those along the coast of Japan or Southeastern Australia are ideal environments in which these effects can develop. Boats in harbours are particularly vulnerable and should put out to sea and deeper water following any tsunami warning.

Fifth, treat rivers exactly like long harbours. When a tsunami gets into a tidal river or estuary where water depths can still be tens of metres deep, the wave can

travel easily up the river to the tidal limits or beyond. Along some coasts, tide limits may be tens of kilometres upriver, and residents living along the riverbanks
may be totally unaware that a threat from tsunami exists. If the river is deep and
allows the penetration of the wave upstream, the height of a long wave can rapidly
amplify where depth shoals or the river narrows. At these locations, water can spill
over levees and banks, flooding any low-lying topography. In its publication
"Tsunami! The Great Waves", NOAA likewise warns, "Stay away from rivers and
streams that lead to the ocean as you would stay away from the beach and ocean if
there is a tsunami."

Sixth, tsunami have an affinity for headlands that stick out into the ocean,
mainly because wave energy is concentrated here by wave refraction. Storm waves
can increase in amplitude on headlands two- or threefold relative to an adjacent
embayed beach. Tsunami are no different. Seventh, if headlands concentrate
tsunami energy because of refraction, then gullies do the same because of funnelling. The highest run-up measured during the Hokkaido Nansei–Oki tsunami of
12 July 1993 was 31.7 m in a narrow gully. On the adjacent coastline the wave did
not reach more than 10 m above sea level. It is safer to climb as far as you can up a
steep slope rather than flee from a tsunami by running up a gully – even one that
appears sheltered because it is hidden from the ocean.

Eighth, tsunami are not blocked by cliffs. Compared to the long wavelengths of
a tsunami, which can still have a wavelength of 12 km at the base of a cliff dropping 20 m into deep water, the height of a cliff is minuscule. Steep slopes are similar to cliffs. Tsunami waves 1–2 m in height have historically surged up cliffs or
steep slopes to heights of 30 m or more above sea level. If one has any doubt of this
then turn to Figure 3.5 and look at the limit of run-up in the background of the
photograph. This photograph was taken at Riang–Kroko following the 12 December 1992 tsunami. While the wave had a height of only a couple of metres
approaching the coast and was stopped on gentle slopes by forest, it ran up to a
height of 26.2 m above sea level on steeper slopes and bulldozed slopes clear of
vegetation. The view from cliffs is great, but anyone standing there during a
tsunami may have a unique life experience. Never do what 10,000 people did at
San Francisco following the Great Alaskan Earthquake of 1964. When they heard
that a tsunami was coming, they raced down to vantage points on cliffs to watch it
come in. Fortunately, the tsunami was a fizzler along this part of the Californian
Coast. However, it killed eleven people at Crescent City to the north.

Finally, tsunami are enhanced in the lee of circular-shaped islands. Not only do
they travel faster here, the height of their run-up can also be greater, especially if
the initial wave is large. Two examples of this effect were presented in this text in
Chapter 2. The 12 December 1992 tsunami along the North Coast of Flores Island,
Indonesia, devastated two villages in the lee of Babi, a small coastal island lying 5
km offshore of the main island. Wave heights actually increased from 2 to 7 m
around the island. Similarly, the 12 July 1993 tsunami in the Sea of Japan destroyed

the town of Hamatsumae lying on a sheltered part of Okusihir Island. The tsunami ran up 30 m above sea level, – more than three times the elevation recorded at some communities fronting the wave on the more exposed coast. Over 800 people were killed in the first instance and 300 people in the latter. Lee sides of islands are particularly vulnerable to tsunami because long waves wrap around these small obstructions as solitary waves, becoming trapped and increasing in amplitude.

WARNING SYSTEMS

The Pacific Tsunami Warning Centre
(Bryant, 1991; International Tsunami Information Center, 1998)

As shown in Chapter 5, the most devastating oceanwide tsunami of the past two centuries have occurred in the Pacific Ocean. For that reason, tsunami warning is best developed in this region. Surprisingly, a coherent Pacific-wide warning system was only introduced following the Chilean tsunami of 1960. To date that system still has flaws. These flaws will be discussed later. The lead time for warnings in the Pacific is the best of any ocean, anywhere up to 24 hours depending upon the location of sites relative to an earthquake epicentre.

Following the Alaskan tsunami of 1946, the U.S. government established tsunami warning in the Pacific Ocean under the auspices of the Seismic Sea Wave Warning System. In 1948, this system evolved into the Pacific Tsunami Warning Centre (PTWC). Warnings were initially issued for the United States and Hawaiian areas, but following the 1960 Chilean earthquake, the scheme was extended to all countries bordering the Pacific Ocean. Japan up until 1960 had its own warning network, believing at the time that all tsunami affecting Japan originated locally. The 1960 Chilean tsunami proved that any submarine earthquake in the Pacific Ocean region could spread oceanwide. The Pacific Warning System was significantly tested following the Alaskan earthquake of 1964. Within 46 minutes of that earthquake, a Pacific-wide tsunami warning was issued. This earthquake also precipitated the need for an International Tsunami Warning System (ITWS) for the Pacific that was established by the Intergovernmental Oceanographic Commission (IOC) of UNESCO at Ewa Beach, Oahu, Hawaii, in 1965. At the same time, other UNESCO/IOC member countries integrated their existing facilities and communications into the system. The United States National Weather Service currently maintains the Center. As of 1999, twenty-five countries cooperate in the Pacific Tsunami Warning System, in one of the most successful disaster mitigation programs in existence. These countries include Canada, the United States and its dependencies, Mexico, Guatemala, Nicaragua, Colombia, Ecuador, Peru, Chile, Tahiti, Cook Island, Western Samoa, Fiji, New Caledonia, New Zealand, Australia, Indonesia, Philippines, Hong Kong, Peoples Republic of China, Taiwan, Democratic Peoples Republic of Korea, Republic of Korea, Japan, and the Russian Federation. An additional ten countries or dependencies receive PTWC warnings. Many of the primary countries also operate national tsunami warning centres, providing warning services for their local area.

The objective of the International Tsunami Warning System is to detect, locate, and determine the magnitude of potentially tsunamigenic earthquakes occurring in the Pacific Basin or its immediate margins. The warning system relies on the detection of any earthquake 6.5 or greater on the Richter scale registering on one of thirty-one seismographs outside the shadow zones of any P or S waves originating in the Pacific region (Figure 9.6). These stations are operated by the Center itself, the Alaskan Tsunami Warning Center, the United States Geological Survey's National Earthquake Information Center, and various international agencies. Once a suspect earthquake has been detected, information is relayed to Honolulu, where requests for fluctuations in sea level on tide gauges are issued to member countries operating sixty tide gauges scattered throughout the Pacific. These gauges can be polled in real time so that warnings can be distributed to a hundred dissemination points with three hours' advance notice of the arrival of a tsunami. The warnings are distribute to local, state, national, and international centres for any earthquake with a surface magnitude, M_s, of 7 or larger. A watch may also be disseminated by the International Tsunami Information Center (ITIC) at Ewa Beach for potential regional tsunami earthquakes with surface magnitudes, M_s, of less than 7.5. Administrators, in turn, disseminate this information to the public, generally over commercial radio and television channels. The National Oceanic and Atmospheric Administration (NOAA) Weather Radio system provides direct broadcast of tsunami information to the public via VHF transmission. The U.S. Coast Guard also broadcasts urgent marine warnings on medium frequency (MF) and very high frequency (VHF) marine radios. Anyone can also receive these tsunami warnings direct via e-mail by subscribing to the PTWC listserver at **TSUNAMI@ITIC.NOAA.GOV.** Local authorities and emergency managers are responsible for formulating and executing evacuation plans for areas under a tsunami warning. If no tsunami of significance is detected at tide gauges closest to the epicentre, the ITIC issues a cancellation. Once a significant tsunami has been detected, its path is then monitored to obtain information on wave periods and heights. These data are then used to define travel paths using refraction–diffraction diagrams, calculated beforehand for any possible tsunami originating in any part of the Pacific region.

The ITIC also gathers and disseminates general information about tsunami, provides technical advice on the equipment required for an effective warning system, checks existing systems to ensure that they are up to standard, aids the establishment of national warning systems, fosters tsunami research, and conducts postdisaster surveys for the purpose of documentation and understanding of tsunami disasters. As part of its research mandate, the ITIC maintains a complete library of publications and a database related to tsunami. Research also involves

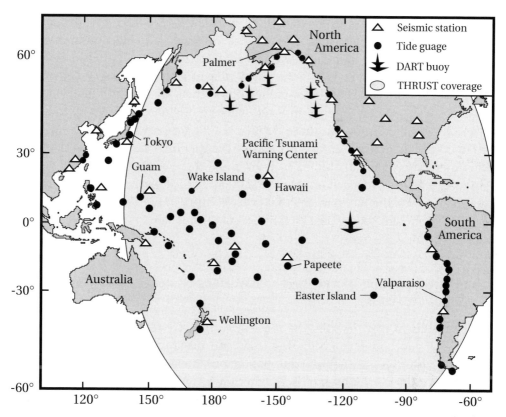

9.6 Location of seismic stations and tide gauges making up the Pacific Tsunami Warning System, buoys forming NOAA's DART network, and area of possible coverage of the THRUST satellite warning system. Sources: Bernard (1991) and González (1999); **http://vishnu.glg.nau.edu/wsspc/tsunami/HI/Waves/waves00.html.**

the construction of mathematical models of tsunami travel times, height information, and extent of expected inundation for any coast. Planners and policy makers use results from these models to assess risk and to establish criteria for evacuation. The ITIC trains scientists of member states who, upon returning to their respective countries, train and educate others on tsunami programs and procedures, thus ensuring the continuity and success of the program. The Center also organises and conducts scientific workshops and educational seminars aimed towards tsunami disaster education and preparedness. In recent years, emphasis has been placed on the preparation of educational materials such as textbooks for children, instruction manuals for teachers, and videos for the lay public. Finally, the ITIC publishes an information and education newsletter on a regular basis. This newsletter is distributed to interested individuals, scientists, and institutions in approximately seventy countries.

The Pacific Warning Tsunami System is being updated to ensure that false alarms are not issued, and that all tsunami are detected. Satellite communications now speed up data collation and warning broadcasts. Other methods of detection are being investigated, including the positioning of sensitive pressure detectors on the ocean bottom adjacent to remote regions such as Alaska and Chile, where the highest-magnitude earthquake-induced tsunami originate. Shorter-period ocean swell will not be detected at these depths, but longer-period tsunami will. Today there is little chance that areas surrounding the Pacific Ocean should suffer loss of life from teleseismic tsunami. As of 1999, about three or four warnings a year are issued for the Pacific Ocean region. In the majority of cases, no tsunami ever eventuates; however, the fact remains that every country in the Pacific Ocean region – even ones not affected by past tsunami events – must always take these warnings seriously.

The West Coast and Alaska Tsunami Warning System
(Sokolowski, 1999b)

The tsunami that followed the Alaskan earthquake of 27 March 1964 were of three types: localised, landslide-induced, and oceanwide. The Pacific Tsunami Warning System is only equipped to handle oceanwide phenomena. Not only did warnings from Honolulu reach Alaska after the arrival of all three types of tsunami, they also went through a process that delayed dissemination to the public along the West Coast of the United States. The West Coast/Alaska Tsunami Warning Center (WC/ATWC) was established in Palmer, Alaska, in 1967 to provide timely and effective tsunami warnings and information for the coastal areas of Alaska. In 1982, the Center's mandate was extended to include the coasts of California, Oregon, Washington, and British Columbia. Finally, in 1996, the Center's responsibility was expanded to include all Pacific-wide tsunamigenic sources that could affect these coasts.

The objectives of the West Coast and Alaska Tsunami Warning Center are to provide immediate warning of earthquakes in the region to government agencies, the media, and the general public, and to accelerate the broadcast of warnings to the wider community along the West Coasts of Alaska, Canada, and the United States. Because tsunamigenic earthquakes can occur anytime, the Center operates continuously twenty-four hours a day throughout the year. To achieve this objective and to reduce labour costs, the Center has been automated with state-of-the-art computers and earthquake-detecting software. Alarms are triggered by any sustained, large earthquake monitored at eight seismometers positioned along the West Coast of North America, and whenever any earthquake in the Pacific basin exceeds a predetermined magnitude. The Center also conducts community preparedness programs to educate the public on how to avoid tsunami if they are caught in the middle of a violent earthquake. Follow-up visits are made to the communities that have experienced a false alarm. The purpose of these visitations is to explain why a warning was issued and to stress the continued need to respond to emergency tsunami warnings.

The West Coast and Alaska Tsunami Warning Center operates a real-time data network of twenty-three short- and long-period seismometers in Alaska linked to the Center using a local digital broadband station. The broadband system allows over eighty dedicated channels of vertical short-period and long-period seismic data to be recorded and processed. Tsunami earthquakes generated by long period seismic waves can automatically be detected in this system. These data can be obtained in real time by the Center's own network or by data acquisition systems maintained by the National Earthquake Information Center, the University of Alaska, the Alaska Volcano Observatory, the Pacific Tsunami Warning Center, the University of Washington, the Incorporated Research Institutions for Seismology, the Geological Survey of Canada, and the United States Geological Survey at Menlo Park Observatory. Tsunamigenic earthquakes can be identified immediately from seismic data using an algorithm that detects P waves. The algorithm then automatically determines the initial magnitude and location of the earthquake using all seismic stations in the network. Once an earthquake's parameters have been determined, a Geographical Information System displays its epicentre on both global and local maps. The global map supports overlays of historical seismicity and tsunami, plate boundaries, major cities, seismometers, tide gauges, modelling results, and tsunami travel time contours. The localised map can also be overlain with the location of urban areas, roads, topography, pipelines, power lines, tide gauges, seismometers, airports, railroads, place names, and past tsunami inundation limits. Tsunami warning messages, based on earthquake location and magnitude, are constructed automatically by the computers. The estimated arrival time of the tsunami at twenty-four sites along the West Coast of North America is also included in the message. Messages are disseminated over the National Warning System, the NOAA satellite system, and a dedicated Federal Aviation Administration Teletype system. Warnings are also issued by e-mail, posted on the Alaska Tsunami Warning Center's web page at **www.alaska.net/~atwc**, and phoned to a number of people locally. Finally, a hard copy of the message is sent to state and provincial emergency services, and to communication systems operated by the U.S. National Weather Service.

Warnings issued by the West Coast and Alaska Tsunami Warning Center are of two types: regional warnings for tsunami produced along the West Coast of North America and ones for tsunami generated outside this area. Regional warnings are issued within 15 minutes of the occurrence of any earthquake having a surface wave magnitude, M_s, greater than 7. Warnings of teleseismic tsunami are issued in coordination with the Pacific Tsunami Warning Center at Ewa Beach, Hawaii. Between 1981 and 1999, tsunami warnings were issued by the Center for eleven regional earthquakes. These warnings were sent within 8–14 minutes of each earthquake. Only two of these earthquakes occurred during normal working hours. Once a warning has been issued, the nearest tide gauges are monitored to confirm the existence of a tsunami, and its degree of severity. Over ninety tide gauges throughout the Pacific Basin are used for this purpose. Approximately 75% of these sites are maintained by NOAA's National Ocean Survey, while the Pacific Tsunami Warning Center, the Japanese Meteorological Agency, and others operate the remainder. In order to provide adequate warning in Alaska, the Center also maintains eight tide gauges in Alaska that can be polled remotely for data via dedicated circuits every 15 seconds. These data are compared to the expected arrival time of any tsunami. Many of the National Ocean Survey's tide gauges have automated triggering software that enables immediate transmission of water level data if a tsunami has been detected. Currently, there is a one- to three-hour delay in the transmission of these tidal data; however, by the year 2000, real-time data will be available. A tsunami warning is cancelled or extended based on this information, historic tsunami records, and precomputed tsunami models. These models have been constructed for earthquakes having moment magnitudes of 7.5, 8.2, and 9.0 along the Pacific Plate boundary from Honshu, Japan, to the Cascadia subduction zone. In the near future, these models and the observed height of tsunami on tide gauges will be used to predict expected tsunami heights along the West Coast of North America following any earthquake.

Flaws in Regional Warning Systems

(Bryant, 1991; Walker, 1995; González, 1999)

The Pacific Warning Tsunami System is not flawless. Not all countries or dependencies in the Pacific Ocean are part of the Pacific Tsunami Warning System. For example, Wallis and Futuna, Kiribati, the Solomon Islands, Tonga, and Vanuatu are not members. The risk still exists in Japan and other island archipelagos along the western rim of the Pacific for local earthquakes to generate tsunami too close to shore to permit advance warning. Almost 99% of the deaths from tsunami in the Pacific Ocean over the past century have occurred in areas where the tsunami reached shore within 30 minutes of being generated. For example, the 7.8 surface magnitude earthquake that struck in the Moro Gulf on the southwest part of the island of Mindanao, the Philippines, on 17 August 1976, generated a 3.0- to 4.5-m-high local tsunami. The event was virtually unpredictable because the earthquake occurred within 20 km of a populated coastline. The International Tsunami Information Center in Honolulu put the death toll at 8,000 people with 90,000 left homeless; however, the actual loss of life was probably between 400 and 500 people with 12,000 homeless. While this tsunami is generally believed to be the last major under-predicted tsunami in recent years, three obscure tsunami killed between 539–700 people on the island of Lembata, Indonesia, in 1979. All of these tsunami are obscure because the events responsible for their generation are unknown. Not only were these latter events poorly documented, they also were totally unpredicted. Then there was the PNG event of 17 July 1978 (Figures 5.20–5.22). Tsunami around the islands of the West Pacific Ocean are underrated.

The accuracy of any warning system does not rely upon the number of tsunami, but upon the number that actually arrive at a coast. False alarms weaken the credibility of any warning system. Although tide gauges can detect tsunami close to shore, they cannot predict run-up heights accurately. Consequently, 75% of tsunami warnings since 1950 have resulted in erroneous alarms. For example, on 7 May 1986, following an earthquake in the Aleutian Islands, and again in 1994 after an earthquake north of Japan, Pacific-wide tsunami warnings were issued for tsunami that never eventuated. Both events cost 30 million dollars in lost salaries and business revenues in Hawaii, where evacuations were ordered. The people who distribute such warnings are only human. Each time a false warning is issued, it weakens their confidence in predicting future tsunami, especially if the tsunami have originated from less well-known source regions. Worse than a false alarm is one that is realistic, but where the time has been underestimated. Tsunami travel charts have been constructed for tsunami originating in many locations around the Pacific Ocean. Many of these charts are inaccurate, with tsunami travelling faster than predicted. Before 1988, about 70% of the Pacific Ocean did not have publicly accessible bathymetry to permit accurate tsunami travel-time forecasting. Fortunately, since the end of the Cold War, these data have become more available.

Improvements have also been made in detecting teleseismic tsunami in the North Pacific. Seabed transducers have been installed and linked to satellites for rapid communication. This networking can also be used to forewarn of local tsunami, overcoming the necessity for long cable connections to shore that the Japanese experimented with unsatisfactorily in the earlier 1980s. NOAA is deploying six of these deep ocean buoys in a project known as Deep-Ocean Assessment and Reporting of Tsunami (DART). These buoys operate under the principle that long-wave motion can be sensed on the deepest sea floor (Equation 2.2). The bottom transducers can detect a tsunami with heights of only 1 cm in water depths of 6,000 m. By the year 2000, five of these buoys will be deployed across the Northeast Pacific Ocean to intercept tsunami originating from the Alaska–Aleutian seismic corridor, and from the West Coast of North America. One buoy will also be positioned on the equator to detect tsunami originating off the South American Coast (Figure 9.6).

Earthquakes do not cause all tsunami. A relatively small earthquake can trigger a submarine landslide that then generates a much bigger tsunami. Nor is the size of an earthquake necessarily a good indicator of the size of the resulting tsunami. The 17 July 1998 tsunami along the Aitape Coast of Papua New Guinea illustrated this fact. The earthquake that generated this event only registered 7.1 on the Richter scale, yet the resulting tsunami at shore was up to 15 m high. As described in Chapter 5, such tsunami earthquakes are common. For example, the 1 April 1946 Alaskan earthquake had a surface magnitude of 7.2 on the Richter scale, but it generated run-ups of 16.7 m as far away as Hawaii (Figure 2.13). On 15 June 1896, an earthquake that was scarcely felt along the coast of Japan generated the Sanriku tsunami that produced run-ups of 38.2 m above sea level and killed 22,000 people.

Finally, our knowledge of tsunami is rudimentary for many countries and regions in the Pacific Ocean. While travel time maps have been drawn up for source regions historically generating tsunami – for example the coasts of South America, Alaska, and the Kamchatka Peninsula – not all of the coastline around the Pacific Rim has been studied. This was made apparent on 25 March 1998 when an earthquake with a surface magnitude, M_s, of 8.8 occurred in the Balleny Islands region of the Antarctic directly south of Tasmania, Australia. Because of the size of the earthquake, a tsunami warning was issued, but no one knew what the consequences would be. The closest tide gauges were located on the South Coast of New Zealand and Australia. Forecasters at the ITIC in Hawaii had to fly by the seat of their pants and wait to see if any of these gauges reported a tsunami before they issued warnings further afield. While that may have helped residents in the United States or Japan, it certainly was little comfort to residents living along coastlines facing the Antarctic in the Antipodes. In cities such as Adelaide, Melbourne, Hobart, and Sydney emergency hazard personnel knew they were the "mine canaries" in the warning system. Fortunately, the Antarctic earthquake was not conducive to tsunami, and no major wave propagated into the Pacific Ocean.

Localised Tsunami Warning Systems
(Bernard, 1991; Okal et al., 1991; Shuto et al., 1991; Reymond et al., 1993; Schindelé et al., 1995; Furumoto et al., 1999; González, 1999)

A tsunami originates in, or near, the area of the earthquake that creates it. It propagates outwards in all directions at a speed that depends upon ocean depth. In the deep ocean, this speed may exceed 600 kms^{-1}. In these circumstances, the need for rapid data handling and communication becomes obvious if warnings are to be issued in sufficient time for local evacuation. Because of the time spent in collecting seismic and tidal data, the warnings issued by the PTWC cannot protect areas against local tsunami in the first hour after generation. For this purpose, regional warning systems such as the Alaskan Tsunami Warning System have been established. Local systems generally have data from a number of seismic and tidal stations telemetered to a central headquarters. Nearby earthquakes have to be detected within 15 minutes or less, and a warning issued soon afterwards to be of any benefit to the nearby population. Because warnings are based solely upon a seismic signature, false warnings are common. At present, warning systems tend to err on the side of caution to the detriment of human life.

Not all of the tsunamigenic source areas in the Pacific Ocean have developed localised warning systems. Besides the Alaska Tsunami Warning Center, a regional centre was established in 1975 for the Hawaiian Islands. Separate warning systems also exist for Russia (formerly the USSR), French Polynesia, Japan, and Chile. The Russian warning system was developed for the Kuril–Kamchatka region of North-eastern Russia following the devastating Kamchatka tsunami of 1952. This system is geared towards the rapid detection of the epicentre of coastal tsunamigenic earthquakes because some tsunami here only take 20–30 minutes to reach shore. In French Polynesia, a system was developed in 1987, for both near- and far-field tsunami, by the Polynesian Tsunami Warning Center at Papeete, Tahiti. The system uses the automated algorithm TREMORS described in Chapter 5. This system calculates the seismic moment, M_o, which is then used to compute the expected tsunami height using the following formula:

$$H_o = 0.3\, M_o\, (\sin \Delta)^{0.5}\, (90\, \Delta^{-1})^{0.5} \qquad\qquad 9.1$$

where H_o = crest-to-trough amplitude of the tsunami in the open ocean
$\quad\quad\ M_o$ = seismic moment (equated to M_m, Eq 5.1)
$\quad\quad\ \Delta$ = distance measured in degrees to the earthquake epicentre

The tsunami wave can be corrected for distance from the epicentre and for inshore bathymetric effects. This simple method gave surprisingly good forecasts of tsunami wave heights for seventeen tsunami that reached Papeete between 1958 and 1986, including the Chilean tsunami of 1960 (Table 9.1). Because the TREMORS system is not site-specific and the underlying equipment is inexpen-

TABLE 9.1   Tsunamic Warning Framework for Tahiti Using Mantle Moments

Mantle Moment (M_m)	Tsunami Risk	Action to Be Taken
<7	Small	No tsunami risk
7-8	Very large tsunami improbable	Tsunami earthquake cannot be ruled out Await further analysis
8–8.7	Tsunami up to 1 m in height	Await further information
8.7–9.3	Substantial risk	If earthquake in the Samoa–Tonga–Kermadec region, immediate Tsunami watch
>9.3	Very destructive tsunami	If earthquake in the Samoa–Tonga–Kermadec region, tsunami alarm immediate tsunami watch issued for other source regions

Source: Based on Okal et al., 1991.

sive, there is no reason why the system could not be installed in any country bordering the Pacific Ocean.

In Japan, a number of systems are used for local tsunami prediction. Tsunami warning began in Japan in 1941 under the auspices of the Japan Meteorological Agency. Originally, coverage was only for the northeast Pacific Ocean Coast, but this was extended nationwide in 1952. The Japan Meteorological Agency has a national office in Tokyo and five regional observatories, each responsible for local tsunami warnings. Following the Chilean tsunami of 1960, communities threatened by tsunami in Japan were identified and protective seawalls built (Figure 9.7). However, large tsunami still require evacuation. Near-field tsunami are a threat in Japan, especially along the Sanriku coastline of Northeastern Honshu, where only 25–30 minutes of lapse time exists between the beginning of an earthquake and the arrival at shore of the resulting tsunami. If it is assumed that most people can be evacuated within 15 minutes of a warning, then tsunamigenic earthquakes here must be detected within the first 10 minutes. The P wave for any local earthquake can be detected within seconds using an extensive network of high frequency and low magnification seismometers. The Japanese Warning System also utilises satellite dissemination in case the ground base network is destroyed in a tsunamigenic earthquake. Algorithms have been written to estimate the seismic moment of an earthquake using as little as 5 minutes of record. Within the next 2 minutes the height of the tsunami along the adjacent coastline can be predicted using graphic forecasting models based upon the size of prior earthquakes and the distance to their epicentres.

9.7 Ryoishi, a typical town along the Sanriku Coast of Japan, protected against tsunami by 4.5-m high walls. These walls are now common around the Japanese Coast; however, they do not offer protect against tsunami having historical run-ups. Source: Fukuchi and Mitsuhashi (1983).

Based upon this forecast a tsunami bulletin is issued as a warning, watch, or "no danger" advisory. Warnings are passed through central government offices, which include the Maritime Safety Agency, which transmits warnings to harbour authorities, fishing fleets, and fishermen, and the Nippon Broadcasting Corporation, which broadcasts warnings nationally on radio and television. At the same time warnings are transmitted to prefectures and to local authorities via L-ADESS, the Local Automatic Data Editing and Switching System. At the local level warnings are then issued via the Simultaneous Announcement Wireless System (SAWS). This system can switch on sirens and bells, and even radios in individual homes. Mobile loudspeakers mounted on fire trucks will also cruise the area broadcasting the warning. In extreme cases, a network of individual contacts has been established and the tsunami warning can be transmitted by word of mouth or over the telephone. Warnings issued within 15 minutes may not be good enough to save lives. Local authorities may hesitate to initiate SAWS and wait for confirmation of a tsunami warning for their particular coast to appear in map form on television. These maps take time to be drawn and do not appear as part of the initial warning. Even where a direct warning is heeded, it may be insufficient. In the Sea of Japan, the lapse time between the beginning of an earthquake and the arrival at shore of the resulting tsunami can be as low as 5 minutes. For example, a tsunami warning was broadcasted directly to the public, via television and radio, within 5 minutes of the Okushiri, Sea of Japan, earthquake of 12 July 1993. However, by then, the tsunami had already reached shore and was taking lives.

All of the warning systems in the Pacific assume that a teleseismic tsunami originates in some underpopulated country far, far away. Chile is one of those far-away countries, but with a significant coastal population. As shown in Chapters 4 and 5, tsunamigenic earthquakes here have tended to occur every fifty years with a deadly impact. Chile does not have the privilege of being able to rely upon the Pacific Tsunami Warning System, because what is a distant earthquake to the PTWS can well be a localised earthquake in Chile. Project THRUST (Tsunami Hazards Reduction Utilizing Systems Technology) was established offshore from Valparaiso, Chile, in 1986 to provide advance warning of locally generated tsunami along this coastline within 2 minutes. When a sensor placed on the seabed detects a seismic wave above a certain threshold, it transmits a signal to the GEOS geostationary satellite, which then relays a message to ground stations. The signal is processed, and another signal is transmitted via the satellite to a low-cost receiver and antenna, operating twenty-four hours a day, located along a threatened coastline. This designated station can be preprogrammed to activate lights and acoustic alarms, and to dial telephones and other emergency response apparatus when it receives a signal. The GOES satellite also alerts tide gauges near the earthquake to begin sending data, via satellite, both to local authorities and to the Pacific Tsunami Warning Center to confirm the presence of a tsunami. For a cost of

$15,000, a life-saving tsunami warning can be issued to a remote location within 2 minutes of a tsunamigenic earthquake. The warning system is independent of any infrastructure that could be destroyed during the earthquake. In August 1989, the THRUST system was integrated into the Chilean Tsunami Warning System with a response time of 17–88 seconds. The system provides coverage of all but the southernmost tip of the South American continent (Figure 9.6). Response times of 5–10 seconds are now technically possible. Potentially a GEOS Satellite warning system could be installed for any coastline in the Pacific Ocean except Eastern Asia.

HOW LONG HAVE YOU GOT?

Distant or teleseismic tsunami in the Pacific Ocean leave a signature that provides sufficient lead time for dissemination of a warning and evacuation. For example, the Hawaiian Islands will get more than six hours' warning of any tsunami generated around the Pacific Rim, while the West Coast of United States receives more than four hours' notice of tsunami originating from either Alaska or Chile. The real concern is the potential warning time, or margin of safety, if a tsunami originates near the edge of the continental shelf, off the Hawaiian Islands or the continental shelf of Washington State. There are two possible scenarios for locally generated tsunami. In the first scenario, a tsunamigenic earthquake is responsible for the tsunami. The earthquake can occur at the shelf break or in deeper water offshore. In either case, once the wave begins to cross the continental shelf, the depth of water determines its velocity. Hence the slope and width of the shelf dictate the tsunami's travel time. In the second scenario, the earthquake generates a submarine landslide on the shelf slope. In this case, the longer it takes for a submarine slide to develop, the further it has moved from shore, and the longer it takes for the resulting tsunami to propagate to the coast.

A crude approximation of the time it takes tsunami spawned by these processes to cross a shelf can be determined by dividing the shelf into segments, and calculating the time it takes the wave to pass through each segment using Equation 2.2. The calculations are simplified if the shelf is assumed to have a linear slope. These results are presented in Figure 9.8 for different shelf slopes and widths. These relationships should be treated cautiously because they are based upon simplified assumptions. For example, the Grand Banks earthquake of 18 November 1929 occurred at the edge of the continental shelf, 300 km south of the Burin Peninsula of Newfoundland that was eventually struck by the resulting tsunami. This tsunami arrived two and a half hours after the earthquake – well within the four hours indicated in Figure 9.8. Tsunami induced by submarine slides may travel as fast as 1,500 km hr^{-1} – much faster than linear theory would suggest. If anything, the margin of safety shown in Figure 9.8 is too lenient.

Figure 9.8 shows that there is a log-linear relationship between travel time and the distance to the shelf break. This relationship holds for shelf widths as narrow

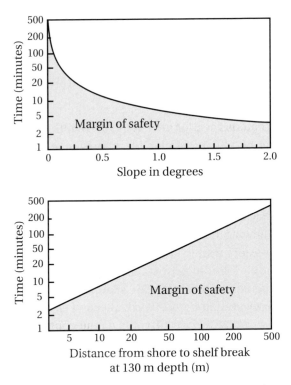

9.8 Travel time for tsunami moving across a continental shelf. **A.** For various slopes, **B.** for various shelf widths.

as 2 km and as wide as 500 km. The figure also indicates that the travel time for a tsunami asymptotically approaches 3.25 minutes for the steepest shelf slopes. These relationships can be put into a more familiar context using two examples, both of which have already been discussed in this text. In the first example – that of the East Coast of the United States – the shelf break lies more than 165 km from shore. Here, a tsunami generated at the edge of the shelf would take over 135 minutes or 2.2 hour to reach the closest point at shore. This does not seem like much when compared to the time that residents along the West Coast of the United States have for tsunami generated in Alaska or Chile, but it is more than sufficient when compared to the second example, that off Sydney, Australia, where the shelf is steep, being only 12–14 km wide. Unlike the East Coast of the United States, substantial evidence has been found along this coast for the impact of mega-tsunami. Here, a tsunami generated on the continental slope would take only 10–12 minutes to reach shore. Within this time, one would be hard-pressed to reach safety if sunbathing on a local beach, or even worse, surfing off one of the headlands. At Wollongong, south of Sydney, the seabed also shows geological evidence for a submarine landslide measuring 20 km long and ten kilometres wide positioned 50 km offshore. A tsunami generated by this slide would only take 40 minutes to reach shore.

WHERE SHOULD YOU GO IF THERE IS A TSUNAMI WARNING?
(Wiegel, 1970; Shuto, 1993)

While it may seem obvious from the previous sections, there is more to this question than meets the eye. Obviously one shouldn't rush to cliffs, take to boats inside harbours, or decide it is a good time to have lunch at your favourite quayside cafe. Don't do what the residents of San Francisco did during the Alaskan tsunami of 1964 and flock to the coast to see such a rare event. And don't do what the residents of Hilo, Hawaii, did during the Alaskan tsunami event of 1 April 1946 (Figure 9.4), and hurry back to the coast following the arrival of the first few tsunami waves. Here, people returned to the coastal business area to see what damage had occurred, only to be swamped by the third and biggest wave. Big waves later in a wave train are more common than generally believed. For example, the eighth wave during the 1 April event was the biggest along the north shore of Oahu. Figure 2.3 also shows that the biggest wave can occur after the initial one. Under extreme conditions on coastlines where tsunami are recurrent, for example the Sanriku Coast of Northeastern Honshu Island, Japan, the government has gone to extreme lengths to build seawalls behind beaches to protect towns against tsunami. Figure 9.7 shows the walls protecting the town of Ryoishi against a tsunami 4.5 m high. Similar walls have been constructed in and around Tokyo and other metropolitan areas in Japan. Such walls offer a false sense of security. The southernmost part of Aonae, Okushiri Island, was completely surrounded by such a wall but was destroyed by the Hokkaido Nansei–Oki tsunami of 12 July 1993, which had a run-up height of 7–10 m. Events of this magnitude are common in Japan. The worse scenario for the town of Ryoishi shown in Figure 9.7 would be for the tsunami to overtop the seawalls. In this case, residents would be trapped against the barrier by the backwash.

Most people can escape to safety with as little as 10 minutes' warning of a tsunami. Along the northern coastline of Papua New Guinea where the July 1998 tsunami had such an impact, people have been encouraged to adopt a tree. In Chapter 1, people who were stranded on the Sissano barrier with nowhere to flee did have an option. As shown in Figure 5.22, a substantial number of trees withstood the impact of the tsunami even though it was 15 m high and moved at a velocity of 10–15 m s^{-1}. Notches can be cut into trees as toeholds, and people can easily climb a tree and lash themselves to the trunk in a matter of minutes. Urban dwellers may not have the opportunity to be as resourceful because of the lack of trees (Figure 9.1). It is an interesting exercise to stand with a group of people on an urban beach and say, "Where would you go if an earthquake just occurred and a tsunami will arrive in ten minutes?" Most people soon realise that they should run to the nearest hill, preferably to the sides of the beach and away from the coast. However, in a suburb such as that shown in Figure 9.1, this option may be neither obvious nor feasible. The only choice may be to seek safety in buildings. Personally I would look for the closest and tallest concrete building, preferably an office building (apartment buildings have secured access), run to the lobby, push the elevator button, and go to the top floor. Hopefully, the tsunami would not repeat the scene of the Scotch Cap lighthouse, which the 1 April 1946 tsunami wrecked (Figures 2.1 and 2.12).

Research has investigated the ability of buildings to withstand the force of a tsunami. Damage to structures by tsunami results from five effects. First, water pressure exerts a buoyant or lift force wherever water partially or totally submerges an object. This force tends to lift objects off their foundations. It is also responsible for entraining individual boulders. Second, the initial impact of the wave carries objects forward. The impact forces can be aided by debris entrained in the flow or, in temperate latitudes, by floating ice. For these reasons, litter often defines the swash limit of tsunami waves. Third, surging at the leading edge of a wave can exert a rapidly increasing force that can dislodge any object initially resisting movement. Fourth, if the object still resists movement, then drag forces can be generated by high velocities around the edge of the object, leading to scouring. Finally, hydrostatic forces are produced on partially submerged objects. These forces can crush buildings and collapse walls. All of these forces are enhanced by backwash that tends to channelise water, moving it faster seaward.

Various building types and their ability to withstand tsunami are summarised in Figure 9.9. The data come from the 1883 Krakatau, 1908 Messina, 1933 Sanriku, 1946 Alaskan, and 1960 Chilean tsunami. Lines on this figure separate undamaged, damaged, and destroyed buildings. Wood buildings offer no refuge from tsunami. Fast-moving water greater than 1 m in depth will destroy any such structures unless they are perched on cross-linked iron struts sunk into the ground. Stone, brick, or concrete block buildings will withstand flow depths of 1–2 m but are destroyed by greater flows. The Nicaraguan tsunami of 2 September 1992 destroyed all such buildings wherever the wave ran up more than 2 m (Figure 5.16). Even concrete pads that require significant force to be moved can be swept away by such flows. The NOAA (1998) in its publication "Tsunami! The Great Waves" states, "Homes and small buildings located in low-lying coastal areas are not designed to withstand tsunami impacts. Do not stay in these structures should there be a tsunami warning". Reinforced concrete buildings will withstand flow depths of up to 5 m. Such depths have only occurred during the severest tsunami, and then only along isolated sections of coastline. If there is no escape, the safest option is to shelter in a reinforced concrete building, preferably in the first instance above the ground floor level. The NOAA publication also states, "High, multi-story, reinforced concrete hotels are located in many low-lying coastal areas. The upper floors of these hotels can

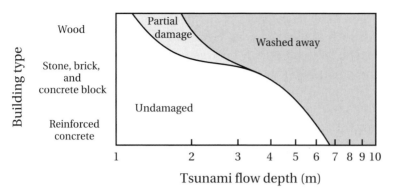

9.9   The degree of damage for different housing types produced by varying tsunami flow depths (based on Shuto, 1993).

provide a safe place to find refuge should there be a tsunami warning and you cannot move quickly inland to higher ground".

WHAT IF IT IS AN ASTEROID OR COMET?
(Verschuur, 1996; Ward and Asphaug, 2000)

Despite the image conveyed by recent disaster movies such as *Deep Impact* and *Armageddon,* the risk from large asteroids or comets is minor. The main perceived threat comes from objects 1–10 km in size that, before the 1990s, had escaped detection. As of 1998, over 600 Near Earth Asteroids (NEAs) have been discovered, consisting dominantly of Apollo objects. The largest of these objects is 8 km in diameter. By 1999, scientists estimate that they had detected only 10% of the meteors intersecting the Earth's orbit. However, NASA has since instituted a dedicated program called Spaceguard to detect 90% of objects greater than 1 km in size by the year 2010. Between December 1995 and August 1998 the Near Earth Tracking Program (NEAT) at the Jet Propulsion Laboratory has detected forty-nine NEAs, of which half had a diameter greater than 1 km. This threat from large asteroids may be illusionary when compared to that posed by objects between 200 and 1,000 m in size. Any of these objects can still generate devastating basinwide tsunami. Statistics on the frequency of these smaller Near Earth Objects (NEDs) are summarised in Table 9.2. There may be up to 2,000 NEOs greater than 1 km in diameter, 10,000 greater than 0.5 km in diameter, 300,000 greater than 100 m in diameter, and 150 million greater than 10 m in size. The error bar on these numbers is approximately 50%. Unless any of the larger objects are detected well in advance, the present threat is still basically random and unpreventable. As of the year 2000, no feasible program had been developed to mitigate the threat from cosmogenic tsunami.

The randomness of meteorites impacting into the world's oceans can be used to calculate the risk of tsunami for any coastline. The probabilities of any point on

TABLE 9.2 Estimated Number of Near Earth Objects (NEOs) by Size, Return Interval, and Probability of Discovery by A.D. 2050

Diameter (m)	Possible Number of Objects Greater in Size	Return Interval in Years	Probability of Object Impacting in Next 50 years (percentage)	Percentage of Objects Discovered by A.D. 2010	Chance of a Known Object Impacting by A.D. 2050 (percentage)
10	150,000,000	2	100.00	<0.01	<0.010
50	3,000,000	100	50.00	0.50	0.25
100	500,000	1,000	5.00	2.00	0.10
200	100,000	5,000	1.00	20.00	0.20
500	10,000	40,000	0.12	50.00	0.06
1,000	2,000	100,000	0.05	90.00	0.05

Source: Based on Figure 8.2 and the Web page of Michael Paine at http://www1.tpgi.com.au.users/tpsseti/Impact.jpg.

Earth being hit directly by stony meteorites with a radii of 2 m, 5 m, 10 m, 15 m, and 25 m within a millennium timespan are 0.049%, 0.00538%, 0.00249%, 0.00159%, and 0.0009% respectively. These probabilities can be integrated at any point along a coastline. For example, any coastal city has a 7.1%, 2.9%, and 0.29% probability of at least one cosmogenic tsunami greater than 2 m, 5 m, and 25 m respectively in height within the next thousand years. This calculation is based upon the middle range of modelled tsunami wave heights and includes atmospheric ablation effects. These probabilities can also be related to the size of the meteorite. Meteorites less than 1 km in diameter will generate over 90% of all cosmically induced tsunami greater than 10 m in height. Larger meteorites contribute little to the cosmogenic tsunami hazard. This result indicates that the present Spaceguard program of asteroid detection is being directed towards the wrong size of asteroids. The program should be concentrating upon much smaller objects. It also indicates that small meteorites, and hence catastrophic tsunami, can be very common along coastlines. This fact conceivably accounts for the evidence of repetitive mega-tsunami along the South Coast of New South Wales over the past 7,000 years. Other stable coastlines should show similar evidence.

Finally, it is possible to calculate the probability of variously sized cosmogenic tsunami for any coastal location. The results for six coastal cities – San Francisco, New York, Tokyo, Hilo, Perth, and Sydney – are presented in Figure 9.10 for stony meteorites. This figure takes into account the effects of atmospheric ablation. San Francisco has the greatest exposure, facing 1.7×10^8 km^2 of ocean that plausibly could be struck by a meteorite impact. New York has the least exposure, facing only 0.64×10^8 km^2 of ocean. However, Hilo, Hawaii, has the highest probability of being struck by a cosmogenic tsunami. Here, the probabilities are 15.3% and 5.8% for a

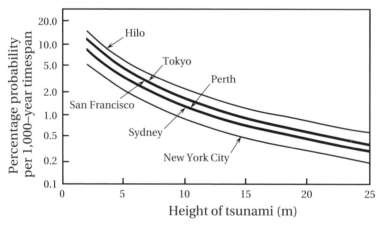

9.10 Probability of meteoritic tsunami of various heights striking a selection of world cities (based on Ward and Asphaug, 2000). Values are for stony meteorites having a density of 3 g cm^{-3}. The results take into account the effects of atmospheric ablation upon the meteorites.

wave of 2 and 5 m in height respectively occurring within the next millennium. At San Francisco, the probabilities for similarly sized waves are 12.0% and 4.1% respectively. Sydney, which is situated on a 400-km stretch of coastline displaying the best evidence for catastrophic tsunami yet identified, has probabilities of 8.8% and 3.2% respectively for these two wave heights. Interestingly, the probabilities for meteorite impacts in the ocean are higher adjacent to Perth than to Sydney. Our recent fieldwork suggests that catastrophic tsunami have been as frequent along the Southwest Coast of Western Australia as along that of New South Wales (Figure 8.13). Note that, of the six cities, Sydney ranks fifth in terms of the risk from cosmogenic tsunami. However, to date it has the best-defined regional evidence of catastrophic tsunami. The challenge is now to find similar evidence for other coastlines and to unravel the chronology associated with it.

IS IT ALL THAT BAD? THE CASE OF SYDNEY

It is possible to assess the risk of tsunami to human life in urban areas. In the case of Sydney, which has a population of over 4 million people, very few people would even witness a large tsunami event occurring, let alone become a casualty of one. The evidence for mega-tsunami along the coast of New South Wales shown in Chapters 3 and 4 extends along the Sydney coastline. Detailed maps of two sections along this coast are shown in Figure 9.11. Boulders similar in size and imbrication to those at Gum Getters Inlet (Figure 8.11) are piled along the cliffs north of Little Bay. Many have been moved with ease in suspension onto ledges and clifftops (Figure 9.12). Large imbricated boulders trail around the cliffs into Sydney Harbour at South Head. Cannae Point on the north side of the harbour's entrance has all the appearances of a toothbrush-shaped headland. Gravelly sands were dumped in a sandsheet downslope from The Gap towards the harbour after being swept over cliffs 22 m high. The first impression is one of devastating damage were such an event to recur. However, this is not necessarily the case.

In each of the detailed maps, the extent of residential buildings is also shown. In the first case, that of Little Bay where boulders were transported by mega-tsunami and stacked against cliffs 20 m high (Figure 9.12), remarkably little damage to buildings would occur because houses have been set back from a coast that is fringed by golf courses and a military reserve. Certainly, the residents of houses at the head of the gully draining into Little Bay would get an impressive view of the tsunami racing towards them up the coast. However, the hinterland behind this bay is relatively sheltered from the coast. The mega-tsunami event that occurred around A.D. 1500, while overtopping some of the cliffs, stopped just short of the houses, which lie 30 m above sea-level and 600 m from the shore. Farther north at The Gap, where the evidence for mega-tsunami is just as impressive, the impact would be similar. Very little development exists on the headlands bracketing the entrance to Sydney Harbour. The mega-tsunami of A.D. 1500 does not appear to have overtopped the

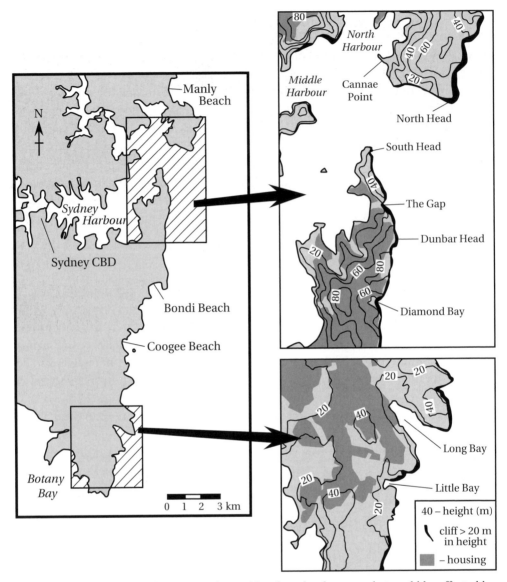

9.11 A sampling of coastal areas in Sydney with urban development that could be affected by large tsunami. **A.** The Gap; **B.** Little Bay.

cliffs at Dunbar Head, which are 70 m high, because the wave was travelling north-wards at an angle to the coast. The tsunami, however, did overtop the cliffs on North Head, which lay directly in the path of the wave. These latter clifftops are uninhabited. The wave would also have funnelled up the gullies leading inland from embayments such as Long Bay, Cogee Beach, Bondi Beach, and Diamond Bay. These gullies are densely urbanised with single- and multistoreyed dwellings.

9.12 A boulder transported by tsunami at the front of cliffs at Little Bay, Sydney. The boulder lies 7–8 m above sea level and 50 m from the ocean. Each corner of the boulder rests on a smaller boulder or on bedrock. The contacts are clean, without any evidence of the fracturing or crushing that should have occurred if the boulder had been tossed up onto the ledge. Instead, the boulder was transported in suspension and gently settled from turbid flow. Note the imbricated boulders at the base of the cliff in the background.

However, no more than 10,000 people would be threatened, and most of these could be evacuated to safety with sufficient warning. If a submarine landslide on the adjacent continental shelf caused the tsunami, the lead time could be as little as 15 minutes with a resulting high death toll. The warning time would be insufficient. However, if an earthquake or a meteorite impact in the south Tasman Sea caused the tsunami, people would have several hours warning to evacuate threatened areas. The latter scenario of course assumes that the threatened areas have been identified, and that the State Emergency Service has drawn up adequate evacuation plans.

CHAPTER TEN

Epilogue

This book began with five stories that were based upon legends and historical records of tsunami. These stories formed the basis of subsequent description and discussion about tsunami in the text. It is perhaps appropriate to end this book with five stories that foreshadow or prophesy about the nature of tsunami events in the near future. Each of these stories centres upon one of the underrated aspects about tsunami dealing either with their mechanism of formation or location.

FIVE STORIES

1 An Unsuspected Earthquake

Charleston – dances, southern nights, earthquakes – is home to the event that everyone ignores but that crops up in spreadsheets and Internet databases as one of the worst earthquakes in U.S. history. The city lies in a broad embayment along the Southeast Coast (Figure 10.1). The earthquake of 31 August 1886 was considered a rarity, a whim, an isolated stochastic perturbation of the restless Earth, but never a harbinger of things to come. In 1886, the earthquake began as a barely perceptible tremor at 9:51 P.M. and grew to a sound like a heavy metal body rolling over gravel and then into a roar. It was felt over an area of 5 million km², as far as New York City, Milwaukee in the midwest, Havana, and Bermuda. Seven additional shocks followed during the next twenty-four hours. The earthquake's recurrence interval was evaluated as once in a millennium, maybe twice. The warning signs of a similar event in the future were always present – 1903, 1907, 1912, 1914, 1924, 1945, 1952, 1959, 1960, 1964, and 1967. The epicentre for the 12 March 1960 event was off the coast of South Carolina. Further tremors occurred throughout the 1970s and then nothing – not until that July 4th weekend in the early part of the twentieth-first century.

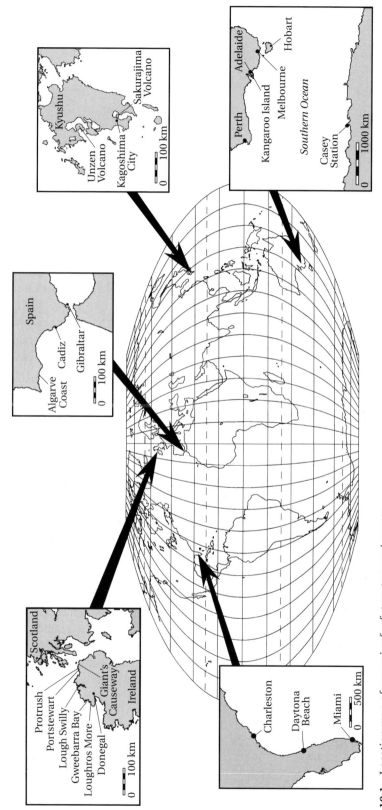

10.1 Location map of scenarios for future tsunami events.

It began as a low rumble around 10:00 A.M. Some describe it as a locomotive coming in from the ocean. The ground shook so violently that five- to nine-storey buildings immediately collapsed. The shaking went on forever – 10 seconds, 30 seconds, a minute, two. Puffs of sand issued from hundreds of sand fountains on the beaches along the coastline before they slowly sank into the Atlantic Ocean. Some of the lagoonal sediments also lost their bearing strength, and houses and apartments built during the housing boom of the late twentieth century sank at weird angles into the substratum. It was a catastrophic earthquake, and within 30 minutes the region was affected by many aftershocks registering over 7.0 on the Richter scale. Silently, 200 km offshore, turbidity currents slithered off the edge of the continental shelf and into the abyss. Hundreds merged into tens and then into one stream of immense proportions. Anyone who had been diving at these depths would have noticed the strong undertow and, more important, the slow oscillatory rotation that was taking place throughout the water column. A tsunami was being generated, fed by a myriad of streams of coalescing debris on the seabed and propagating shoreward.

Ninety minutes later the water withdrew from over 200 km of coastline, from Brunswick, Georgia, in the south to Wilmington, North Carolina, in the north. Water flowed out of estuaries, leaving peats exposed above sea level for the first time in thousands of years. Boats drifted on their anchors; some broke free and surged into the Atlantic through tidal inlets. Then the sea came back. It was preceded by a change in wind ever so slight. A hump on the horizon grew into a wave, and then a wall began to form closer to shore. The sound became deafening, like hundreds of jet planes landing at once along the whole coast. The drifting boats became ballistic missiles, dragged upright on their anchors and then shot into the air like champagne corks at the millennium celebrations as they broke free. The wave was 15 m high when it reached the coastline directly off Charleston, 6 m high opposite Savannah and 8 m high northwards along the barrier islands leading to Cape Hatteras. Not since the Great Lisbon Earthquake of 1 November 1755 had the Atlantic Ocean experienced such a big wave. The wave raced over barriers, through tidal inlets, and 40 km up estuaries. When it appeared that the full force of the wave was expended, a second and then a third wave roared into the coast, obliterating all surviving signs of human occupation. Within half an hour the waves spread outwards from Charleston and began to rise in amplitude along the Florida Coast. Because the coastline southward swept into the Atlantic, the wave grew in intensity and amplitude towards Daytona Beach, Cape Canaveral, West Palm Beach, Fort Lauderdale, and Miami. The beach ridges on promontories along the coast were eroded and then laid down again as bedforms under the swamping waves. Houseboats, yachts, tourists sunbathing on beaches all were inundated without any warning. Ten minutes after the first wave had obliterated the coast, a single curt message was issued by the National Weather Service,

This is a tsunami information message, no action required . . .
An earthquake, preliminary magnitude 8.2, occurred at 1458 UTC
4 Jul 2011, located near latitude 32 N longitude 79 W in the vicinity
of Charleston S Carolina.

Evaluation: no destructive Atlantic-wide tsunami threat exists.

This will be the only bulletin issued unless additional information
becomes available.

2 An Unassuming Earthquake

Cádiz on the Southwest Coast of Spain is the last place in the world that
should be struck by an earthquake (Figure 10.1). Actually the earthquake wasn't
centred here but offshore along the extension of the plate boundary that had
given rise to the Great Lisbon Earthquake of 1 November 1755. The earthquake
wasn't big. It only had a surface magnitude of 7.0. Hardly anyone in the city felt it,
which was unusual because it was siesta time and if an earthquake was going to
be noticed at all, it would be noticed while people were resting. Some of the fishermen were suspicious. For the past two weeks they had seen dead fish floating
offshore, and one had even reported seeing the ocean bubbling around his boat.
It wasn't a good day at all. It was grey and drizzly, and the horizon was bumpy
from the heavy swell running along the coast after the storm of the past two days.
Even that storm was unusual. It probably had something to do with greenhouse
warming.

One or two of the fishermen wandered down to the breakwall and casually
scanned the ocean as they talked about their run of bad luck. The bumps had
moved. They were closer to shore now and appeared to be growing in height.
They were! Within 30 seconds a coherent wall of water formed and increased
to a height of 10 m before it slammed into the coastline. There was no time
for the fishermen to flee. They were picked up by the wave and swept into
the harbour. The wave washed across the adjacent beach, splashed against the
15-storey hotels and apartments that lined the backshore, and squeezed between
the buildings and into the streets behind. North of the beach, it ran into the
bay and along the harbour foreshores. Docks were swamped, and boats were
picked up and ripped from their moorings or sunk on the spot. The wave
surged across the bay and up the Guadelete River. Within 10 minutes, another
wave struck the coast and finally a third came ashore. It was all over in 30
minutes. One of the most picturesque cities on the Spanish West Coast had just
experience a tsunami earthquake, which supposedly only occurs in the Pacific
Ocean.

3   A Submarine Landslide

The signs were ominous. There were the small earthquakes with surface magnitudes registering 3–4 on the Richter scale. They had increased in frequency to
the extent that the Norwegian government instituted tsunami evacuation drills in
the major cities along the coast – at Bergen, Stavenger, and more than a dozen
smaller communities. The Storegga slides had occurred here more than 7,000
years ago, but the cause of the slides and their resulting tsunami were still a heated
point of debate. The devolved government in Scotland took particular note of the
events because the most widespread evidence of the tsunami from the third
Storegga slide existed along its East Coast. The debate went no further because
geological events were not political ones and Scotland didn't count any more.
Everyone was wrong. What wasn't noticed were the smaller earthquakes along the
continental shelf edge off the coast of Ireland. While the Storegga Coast off Norway
had been the source for three major submarine slides, eleven others had occurred
over the same time span along the step continental slope off the coast of Northwestern Europe. At least seven of these had occurred along the coast of Ireland
within a few hundred kilometres of the coast.

At 4:58 A.M. on that Sunday morning in April, the shelf slope finally succumbed
to the enormous pressures that had been building up over the last 5,000 years of
higher sea levels during the Holocene. The triggering earthquake was minor, and
because it was a Sunday, the whole event went relatively unnoticed. Surveys afterwards found it difficult finding anyone who had felt the earthquake; however,
everyone had stories to tell of the consequences. Slowly the tsunami from the slide
built up, and within two hours the first communities along the West Coast of
County Donegal were witnessing its effects. Headlands along 350 km of rugged
coastline were swamped, while flat pocket beaches in sheltered embayments were
totally eroded. The wave was amplified by funnelling in embayments such as
Loughros More and Gweebarra Bays, and into Lough Swilly running down to Let-
terkenny (Figure 10.1). Here the wave reared from 8 m along the open coast to over
15 m inside embayments. The tsunami had its most dramatic effect to the east.
Where the shelf shallowed and the coastlines of Ireland and Scotland came closer
together, the wave not only maintained its height but also underwent amplification. At the Giant's Causeway on the northern tip of Ulster, it broke again over the
knob of basalt columns and deepened the canyon that formed the toothbrush-
shaped headland. Bedrock sculpturing effects were unmistakable along the whole
of the coast of Northern Ireland.

Unfortunately, the death toll was high. Unlike the Grand Banks tsunami of
November 1929 – which struck a similarly shaped but sparsely inhabited coastline –
this tsunami hit a more densely populated coastline. University students at Col-
eraine living in the coastal communities of Portstewart and Protrush succumbed to
the waves. The countryside around the towns of Gweedora and Donegal was partic-

ularly hard hit. The number of dead will never be known because rural marginalisation along one of the most isolated coastlines in Western Europe ensured that many victims had no community contacts and hence went missing without being noticed. Experts afterwards stated that the whole event was an abnormality. Meanwhile the offshore slopes continued to build up pressure.

4 A Volcanic Eruption

Sakurajima Volcano on the Island of Kyushu, Japan is not a well-known one (Figure 10.1). It was overshadowed further north by Unzen on the Shimabara Peninsula, which during its eruption on 21 May 1792, caused a tsunami with a maximum run-up of 55 m. That wave killed over 14,000 people. Sakurajima had never had eruptions like this; however, people lived closer to it. About 7,000 people lived at the foot of the volcano and half a million people lived in Kagoshima City 10 km to the west. The oldest documented eruption had taken place in A.D. 764. Since then, five major eruptions had occurred – each generating pyroclastic ash and lava flows that had reached the ocean. On 9 September 1780, one of the ash flows had produced a 6-m-high tsunami. Since 1955, Sakurajima had burst into life. That would not have been so bad, but the location of the volcano was dangerous. So was its height. Unlike many other explosive Japanese volcanoes, Sakurajima was high, rising to over 1,000 m above sea level. It also protruded like a hernia into Kagoshima-wan Bay on the southern end of Kyushu Island. It was a disaster waiting to happen.

The eruptions since 1955 just seemed to go on and on. They were cyclic and intense, and should have warned authorities that all was not well. Towards autumn in the early part of the twenty-first century the seismic tremors and eruptions became more frequent. The local officials even thought about advertising the eruptions, because unlike others, these could be viewed from the relative safety of Kagoshima across the bay. On that sunny morning, with the latest eruption sending ash high into the sky towards the east, the unthinkable happened. One last major earthquake shook the region and the oversteepened slope of the volcano collapsed to the west. Before the slope could disintegrate into the ocean, the volcano blasted through its flank in the largest basal surge since Mt. St. Helens in May 1981. The severity of the situation became immediately apparent to all who were watching the eruption from the city.

The authorities and subsequent investigations could never define the main cause of death for the 20,000 people that died that day. To allay fears in other communities – such as those around Unzen Volcano to the north – the experts said that the wave was not a tsunami, that the water came from the volcano, that the death toll was due to the lateral pyroclastic flow, and that the blast was a freak of nature – never recorded before in the history of Japanese eruptions. Certainly part of the basal surge had spread across the ocean's surface and swept through the city. The melting of glass and metal in the path of the blast confirmed that. However, a few witnesses implicated a tsunami. The bank manager, who saw the eruption and then ducked into the vault and shut it, swore that the ash cloud had sunk below the ocean. He had paused in his retreat because he saw the ocean heaving erratically like a cat crawling under a carpet. Then further down the bay there were the fishermen who actually saw the floor of the bay exposed in the 10-km gap between the base of the volcano and the city. One of them, before cutting the anchor of his boat and deciding to ride the wave out, said that the whole bay splashed in the air over the city. Certainly, there was plenty of evidence of water swamping the city. A camera attached to the Internet, and updated every minute, even showed a fussy picture of a 50-m-high wave in the last of its frames. While many said that the debris deposited in the city originated from the volcano, the presence of rounded boulders, marine mud, and shell left little doubt that the seabed had been swept clean. The final details of the disaster may never be known, but over subsequent years people slowly moved away from the foreshores of Kagoshima-wan Bay and the waters surrounding Unzen volcano further to the north. The seas were perceived as being too dangerous.

5   A Meteorite Impact with the Ocean

It hurtled around the Sun as it had thousands of times before, spinning, dark, ominous, 65 million tonnes of stony conglomerate formed in the birth of the solar system 5 billion years ago. As it swung from behind the Sun it was silhouetted against a distant pale blue speck, the planet Earth, with which it would rendezvous ten weeks later. Closing in on the Earth, it should have missed, but this time deviated ever so slightly from its path because of the gravitational attraction of the Earth and its moon. The meteorite's fate was sealed. It spun through the Earth's atmosphere at 25 km s^{-1} on a low southeast trajectory. It began to heat up, and just before striking the ocean, it fragmented covering an area four times larger than its original 250-m diameter. Along the South Coast of Australia, late on a clear, warm summer's day, a few residents who happened to look south noticed a dull glow hanging over the horizon. Some even said that they could read by the light as night fell. Within seconds, the Australian Antarctic Division in Hobart lost contact with Casey Station in the Antarctic (Figure 10.1). Such blackouts were common, but this one was permanent. The crew on the supply ship standing off Casey never knew what hit them. It was all over within 10 seconds as the meteorite struck the ocean less 100 km away in a blast equivalent to more than 3,000 megatons of TNT. In that interval, billions of tonnes of water were thrown at the speed of sound into the atmosphere and vapourised. The vapour – heated to 5,000° C – struck the ship and instantly incinerated it.

Within 10 minutes a tsunami had propagated away from the centre of impact and was approaching the ice cap. When it reached shore the wave was almost

100 m high. It sloshed over the ice and then ran back into the ocean together with millions of tonnes of melted ice and water that had condensed out of the atmosphere. After five hours the lead wave from the tsunami generated by the impact had crossed the Southern Ocean and was approaching the first tide gauge of any note – Adelaide. This wave was followed by a larger one generated by the slosh from the ice cap. In the late evening, the wave approached Kangaroo Island, which protected the mouth of the Gulf of St. Vincent leading to the city. The waves surged over the rocky coastline as effortlessly as a previous tsunami that had wiped out Aboriginal culture on the island 500 years before. The island absorbed the brunt of the wave; however, the tsunami refracted around the ends of the island and ran in a crisscross fashion up the funnel-shaped Gulf. The wavelets increased in amplitude from 5 m to 10 as they impinged upon the western shore of the mainland. The Adelaide tide gauge never registered a thing. It was instantly obliterated as waves surged up the Torrens River and through the Central Business District of the city. In the flatter coastal suburbs, successive waves smashed up to 2 km inland, leaving a mass of demolished houses and shops stacked up at the limit of run-up. The whole scene was broadcast live nationwide from the Goodyear Blimp hovering over the Adelaide Cricket Ground for the day–night match between Australia and India. Viewers sat stunned in front of their TV sets as the wave crashed through the outer stands of the cricket ground. In Melbourne, frantic activity could be seen in one or two houses as their occupants prepared to flee. Only they knew that the waves were minutes away from that city.

CONCLUDING COMMENTS

The above scenarios have been deliberately contrived to highlight the fact that, while earthquakes are commonly thought as the cause of tsunami, tsunami have many sources. None of the stories should be viewed as unbelievable. In fact, the tone, voice, and storyline of each deliberately matches those of the historical accounts and legends presented in Chapter 1. If these concluding stories are perceived as tall tales, then the reader is likely to deny the magnitude of past historical events such as the tsunami generated by the Lisbon earthquake of 1755 or by the volcanic eruption of Krakatau in 1883. If the run-up heights of 40 m generated by these two events are trivialised, then bigger events in isolated parts of the globe are more likely to be ignored. We then are prone to mock the descriptions of tsunami present in ancient historical writings. Finally, we unashamedly convert history into legends and legends into myths. It is human nature to minimise hazards, and that is why tsunami are so underrated. Unlike any past civilisation, Western Civilisation is unique in its settlement of the shoreline and its development of great coastal cities. We develop ever larger ports in harbours and along the open coast, establish retirement villages on coastal marshes and barrier islands, talk up the value of coastal real estate, and glamorise seaside holidays. Our civilisation is so dependent upon the coastline and marine trade that it in turn plays down marine hazards. We then marvel at devastating hurricanes and attribute them to phenomena such as global warming, and farewell sporting seamen on ocean races who then die in storms that we term abnormal. The purpose of this textbook has been to make readers aware that tsunami are ubiquitous along our shorelines. The only guarantee or prediction is that they will happen again, sometime soon, on a coastline near you – on a reservoir, a lake, a sheltered sea, inside a coral barrier, in the lee of an island, or along an open coastline. Our present knowledge about marine hazards is biased. Tsunami are very much an underrated, widespread hazard. Any coast is at risk.

References

Aalto, K. R., Aalto, R., Garrison-Laney, C. E., and Abramson, H. F. 1999. Tsunami (?) sculpturing of the Pebble Beach wave-cut platform, Crescent City area, California. *Journal of Geology* v. 107, pp. 607–622.

Abe, K. 1979. Size of great earthquakes of 1837–1974 inferred from tsunami data. *Journal of Geophysical Research* v. 84, pp. 1561–1568.

1983. A new scale of tsunami magnitude, M_t. *In* Iida, K., and Iwasaki, T. (Eds.) *Tsunamis – Their Science and Engineering.* Terra Scientific Publishing, Tokyo, pp. 91–101.

Abe, K., Tsuji, Y., Imamura, F., Katao, H., Iio, Y., Satake, K., Bourgeois, J., Noguera, E., and Estrada, F. 1993. Field survey of the Nicaragua earthquake and tsunami of September 2, 1992. *Bulletin of the Earthquake Research Institute* v. 68, pp. 23–70.

Allen, J. R. L. 1984. *Sedimentary Structures: Their Character and Physical Basis:* v. 1, Amsterdam, Elsevier.

Alexander, H. S. 1932. Pothole erosion. *Journal of Geology* v. 40, pp. 305–337.

Altinok, Y., Alpar, B., Ersoy, S., and Yalciner, A. C. 1999. Tsunami generation of the Kocaeli earthquake (August 17th 1999) in the Izmit Bay; coastal observations, bathymetry and seismic data. *Turkish Journal of Marine Sciences* v. 5, pp. 131–148.

Alvarez, W. 1997. *T. Rex and the Crater of Doom.* Princeton University Press, Princeton.

Andrade, C. 1992. Tsunami generated forms in the Algarve barrier islands (South Portugal). *Science of Tsunami Hazards* v. 10, pp. 21–33.

Asher, D. J., Clube, S. V. M., Napier, W. M., and Steel, D. I. 1994. Coherent catastrophism. *Vistas in Astronomy* v. 38, pp. 1–27.

Atwater, B. F. 1987. Evidence for great Holocene earthquakes along the outer coast of Washington State. *Science* v. 236, pp. 942–944.

Bak, P. 1997. *How Nature Works.* Oxford University Press, Oxford.

Baker, V. R. 1978. Paleohydraulics and hydrodynamics of scabland floods *In* Baker, V. R., and Nummedal, D. (Eds.) *The channeled scabland.* NASA Office of Space Science, Planetary Geology Program, pp. 59–79.

(Ed.) 1981. *Catastrophic flooding: The origin of the channeled scabland.* Stroudsburg, Dowden Hutchinson & Ross, 360 pp.

Baptista, M. A., Miranda, P. M. A., Miranda, J. M., and Mendes Victor, L. 1996. Rupture extent of the 1755 Lisbon earthquake inferred from numerical modeling of tsunami data. *Physics and Chemistry of the Earth* v. 21, pp. 65–70.

Bascom, W. L. 1959. *Ocean Waves*. Scientific American Reprint.

Ben-Menahem, A., and Rosenman, M. 1972. Amplitude patterns of tsunami waves from submarine earthquakes. *Journal of Geophysical Research* v. 77, pp. 3097–3128.

Bernard, E. N. 1991. Assessment of Project THRUST: past, present, future. *Proceedings of the 2nd UJNR Tsunami Workshop*, Honolulu, Hawaii, 5–6 November 1990, National Geophysical Data Center, Boulder, pp. 247–255.

Beroza, G. C. 1995. Seismic source modeling. U.S. National Report to International Union of Geodesy and Geophysics 1991–1994, *Review of Geophysics Supplement* v. 33 **http://earth.agu.org/revgeophys/beroza01/ beroza01.html**.

Blong, R. J. 1984. *Volcanic Hazards: A Sourcebook on the Effects of Eruptions*. Academic Press, Sydney.

Bohor, B. F. 1996. A sediment gravity flow hypothesis for siliciclastic units at the K/T boundary, northeastern Mexico. *Geological Society of America Special Paper* No. 307, pp. 183–195.

Bolt, B. A., Horn, W. L., MacDonald, G. A., and Scott, R. F. 1975. *Geological hazards*. Springer-Verlag, Berlin.

Bondevik, S., Svendsen, J. I., Johnsen, G., Mangerud, J., and Kaland, P. E. 1997a. The Storegga tsunami along the Norwegian coast, its age and run-up. *Boreas* v. 26, pp. 29–53.

Bondevik, S., Svendsen, J. I., and Mangerud, J. 1997b. Tsunami sedimentary facies deposited by the Storegga tsunami in shallow marine basins and coastal lakes, western Norway. *Sedimentology* v. 44, pp. 1115–1131.

Bouma, A. H., and Brouwer, A. (Eds.) 1964. *Turbidites*. Elsevier, Amsterdam.

Bourgeois, J., and Leithold, E. L. 1984. Wave-worked conglomerates–depositional processes and criteria for recognition *In* Koster, E. H., and Steel, R. J. (Eds.) Sedimentology of Gravels and Conglomerates. *Canadian Society of Petroleum Geologists Memoir* No. 10, pp. 331–343.

Bourgeois, J., Hansen, T. A., Wiberg, P. L., and Kauffman, E. G. 1988. A tsunami deposit at the Cretaceous–Tertiary boundary in Texas. *Science* v. 241, pp. 567–570.

Bourrouilh-Le Jan, F. G., and Talandier, J. 1985. Sédimentation et fracturation de haute énergie en milieu récifal: Tsunamis, ouragans et cyclones et leurs effets sur la sédimentologie et la géomorphologie d'un atoll: Motu et hoa, à Rangiroa, Tuamotu, Pacifique SE. *Marine Geology* v. 67, pp. 263–333.

Branney, M., and Zalasiewicz, J. 1999. Burning clouds. *New Scientist* 17 July, pp. 36–41.

Briggs, M. J., Synolakis, C. E., Harkins, G. S., and Green, D. R. 1995. Laboratory experiments of tsunami runup on a circular island. *Pure and Applied Geophysics* v. 144, pp. 569–593.

Bryant, E. 1991. *Natural Hazards*. Cambridge University Press, Cambridge, UK.

Bryant, E., and Nott, J. 2001. Geological indicators of large tsunami in Australia. *Natural Hazards* (in press).

Bryant, E., and Young, R. W. 1996. Bedrock-sculpturing by tsunami, South Coast New South Wales, Australia. *Journal of Geology* v. 104, pp. 565–582.

Bryant, E., Young, R. W., and Price, D. M. 1992. Evidence of tsunami sedimentation on the southeastern coast of Australia. *Journal of Geology* v. 100, pp. 753–765.

Bryant, E., Young, R. W., and Price, D. M. 1996. Tsunami as a major control on coastal evolution, Southeastern Australia. *Journal of Coastal Research* v. 12, pp. 831–840.

Bryant, E., Young, R. W., Price, D. M. Wheeler, D., and Pease, M. I. 1997. The impact of tsunami on the coastline of Jervis Bay, southeastern Australia. *Physical Geography* v. 18, pp. 441–460.

Camfield, F. E. 1994. Tsunami Effects on Coastal Structures. *Journal of Coastal Research* Special Issue No. 12, pp. 177–187.

Carlson, R. R., Karl, H. A., and Edwards, B. D. 1991. Mass sediment failure and transport features revealed by acoustic techniques, Beringian margin, Bering Sea, Alaska. *Marine Geotechnology* v. 10, pp. 33–51.

Carracedo, J. C., Day, S., Guillou, H., Rodríguez Badiola, E., Canas, J. A., and Pérez Torrado, F. J. 1998. Hotspot volcanism close to a passive continental margin: The Canary Islands. *Geological Magazine* v. 135, pp. 591–604.

Chyba, C. F., Thomas, P. J., and Zahnle, K. J. 1993. The 1908 Tunguska explosion: Atmospheric disruption of a stony asteroid. *Nature* v. 361, pp. 40–44.

Cita, M. B., Camerlenghi, A., and Rimoldi, B. 1996. Deep-sea tsunami deposits in the eastern Mediterranean: New evidence and depositional models. *Sedimentary Geology* v. 104, pp. 155–173.

Clague, J. J., and Bobrowsky, P. T. 1994. Evidence for a large earthquake and tsunami 100–400 years ago on Western Vancouver Island, British Columbia. *Quaternary Research* v. 41, pp. 176–184.

Clifton, H. E. 1988. Sedimentologic relevance of convulsive geologic events. *Geological Society of America Bulletin* Special Paper No. 229, pp. 1–5

Coleman, P. J. 1968. Tsunamis as geological agents. *Journal of the Geological Society of Australia* v. 15, pp. 267–273.

Cornell, J. 1976. *The Great International Disaster Book.* Scribner's, New York.

Cox, K. 1994. Sand holds clues to quake of '29. *The Globe and Mail,* Toronto, November 21.

Crawford, D. A., and Mader, C. L. 1998. Modeling asteroid impact and tsunami. *Science of Tsunami Hazards* v. 16, pp. 21–30.

Dahl, R. 1965, Plastically sculptured detail forms on rock surfaces in northern Nordland, Norway. *Geografiska Annaler* v. 47A, pp. 3–140.

Darienzo, M. E., and Peterson, C. D. 1990. Episodic tectonic subsidence of Late Holocene salt marshes, Northern Oregon central Cascadia margin. *Tectonics* v. 9, pp. 1–22.

Dawson, A. G., 1994. Geomorphological effects of tsunami run-up and backwash. *Geomorphology* v. 10, pp. 83–94.

1999. Linking tsunami deposits, submarine slides and offshore earthquakes. *Quaternary International* v. 60, pp. 119–126.

Dawson, A. G., Foster, I. K. L., Shi, S., Smith, D. E., and Long, D. 1991. The identification of tsunami deposits in coastal sediment sequences. *Science of Tsunami Hazards* v. 9, pp. 73–82.

Dawson, A. G., Hindson, R., Andrade, C., Freitas, C., Parish, R., and Bateman, M. 1995. Tsunami sedimentation associated with the Lisbon earthquake of 1 November AD 1755: Boca do Rio, Algarve, Portugal. *The Holocene* v. 5, pp. 209–215.

Dawson, A. G., Long, D., and Smith, D. E. 1988. The Storegga slides: Evidence from eastern Scotland for a possible tsunami. *Marine Geology* v. 82, pp. 271–276.

Dawson, A. G., Smith, D. E., Ruffman, A., and Shi, S. 1996. The diatom biostratigraphy of tsunami sediments: Examples from recent and Middle Holocene events. *Physics and Chemistry of the Earth* v. 21, pp. 87–92.

de Lange, W. P., and Healy, T. R. 1986. New Zealand tsunamis 1840–1982. *New Zealand Journal Geology and Geophysics.* v. 29, pp. 115–134.

Dominey-Howes, D. T. M. 1996. Sedimentary deposits associated with the July 9th 1956 Aegean Sea tsunami. *Physics and Chemistry of the Earth* v. 21, pp. 51–55.

Estensen, M. 1998. *Discovery: The Quest for the Great South Land.* Allen and Unwin, Sydney.

Foster, I. D. L., Albon, A. J., Bardell, K. M., Fletcher, J. L., Jardine, T. C., Mothers, R. J., Pritchard, M. A., and Turner, S. E. 1991. High energy coastal sedimentary deposits: An evaluation of depositional processes in southwest England. *Earth Surface Processes and Landforms* v. 16, pp. 341–356.

Fujita, T. T. 1971. *Proposed Mechanism of Suction Spots Accompanied by Tornadoes.* Preprints, Seventh Conference on Severe Local Storms, American Meteorological Society, Kansas City, pp. 208–213.

Fukuchi, T., and Mitsuhashi, K. 1983. Tsunami countermeasures in fishing villages along the Sanriku coast, Japan. *In* Iida, K., and Iwasaki, T. (Eds.) *Tsunamis – Their Science and Engineering.* Terra Scientific Publishing, Tokyo, pp. 389–396.

Furumoto, A. S., Tatehata, H., and Morioka, C. 1999. Japanese tsunami warning system. *Science of Tsunami Hazards* v. 17, pp. 85–105.

Geist, E. L. 1997a. Native American Legends in the Pacific Northwest. **http://walrus.wr.usgs.gov/ docs/projects/cascadia/tsunami/NAlegends.html**.

   1997b. Local tsunamis and earthquake source parameters. *Advances in Geophysics* v. 39, pp. 117–209.

Gelfenbaum, G., and Jaffe, B. 1998. *Preliminary Analysis of Sedimentary Deposits from the 1998 PNG Tsunami.* United States Geological Survey, **http://walrus.wr.usgs.gov/ tsunami/itst.html**.

Gersonde, R., Kyte, F. T., Bleil, U., Diekmann, B., Flores, J. A., Gohl, K., Grahl, G., Hagen, R., Kuhn, G., Sierro, F. J., Völker, D., Abelmann, A., and Bostwick, J. A. 1997. Geological record and reconstruction of the late Pliocene impact of the Eltanin asteroid in the Southern Ocean. *Nature* v. 390, pp. 357–363.

Goff, J. R., and Chagué-Goff, C. 1999. A Late Holocene record of environmental changes from coastal wetlands: Abel Tasman National Park, New Zealand. *Journal Quaternary International* v. 56, pp. 39–51.

González, F. I. 1999. Tsunami! *Scientific American* May, pp. 44–55.

Goodman, J. 1997. What Is the Greatest Height a Tsunami Could Reach Should a Large Meteorite Strike the Ocean? **http://madsci.wustl.edu/posts/archives /aug97/868937833.Ph.r.html**.

Grazulis, T. P. 1993. *Significant Tornadoes 1680–1991: A Chronology and Analysis of Events.* St. Johnsbury, Environmental Films.

Gusiakov, V. K., and Osipova, A. V. 1993. Historical tsunami database for the Kuril–Kamchatka region. *In* Tinti, S. (Ed.) *Tsunamis in the World.* Kluwer, Dordrecht, pp. 17–30.

Hall, J. 1812. On the revolutions of the earth's surface. *Transactions of the Royal Society of Edinburgh* v. 7, pp. 169–212.

Hamer, M. 1999. Solitary killers. *New Scientist* No. 2201, 28 August, pp. 18–19.

Hansen, W. R. 1965. Effects of the earthquake of March 27, 1964 at Anchorage, Alaska. *United States Geological Survey Professional Paper* No. 542-A.

Harbitz, C. B. 1992. Model simulations of tsunamis generated by the Storegga Slides. *Marine Geology* v. 105, pp. 1–21.

Hasegawa, I. 1992. Historical variation in the meteor flux as found in Chinese and Japanese chronicles. *Celestial Mechanics and Dynamical Astronomy* v. 54, pp. 129–142.

Haslett, S. K., Bryant, E. A., and Curr, R. H. F. 2000. Tracing beach sand provenance and transport using foraminifera: Examples from NW Europe and SE Australia. *In* Foster, I. (Ed.) *Tracers in Geomorphology.* Wiley, Chichester, pp. 437–452.

Hatori, T. 1986. Classification of tsunami magnitude scale. *Bulletin of the Earthquake Institute* v. 61, pp. 503–515 (in Japanese).

Hearty, P. J. 1997. Boulder deposits from large waves during the Last Interglaciation on North Eleuthera Island, Bahamas. *Quaternary Research* v. 48, pp. 326–338.

Hearty, P. J., Neumann, A. C., and Kaufman, D. S. 1998. Chevron ridges and runup deposits in the Bahamas from storms late in Oxygen-Isotope substage 5e. *Quaternary Research* v. 50, pp. 309–322.

Heaton, T. H., and Snavely, P. D. 1985. Possible tsunami along the northwestern coast of the

United States inferred from Indian traditions. *Bulletin of the Seismological Society of America* v. 75 No. 5, pp. 1455–1460.

Heezen, B. C., and Ewing, M. 1952. Turbidity currents and submarine slumps, and the 1929 Grand Banks earthquake. *American Journal of Science* v. 250, pp. 849–873.

Heinrich, Ph., Guibourg, S., and Roche, R. 1996. Numerical modeling of the 1960 Chilean tsunami: Impact on French Polynesia. *Physics and Chemistry of the Earth* v. 21 No. 12, pp. 19–25.

Henry, R. F., and Murty, T. S. 1992. Model studies of the effects of the Storegga slide tsunami. *Science of Tsunami Hazards* v. 10, pp. 51–62.

Hills, J. G., and Mader, C. L. 1997. Tsunami produced by the impacts of small asteroids. *Annals of the New York Academy of Sciences* v. 822, pp. 381–394.

Hindson, R. A., Andrade, C., and Dawson, A. G. 1996. Sedimentary processes associated with the tsunami generated by the 1755 Lisbon earthquake on the Algarve Coast, Portugal. *Physics and Chemistry of the Earth* v. 21, pp. 57–63.

Holcomb, R. T., and Searle, R. C. 1991. Large landslides from oceanic volcanoes. *Marine Geotechnology* v. 10, pp. 19–32.

Horikawa, K., and Shuto, N. 1983. Tsunami disasters and protection measures in Japan. *In* Iida, K., and Iwasaki, T. (Eds.) *Tsunamis: Their Science and Engineering.* Reidel, Dordrecht, pp. 9–22.

Hovland, M. 1999. Gas, fire and water. *EOS Transactions of the American Geophysical Union* v. 80, p. 552.

Howorth, R. 1999. Tsunami – the scourge of the Pacific. *COGEOENVIRONMENT News* No. 14, Commission on Geological Sciences for Environmental Planning, International Union of Geological Sciences, January.

Huggett, R. 1989. *Cataclysms and Earth History.* Clarendon, Oxford, 220p.

Iida, K. 1963. Magnitude, energy and generation of tsunamis, and catalogue of earthquakes associated with tsunamis. Proceedings of Tsunami Meetings Associated with the 10th Pacific Science Congress, *International Union of Geodesy and Geophysics Monograph* No. 24, pp. 7–18.

1983. Some remarks on the occurrence of tsunamigenic earthquakes around the Pacific. *In* Iida, K., and Iwasaki, T. (Eds.) *Tsunamis: Their Science and Engineering.* Reidel, Dordrecht, pp. 61–76.

1985. Activity of tsunamigenic earthquakes around the Pacific. *Proceedings of the International Tsunami Symposium,* August 6–9, Department of Fisheries and Oceans, Sidney, British Columbia, pp. 1–6.

Iida, K., and Iwasaki, T. (Eds.) 1983. *Tsunamis: Their Science and Engineering.* Reidel, Dordrecht.

Intergovernmental Oceanographic Commission 1999. Historical Tsunami Database for the Pacific, 47 B.C.–1998 A.D.. Tsunami Laboratory, Institute of Computational Mathematics and Mathematical Geophysics, Siberian Division Russian Academy of Sciences, Novosibirsk, Russia, **http://tsun.sscc.ru/HTDBPac1/.**

International Tsunami Information Center 1998. **http://www.shoa.cl/oceano/itic/frontpage.html**.

Johnson, D. 1998. Night Skies of Aboriginal Australia: A noctuary. *Oceania Monograph* No. 47, University of Sydney, Sydney.

Johnson, J. M. 1999. Heterogeneous coupling along Alaska–Aleutians as inferred from Tsunami, seismic, and geodetic inversions. *Advances in Geophysics* v. 39, pp. 1–106.

Johnstone, B. 1997. Who killed the Minoans? *New Scientist* 21 June, pp. 36–39.

Jones, A. T., and Mader, C. L. 1996. Wave erosion on the southeastern coast of Australia: Tsunami propagation modelling. *Australian Journal of Earth Sciences* v. 43, pp. 479–483.

Jones, B., and Hunter, I. G. 1992.Very large boulders on the coast of Grand Cayman: The effects of giant waves on rocky coastlines. *Journal of Coastal Research* v. 8, pp. 763–774.

Jones, D., and Donaldson, K. 1989. *The Story of the Falling Star.* Aboriginal Studies Press, Canberra.

Kajiura, K. 1983. Some statistics related to observed tsunami heights along the coast of Japan. *In* Iida, K., and Iwasaki, T. (Eds.) *Tsunamis – Their Science and Engineering.* Terra Scientific Publishing, Tokyo, pp. 131–145.

Kanamori, H., and Kikuchi, M. 1993. The 1992 Nicaragua earthquake: A slow tsunami earthquake associated with subducted sediments. *Nature* v. 361, pp. 714–716.

Kastens, K. A., and Cita, M. B. 1981. Tsunami-induced sediment transport in the abyssal Mediterranean Sea. *Geological Society of America Bulletin* v. 92, pp. 845–857.

Kawana, T., and Nakata, T. 1994. Timing of Late Holocene tsunamis originated around the Southern Ryukyu Islands, Japan, deduced from coralline tsunami deposits. *Japanese Journal of Geography* v. 103, pp. 352–376.

    and Pirazzoli, P. 1990. Re-examination of the Holocene emerged shorelines in Irabu and Shimoji Islands, the South Ryukyus, Japan. *Quaternary Research* v. 28, pp. 419–426.

Kawata, Y., Benson, B. C., Borrero, J. C. Borrero, J. L. Davies, H. L. de Lange, W. P., Imamura, F., Letz, H. Nott, J., and Synolakis, C. E. 1999. Tsunami in Papua New Guinea was as intense as first thought. *EOS Transactions of the American Geophysical Union* v. 80, pp. 101, 104–105.

Keating, B. H. 1998. Side-scan sonar images of submarine landslides on the flanks of atolls and guyots. *Marine Geodesy* v. 21, pp. 124–144.

Keating, B. H., Fryer, P., Batiza, R., and Boehlert, G. W. (Eds.) 1987. Seamounts, Islands and Atolls. *American Geophysical Union* Monograph No. 43, Washington.

Kenyon, N. H. 1987. Mass-wasting features on the continental slope of northwest Europe. *Marine Geology* v. 74, pp. 57–77.

Kikuchi, M., and Kanamori, H. 1995. Source characteristics of the 1992 Nicaragua tsunami earthquake inferred from teleseismic body waves. *Pure and Applied Geophysics* v. 144, pp. 441–453.

Komar, P. D. 1998. *Beach Processes and Sedimentation.* 2nd Ed. Prentice-Hall, Upper Saddle River.

Kor, P. S. G., Shaw, J., and Sharpe, D. R. 1991. Erosion of bedrock by subglacial meltwater, Georgian Bay, Ontario: A regional view. *Canadian Journal of Earth Science* v. 28, pp. 623–642.

Kristan-Tollmann, E., and Tollmann, A. 1992. Der Sintflut-Impakt (The Flood impact). *Mitteilungen Der Österreichischen Geographischen Gesellschaft* v. 84, pp. 1–63.

Kuran, U., and Yalçiner, A. C. 1993. Crack propagations, earthquakes and tsunamis in the vicinity of Anatolia *In* Tinti, S. (Ed.) *Tsunamis in the World.* Kluwer, Dordrecht, pp. 159–175.

LaMoreaux, P. E. 1995. Worldwide environmental impacts from the eruption of Thera. *Environmental Geology* v. 26, pp. 172–181.

Lander, J. F., and Lockridge, P. A. 1989. *United States Tsunamis (Including United States Possessions) 1690–1988.* National Geophysical Data Center, Boulder.

Latter, J. H. 1981. Tsunamis of volcanic origin: Summary of causes, with particular reference to Krakatau, 1883. *Bulletin Volcanologique* v. 44, pp. 467–490.

Lewis, J. S. (1999) *Comet and Asteroid Impact Hazards on a Populated Earth.* Academic Press, San Diego.

Lipman, P. W., Normark, W. R., Moore, J. G., Wilson, J. B., and Gutmacher, C. E. 1988. The giant submarine Alika Debris Slide, Mauna Loa, Hawaii. *Journal of Geophysical Research* v. 93, pp. 4279–4299.

Lockridge, P. A. 1985. *Tsunamis in Peru–Chile.* World Data Center A for Solid Earth Geophysics, National Geophysical Data Center, Boulder, Rpt. SE-39.

1988a. Volcanoes generate devastating waves. *Earthquakes and Volcanoes* v. 20, pp. 190–195.

1988b. Historical tsunamis in the Pacific basin. *In* El-Sabh, M. I., and Murty, T. S. (Eds.) *Natural and Man-Made Hazards.* Reidel, Dordrecht, pp. 171–181.

1990. Nonseismic phenomena in the generation and augmentation of tsunamis. *Natural Hazards* v. 3, pp. 403–412.

Long, D., Smith, D. E., and Dawson, A. G. 1989. A Holocene tsunami deposit in eastern Scotland. *Journal of Quaternary Science* v. 4, pp. 61–66.

Mader, C. L. 1974. Numerical simulation of tsunamis. *Journal of Physical Oceanography* v. 4, pp. 74–82.

1988. *Numerical Modeling of Water Waves.* University of California Press, Berkeley.

1990. Numerical tsunami flooding study: 1. *Science of Tsunami Hazards* v. 8, pp. 79–96.

1998. Modeling the Eltanin Asteroid tsunami. *Science of Tsunami Hazards* v. 16, pp. 21–30.

Masson, D. G. 1996. Catastrophic collapse of the volcanic island of Hierro 15 ka ago and the history of landslides in the Canary Islands. *Geology* v. 24, pp. 231–234.

Masson, D. G., Kenyon, N. Y., and Weaver, P. P. E. 1996. Slides, debris flows, and turbidity currents *In* Summerhayes, C. P., and Thorpe, S. A. (Eds.) *Oceanography: An Illustrated Guide.* Manson Publishing, London, pp. 136–151.

McGlone, M. S., and Wilmshurst, J. M. 1999. Dating initial Maori environmental impact in New Zealand. *Quaternary International* v. 59, pp. 5–16.

McSaveney, M., and Goff, J. 1998. Subsidence identified as trigger of catastrophic tsunami. *Globe, New Zealand Institute of Geological and Nuclear Sciences Newsletter,* December, p. 7.

Miller, D. J., 1960. Giant Waves in Lituya Bay, Alaska. *United States Geological Survey Professional Paper* 354-C, pp. 51–86.

Minoura, K., and Nakaya, S. 1991. Traces of tsunami preserved in inter-tidal lacustrine and marsh deposits: Some examples from northeast Japan. *Journal of Geology* v. 99, pp. 265–287.

Minoura, K., Imamura, F., Takahashi, T., and Shuto, N. 1997. Sequence of sedimentation processes caused by the 1992 Flores tsunami: Evidence from Babi Island. *Geology* v. 25, pp. 523–526.

Minoura, K., Nakaya, S., and Uchida, M. 1994. Tsunami deposits in a lacustrine sequence of the Sanriku Coast, Northeast Japan. *Sedimentary Geology* v. 89, pp. 25–31.

Mooley, B. P. J., Burrows, C. J., Cox, J. E., Johnston, J. A., and Wardle, P. 1963. Distribution of subfossil forest remains Eastern South Island, New Zealand. *New Zealand Journal of Botany* v. 1, pp. 68–77.

Moore, D. G. 1978. Submarine slides. *In* Voight, B. (Ed.) *Rockslides and Avalanches, 1: Natural Phenomena.* Elsevier Scientific, Amsterdam, pp. 563–604.

Moore, G. W., and Moore, J. G. 1988. Large-scale bedforms in boulder gravel produced by giant waves in Hawaii. *Geological Society of America* Special Paper No. 229, pp. 101–110.

Moore, J. G., Bryan, W. B., and Ludwig, K. R. 1994a. Chaotic deposition by a giant wave, Molokai, Hawaii. *Geological Society of America Bulletin* v. 106, pp. 962–967.

Moore, J. G., Clague, D. A., Holcomb, R. T., Lipman, P. W., Normark, W. R., and Torresan, M. E. 1989. Prodigious submarine landslides on the Hawaiian Ridge. *Journal of Geophysical Research* v. 94, pp. 17465–17484.

Moore, J. G., Normark, W. R., and Holcomb, R. T. 1994b. Giant Hawaiian landslides. *Annual Review of Earth and Planetary Sciences* v. 22, pp. 119–144.

Moreira, V. S. 1993. Historical tsunamis in mainland Portugal and Azores – case histories. *In* Tinti, S. (Ed.) *Tsunamis in the World.* Kluwer, Dordrecht, pp. 65–73.

Morton, R. A., 1988. Nearshore responses to great storms. *Geological Society of America,* Special Paper No. 229, pp. 7–22.

Murty, T. S. (Ed.) 1977. Seismic sea waves: Tsunamis. *Bulletin of the Fisheries Research Board of Canada* No. 198.

1984. Storm surges-meteorological ocean tides. *Bulletin of the Fisheries Research Board of Canada* No. 212.

Myles, D. 1985. *The Great Waves.* McGraw-Hill, New York.

National Geophysical Data Center and World Data Center A for Solid Earth Geophysics 1984. *Tsunamis in the Pacific Basin 1900–1983.* United States National Oceanic and Atmospheric Administration, Boulder, Colorado, map 1:17,000,000.

1989. *United States Tsunamis (including United States Possessions) 1690–1988.* United States National Oceanic and Atmospheric Administration, Publication No. 41–2, Boulder, Colorado.

1998. *Tsunami Event Database.* United States National Oceanic and Atmospheric Administration, Boulder, Colorado. **http://www.ngdc.noaa.gov/seg/hazard/tsevsrch.shtml# deaths**.

National Oceanic and Atmospheric Administration (NOAA) 1998. Tsunami! The Great Waves. **http://vishnu.glg.nau.edu/wsspc/tsunami/HI/Waves/waves06.html**.

Nemtchinov, I. V., Loseva, T. V., and Teterev, A. V. 1996. Impacts into oceans and seas. *Earth, Moon and Planets* v. 72, pp. 405–418.

Nomanbhoy, N., and Satake, K. 1995. Numerical computations of tsunamis from the 1883 Krakatau eruption. *Geophysical Research Letters* v. 22, pp. 509–512.

Nott, J. 1997. Extremely high-energy wave deposits inside the Great Barrier Reef, Australia: Determining the cause–tsunami or tropical cyclone. *Marine Geology* v. 141, pp. 193–207.

Oh, I. S., and Rabinovich, A. B. 1994 Manifestation of Hokkaido southwest (Okushiri) tsunami, 12 July 1993, at the coast of Korea: 1. Statistical characteristics, spectral analysis, and energy decay. *Science of Tsunami Hazards* v. 12, pp. 93–116.

Okal, E. A. 1988. Seismic parameters controlling far-field tsunami amplitudes: A review. *Natural Hazards* v. 1, pp. 67–96.

1993. Predicting large tsunamis. *Nature* v. 361, pp. 686–687.

Okal, E. A., Talandier, J., and Reymond, D. 1991. Automatic estimations of tsunami risk following a distant earthquake using the mantle magnitude M_m. *Proceedings of the 2nd UJNR Tsunami Workshop,* Honolulu, Hawaii 5–6 November 1990, National Geophysical Data Center, Boulder, pp. 229–238.

Oliver, J. 1988. Natural hazards. *In* Jeans, D. N. (Ed.) *Australia: A Geography.* Sydney University Press., pp. 283–314.

Ortlieb, L., Goy, J. L., Zazo, C., Hillaire-Marcel, C., and Vargas, G. 1995. *Late Quaternary Coastal Changes in Northern Chile.* International Geological Correlation Program Project 367 Guidebook, ORSTRAM, 23–25 November.

Ota, Y., Pirazzoli, P. A., Kawana, T., and Moriwaki, H. 1985. Late Holocene coastal morphology and sea level records on three small islands, the South Ryukyus, Japan. *Geographical Review of Japan* v. 58B, pp. 185–194.

Paine, M. 1999. Asteroid impacts: The extra hazard due to tsunami. *Science of Tsunami Hazards* v. 17, pp. 155–166.

Pararas-Carayannis, G. 1998a. The 1960 Chilean Tsunami. **http://www.geocities.com/ CapeCanaveral/Lab/1029/Tsunami1960.html**.

1998b. The March 27, 1964 Great Alaska Tsunami. **http://www.geocities.com/CapeCanaveral/ Lab/1029/Tsunami1964GreatGulf.html**.

1998c. The Waves That Destroyed the Minoan Empire (Atlantis) **http://www.geocities.com/ CapeCanaveral/Lab/1029/.html**.

1999. Analysis of Mechanism of The Giant Tsunami Generation in Lituya Bay. Science of Tsunami Hazards v. 17, pp. 193–206. **http://www.geocities.com/CapeCanaveral/Lab/ 1029/Tsunami1958LituyaB.html**.

Parker, K. L. 1978. *Australian Legendary Tales*. Bodley Head, London.

Paskoff, R. 1991. Likely occurrence of a mega-tsunami in the middle Pleistocene, near Coquimbo, Chile. *Revista Geológica de Chile* v. 18, pp. 87–91.

Paskoff, R., Leonard, E. M., Novoa, J. E., Ortlieb, L., Radtke, U., and Wehmiller, J. F. 1995. *Field meeting in the la Serena–Coquimbo Bay Area (Chile)*. Guidebook for a field trip 27–28 November, International Geological Correlation Program Project 367.

Peck, C. W. 1938. *Australian Legends*. Lothian, Melbourne.

Pelinovsky, E. 1996. *Tsunami Wave Hydrodynamics*. Institute of Applied Physics, Nizhny Novgorod (in Russian).

Pelinovsky, E., and Poplavsky, A. 1996. Simplified model of tsunami generation by submarine landslides. *Physics and Chemistry of the Earth* v. 21, pp. 13–17.

Pichler, H., and Friedrich, W. L. 1980. Mechanism of the Minoan eruption of Santorini in Thera and the Aegean World: v. 2. *Proceedings of the 2nd International Scientific Congress,* Santorini, Greece, August 1978, pp. 15–30.

Pickering, K. T., Soh, W., and Taira, A. 1991. Scale of tsunami-generated sedimentary structures in deep water. *Journal of the Geological Society London* v. 148, pp. 211–214.

Pinegina, T., Bazanova, L., Braitseva, O., Gusiakov, V., Melekestsev, I., Starcheus, A. 1996. East Kamchatka palaeotsunami traces. *Proceedings of the International Workshop on Tsunami Mitigation and Risk Assessment,* Petropavlovsk–Kamchatskiy, Russia, August 21–24, **http://omzg.sscc.ru/tsulab/content.html**.

Piper, D. J. W., Cochonat, P., and Morrison, M. L. 1999. The sequence of events around the epicentre of the 1929 Grand Banks earthquake: Initiation of debris flows and turbidity current inferred from sidescan sonar. *Sedimentology* v. 46, pp. 79–97.

Rabinovich, A. B., and Monserrat, S. 1996. Meteorological tsunamis near the Balearic and Kuril Islands: Descriptive and statistical analysis. *Natural Hazards* v. 13, pp. 55–90.

Ranguelov, B., and Gospodinov, D. 1995. Tsunami vulnerability modelling for the Bulgarian Black Sea coast. *Water Science and Technology* v. 32, pp. 47–53.

Rasmussen, K. L. 1991. Historical accretionary events from 800 BC to AD 1750: Evidence for Planetary rings around the Earth? *Quarterly Journal of the Royal Astronomical Society* v. 32, pp. 25–34.

Reid, H. F. 1914. The Lisbon earthquake of November 1, 1755. *Bulletin Seismological Society America* v. 4, pp. 53–80.

Reymond, D., Hyvernaud, O., and Talandier, J. 1993. An integrated system for real time estimation of seismic source parameters and its application to tsunami warning. *In* Tinti, S. (Ed.) *Tsunamis in the World*. Kluwer, Dordrecht, pp. 177–196.

Satake, K. 1995. Linear and nonlinear computations of the 1992 Nicaragua Earthquake Tsunami. *Pure and Applied Geophysics* v. 144, pp. 455–470.

1996. Seismotectonics of tsunami. *Proceedings of the International Workshop on Tsunami Mitigation and Risk Assessment,* Petropavlovsk-Kamchatskiy, Russia, August 21–24, 1996, **http://omzg.sscc.ru/tsulab/paper4.html**.

Satake, K. and Imamura, F. (Eds.) 1995. *Tsunamis 1992–1994: Their Generation, Dynamics and Hazard*. Birkhäuser Verlag, Basel, 890p.

Satake, K., Bourgeois, J., Abe, K., Tsuji, Y., Imamura, F., Iio, Y., Katao, H., Noguera, E., and Estrada, F. 1993. Tsunami field survey of the 1992 Nicaragua earthquake. *EOS Transactions of the American Geophysical Union* v. 74, p. 145.

Satake, K., Shimazaki, K., Tsuji, Y., and Ueda, K. 1996. Time and size of a giant earthquake in Cascadia inferred from Japanese tsunami records of January 1700. *Nature* v. 379, pp. 246–249.

Sato, H., Shimamoto, T., Tsutsumi, A., and Kawamoto, E. 1995. Onshore tsunami deposits caused by the 1993 southwest Hokkaido and 1983 Japan Sea earthquakes. *Pure and Applied Geophysics* v. 144, pp. 693–717.

Schindelé, F., Reymond, D., Gaucher, E., and Okal, E. A. 1995. Analysis and automatic processing in near-field of eight 1992–1994 tsunamigenic earthquakes: Improvements towards real-time tsunami warning. *Pure and Applied Geophysics* v. 144, pp. 381–408.

Schubert, C. 1994. Tsunamis in Venezuela: Some observations on their occurrence. *Journal of Coastal Research Special Issue* No. 12, pp. 189–195.

Scoffin, T. P. 1993. The geological effects of hurricanes on coral reefs and the interpretation of storm deposits. *Coral Reefs* v. 12, pp. 203–221.

Self, S., and Rampino, M. R. 1981. The 1883 eruption of Krakatau. *Nature* v. 294, pp. 699–704.

Shaw, J. 1994. Hairpin erosional marks, horseshoe vortices and subglacial erosion: *Sedimentary Geology* v. 91, pp. 269–283.

Shepard, F. P. 1977. *Geological Oceanography.* University of Queensland Press, St. Lucia.

Shi, S., Dawson, A. G., and Smith, D. E. 1995. Coastal sedimentation associated with the December 12th, 1992 tsunami in Flores, Indonesia. *Pure and Applied Geophysics* v. 144, pp. 525–536.

Shiki, T., and Yamazaki, T. 1996. Tsunami-induced conglomerates in Miocene upper bathyal deposits, Chita peninsula, central Japan. *Sedimentary Geology* v. 104, pp. 175–188.

Shimamoto, T., Tsutsumi, A., Kawamoto, E., Miyawaki, M., and Sato, H. 1995. Field survey report on tsunami disasters caused by the 1993 southwest Hokkaido earthquake. *Pure and Applied Geophysics* v. 144, pp. 665–691.

Shoemaker, E. M. 1983. Asteroid and comet bombardment of the earth. *Annual Reviews of Earth Planetary Science* v. 11, pp. 461–494.

Shuto, N. 1993. Tsunami intensity and disasters. *In* Tinti, S. (Ed.) *Tsunamis in the World.* Kluwer, Dordrecht, pp. 197–216.

Shuto, N., and Matsutomi, H. 1995. Field Survey of the 1993 Hokkaido Nansei-Oki earthquake tsunami. *Pure and Applied Geophysics* v. 144, pp. 649–663.

Shuto, N., Goto, C., and Imamura, F. 1991. Numerical simulation as a means of warning for near-field tsunamis. *Proceedings of the 2nd UJNR Tsunami Workshop,* Honolulu, Hawaii 5–6 November 1990, national Geophysical Data Center, Boulder, pp. 133–153.

Smit, J., Roep. Th. B., Alvarez, W., Montanari, A., Claeys, P., Grajales-Nishimura, J. M., and Bermudez, J. 1996. Coarse-grained, clastic sandstone complex at the K/T boundary around the Gulf of Mexico: Deposition by tsunami waves induced by the Chicxulub impact? *Geological Society of America* Special Paper No. 307, pp. 151–182.

Sokolowski, T. J. 1999a. *The Great Alaskan Earthquake & Tsunamis of 1964.* http://www.tsunami.gov/64quake.htm.

1999b. The U.S. West Coast and Alaska Tsunami Warning Center. *Science of Tsunami Hazards* v. 17, pp. 49–56.

Solomon, S. S., and Forbes, D. L. 1999. Coastal hazards and associated management issues on South Pacific Islands. *Ocean & Coastal Management* v. 42, pp. 523–554.

Steel, D. 1995. *Rogue Asteroids and Doomsday Comets.* Wiley, New York.

and Snow, P. 1992. The Tapanui region of New Zealand: Site of a 'Tunguska' around 800 years ago?. *In* Harris, A., and Bowell, E. (Eds.) *Asteroids, Comets, Meteors 1991.* Lunar and Planetary Institute, Houston, pp. 569–572.

Stoddart, D. R. 1950. The shape of atolls. *Marine Geology* v. 3, pp. 369–383.

Synolakis, C. E. 1987. The run-up of solitary waves. *Journal of Fluid Mechanics* v. 185, pp. 523–545.

1991. Tsunami run-up on steep slopes: How good linear theory really is. *Natural Hazards* v. 4, pp. 221–234.

Tadepalli, S., and Synolakis, C. E. 1994. The run-up of *N*-waves on sloping beaches. *Proceedings of the Royal Society London* v. 445A, pp. 99–112.

Talandier, J., and Bourrouilh-Le Jan, F. 1988. High energy sedimentation in French Polynesia: Cyclone or tsunami?. *In* El-Sabh, M. I., and Murty, T. S. (Eds.) *Natural and Man-Made Hazards.* Reidel, Dordrecht, pp. 193–199.

Tappin, D., Matsumoto, T., Watts, P., Satake, K., McMurty, G., Matsuyama, M., Lafoy, Y., Tsuji, Y., Kanamatsu, T., Lus, W., Iwabuchi, Y., Yeh, H., Matsumotu, Y., Nakamura, M., Mahoi, M., Hill, P., Crook, K., Anton, L., and Walsh, J. P. 1999. Sediment slump likely caused 1998 Papua New Guinea Tsunami. *EOS Transactions of the American Geophysical Union* v. 80., pp. 329, 334, 340.

Tinti, S., and Maramai, A. 1999. Large tsunamis and tsunami hazard from the New Italian Tsunami Catalog. *Physics and Chemistry of the Earth* v. 24A, pp. 151–156.

Titov, V. V., and González, F. I. 1997. Implementation and Testing of the Method of Splitting Tsunami (MOST) Model. NOAA Technical Memorandum ERL PMEL No. 112. **http://www.pmel.noaa.gov/pubs/PDF/tito1927tito1927.pdf**.

Titov, V. V., and Synolakis, C. E. 1997. *Extreme inundation flows during the Hokkaido–Nansei–Oki tsunami.* **http://www.pmel.noaa.gov/tsunami/titov97.html**.

Toon, O. B., Zahnle, K., Morrison, D, Turco, R. P., and Covey, C. 1997. Environmental perturbations caused by the impacts of asteroids and comets. *Reviews of Geophysics* v. 35, pp. 41–78.

Trenhaile, A. S. 1997. *Coastal Dynamics and Landforms.* Oxford University Press, Oxford.

Tsuji, Y. 1991. Decay of the initial crest of the 1960 Chilean tsunami scattering of tsunami waves caused by sea mounts and the effects of dispersion. *Proceeding of the 2nd UJNR Tsunami Workshop,* National Geophysical Data Centre, Boulder, pp. 13–25.

Tsuji, Y., Matsutomi, H., Imamura, F., Takeo, M., Kawata, Y., Matsuyama, M., Takahashi, T., Sunarjo, and Harjadi, P. 1995. Damage to coastal villages due to the 1992 Flores Island earthquake tsunami. *Pure and Applied Geophysics* v. 144, pp. 481–524.

Van Dorn, O. W. 1964. Source mechanism of the tsunami of March 28, 1964. *Proceedings of the 9th Conference on Coastal Engineering,* American Society Civil Engineers, New York, pp. 166–190.

1965. Tsunamis. *Advances in Hydrosciences* v. 2 pp 1–48.

Verbeek, R. D. M. 1884. The Krakatoa eruption. *Nature* v. 30, pp. 10–15.

Verschuur, G. L. 1996. *Impact! The Threat of Comets and Asteroids.* Oxford University Press, Oxford.

von Baeyer, H. C. 1999. Catch the wave. *The Sciences* v. 29 no. 3, pp. 10–13.

Walker, G. 1995. The killer wave. *New Scientist* 25 November, pp. 32–34.

Ward, S. N. 1980. Relationships of tsunami generation and an earthquake source. *Journal of Physics of the Earth* v. 28, pp. 441–474.

2000. Tsunamis. *In* Meyers, R. A. (Ed.) *The Encyclopedia of Physical Science and Technology,* 3rd Edition, Academic Press, (in press).

Ward, S. N., and Asphaug, E. 2000. Asteroid impact tsunami: a probabilistic hazard assessment. *Icarus* v. 145, pp. 64–78.

Watts, P. 1998. Wavemaker curves for tsunamis generated by underwater landslides. *Journal of Waterway, Port, Coastal and Ocean Engineering* v. 124, pp. 127–137.

Whelan, M. 1994. The night the sea smashed Lord's Cove. *Canadian Geographic* November/December, pp. 70–73.

White, K. L., and Price, D. M. 1996b. Fluvial deposition on the Shoalhaven deltaic Plain, southern New South Wales. *Australian Geographer* v. 27, pp. 215–234.

Wiegel, R. L. 1964. *Oceanographical Engineering.* Prentice-Hall, Englewood Cliffs, pp. 95–108.

— 1970. Tsunamis. *In* Wiegel, R. L. (Ed.) *Earthquake Engineering.* Prentice-Hall, Englewood Cliffs, pp. 253–306.

Wilson, B. W., Webb, L. M., and Hendrickson, J. A. 1962. The nature of tsunamis, their generation and dispersion in water of finite depth. In *U.S. Coast and Geodetic Survey, National Engineering Science Company Tech. Rpt.* No. SN 57–2, pp. 43–62.

Wright, C., and Mella, A. 1963. Modifications to the soil pattern of South-Central Chile resulting from seismic and associated phenomena during the period May to August 1960. *Bulletin of the Seismological Society America* v. 53, pp. 1367–1402.

Yanev, P. 1993. Hokkaido Nansei-Oki Earthquake of July 12, 1993. *EQE Review* Fall Issue, **http://www.eqe.com/publications/revf93/hokkaido.htm**.

Yeh, H. H. 1991. Tsunami bore runup. *Natural Hazards* v. 4, pp. 209–220.

Yeh, H. H., Imamura, F., Synolakis, C., Tsuji, Y., Liu, P., and Shi, S. 1993. The Flores Island tsunamis. *EOS Transactions of the American Geophysical Union* v. 74, pp. 369.

Yeh, H. H., Liu, P., Briggs, M., and Synolakis, C. 1994. Propagation and amplification of tsunamis at coastal boundaries. *Nature* v. 372, pp. 353–355.

Yokoyama, I. 1978. The tsunami caused by the prehistoric eruption of Thera. *In* Thera and the Aegean World: 1. *Proceedings of the 2nd International Scientific Congress,* Santorini, Greece, August, pp. 277–283.

Young, R., and Bryant E. 1992. Catastrophic wave erosion on the southeastern coast of Australia: Impact of the Lanai tsunami ca. 105 ka? *Geology* v. 20, pp. 199–202.

Young, R., Bryant E., Price D. M., Wirth, L. M., and Pease, L. M. 1993. Theoretical constraints and chronological evidence of Holocene coastal development in central and southern New South Wales, Australia. *Geomorphology* v. 7, pp. 317–329.

Young, R., Bryant E., Price D. M., and Spassov, E. 1995. The imprint of tsunami in Quaternary coastal sediments of southeastern Australia. *Bulgarian Geophysical Journal* v. 21, pp. 24–31.

Young, R., Bryant E., and Price D. M., 1996a. Catastrophic wave (tsunami?) transport of boulders in southern New South Wales, Australia. *Zeitschrift für Geomorphologie* v. 40, pp. 191–207.

Young, R., Bryant E., Price D. M., Dilek, S. Y., and Wheeler, D. J. 1997. Chronology of Holocene tsunamis on the southeastern coast of Australia. *Transactions of the Japanese Geomorphological Union* v. 18, pp. 1–19.

Index

DATE DUE

DATE DUE			
MAR 1 3 2006			
MAY 2 1 2007			
MAY 0 3 2012			